From Lying to Perjury

Foundations in Language and Law

Editors
Janet Giltrow
Dieter Stein

Volume 3

From Lying to Perjury

Linguistic and Legal Perspectives on Lies and Other Falsehoods

Edited by
Laurence R. Horn

DE GRUYTER
MOUTON

ISBN 978-3-11-135688-4
e-ISBN (PDF) 978-3-11-073373-0
e-ISBN (EPUB) 978-3-11-073381-5
ISSN 2627-3950

Library of Congress Control Number: 2022930993

Bibliographic information published by the Deutsche Nationalbibliothek
The Deutsche Nationalbibliothek lists this publication in the Deutsche Nationalbibliografie;
detailed bibliographic data are available on the Internet at http://dnb.dnb.de.

© 2023 Walter de Gruyter GmbH, Berlin/Boston
This volume is text- and page-identical with the hardback published in 2022.
Cover: kokouu/E+/Getty Images
Typesetting: Integra Software Services Pvt. Ltd.
Printing and binding: CPI books, GmbH, Leck

www.degruyter.com

This book is for my grandchildren Jesse, Levi, and Elise, in the hope that they grow up into a world of fairness, kindness, and (mostly) truth

Contents

Laurence Horn
Introduction: On lying and disleading

What is truth?	– Pontius Pilate, *John* 18:38
Truth is truth	– Shakespeare, *Love's Labour's Lost,* IV.ii
Truth isn't truth	– Rudy Giuliani, Meet the Press, 19 August 2018

> If, like truth, lies had but one face, we would be better off, for we would take as given the opposite of what the liar would say. But the opposite of truth has a hundred thousand faces and an undefined playing field . . . There are a thousand ways to miss the mark, just one to hit it.　　　　　　　　　　– Montaigne (1580), my translation

To tell **the truth** or to tell **a lie**? The truth is singular; there is just one; lies come in packs. But **telling** the truth – much less the whole truth – is no simple matter. Historically, the injunction to tell the truth is parasitic on the injunction to avoid the false, which is perhaps an easier goal to fulfill, or at least an easier one from which to spot deviation:

> The first and most necessary area of philosophy is the one that deals with the application of principles such as "We ought not to lie."　　　　　– Epictetus, *Enchiridion* §52

> Thou shalt not bear false witness against thy neighbour.
> 　　　　　　　　　　　　– 9th Commandment, *Exodus* 20:16

> Do not say what you believe to be false.　　– first maxim of Quality, Grice ([1967] 1989: 27)

But what is it to lie? One persistent challenge in delineating the border between lying and intentionally misleading (or deceiving) is the status of true statements with false implicatures. The traditional view is illustrated by an example from Chisholm & Feehan (1977: 155): when I tell you "My leg isn't bothering me too much today" when it's not bothering me at all I may succeed in misleading or deceiving you but I do not lie. Meibauer (2005, 2014a), who subsumes true assertions that induce false implicatures under his "extended definition" of lying, borrows a couplet from Blake's 1803 "Auguries of Innocence" – *A truth that's told with bad intent/Beats all the lies you can invent* – for the title of a paper on the topic (Meibauer 2014b). For Meibauer (2014a: §4.5.3) this is an instance of "lying while saying the truth".[1] But truths told with bad intent remain truths and invented lies are still lies – which doesn't render the former automatically superior to the latter on ethical or moral

---

1 Meibauer is not alone in seeking to assimilate implied falsehoods to lies; see Mahon (2015): §1.5 and the entries in Meibauer 2019 for surveys of such arguments and Wiegmann & Meibauer 2019 for a review of the somewhat inconclusive experimental results.

Laurence Horn, Yale University, e-mail: laurence.horn@yale.edu

https://doi.org/10.1515/9783110733730-001

grounds (see Saul 2012, Timmerman & Viebahn 2019, and Chapters 1–3 in this volume on the ethical considerations). Indeed, as Blake knew, an ill-intentioned truth can do quite as much damage as any lie. While malicious gossip – e.g. A tells B that B's spouse has been flirting with C – can be either true or false, true gossip is often more effective than false gossip in causing harm, because harder to disprove.

On the traditional view, tracing back to Augustine and Aquinas and forward to contemporary perjury statutes, lying requires the saying of what one disbelieves.[2] Abstracting away from the complicating factors affecting the experimental results, the category of truthful misleading (Sorensen 2016, Reboul 2021) is crucially distinct from that of lying, a dichotomy that correlates with independently established distinctions in the theory of meaning and with established criteria for perjury (Saul 2012, Horn 2017, Weissman & Terkourafi 2019).

But what explains why so many experimental subjects are willing to waive the difference and (perversely?) assess true statements with false implicatures as lies? Nor is it just experimental subjects who have boundary issues. A standard online dictionary (https://www.merriam-webster.com/dictionary/lie) offers two quite distinct senses for the verb *lie:*

lie, v.

(1) to make an untrue statement with intent to deceive

(2) to create a false or misleading impression

Based on the discussion in Mahon 2015 and several chapters in this volume, we can recognize that sense (1) is not entirely uncontroversial, as it stipulates actual rather than believed falsity and rules out bald-faced lies, but our focus here is sense (2). We can view this as a broadening of (1), a case of speaking loosely (supported by our willingness to concede "I wasn't technically/exactly lying, but . . ."), or as an instance of what Lasersohn (1999) calls a pragmatic halo. Alternatively, we can accept this as a case of true polysemy. When Mark Twain (1880) refers to "silent lies" (a.k.a. "lies of omission") or when Clancy Martin (2016) claims that all lovers lie (i.e. hide or disguise their true feelings), they invoke Merriam-Webster's sense (2), as we do whenever we observe that statistics or appearances can lie or, on the nominal side, that one's marriage or one's whole life has been a lie.

---

2 The relevant notion of saying must be broad enough to allow for assertions (and hence lies) via sign language and local communicative conventions but narrow enough to exclude irony, sarcasm, and play-acting. Following Stokke (2013), an assertion warrants the truth of the asserted content (cf. also Carson 2006, 2010) and counts as a proposal to change the common ground.

When we turn to truth, we find another apparent polysemy and an instructive diachrony. The earliest OED cites for *true* (senses 1–3) involve personal and subjective ascriptions with positive affect, glossed variously as 'loyal, faithful, trustworthy, honest, sincere', whence "true love", "true friend", or Polonius's instruction "To thine own self be true". Only later do we find attestations of the epistemic or alethic sense (OED *true*, 4a): 'Of a statement, idea, belief, etc.: in accordance with fact; agreeing with reality; correct'. This meaning shift exemplifies the quasi-universal tendency for epistemic senses to develop out of deontic ones and not vice versa (cf. e.g. Bybee 1985, Shepherd 1982, Traugott 1989).

While *false* allows subjective readings *(false friend, false promise)* along with objective ones, *lie* – in both verbal and nominal use – has invoked negative affect from its inception, and this tendency has only increased over the centuries. Here is the OED (s.v. *lie*, n.):

> In mod[ern] use, the word is normally a violent expression of moral reprobation, which in polite conversation tends to be avoided, the synonyms *falsehood* and *untruth* being often substituted as relatively euphemistic.

Besides *falsehood* and *untruth,* a lie can come pejoratively garbed as an equivocation, a fib, a hoax, a palter, a prevarication, or a whopper, an instance of a misrepresentation or misspeaking, not to speak of fairy tales, fictions, tall tales, urban legends, or yarns. Note the absence of true antonyms for these terms (other than the "true story", which perhaps doth protest overmuch). The euphemistic flavor of these substitutions and circumlocutions attests to the moral dimension of *lie*, which is never far away. While Aquinas allowed for "jocose" lies (ironic or sarcastic utterances, not intended to deceive)[3] and altruistic lies (our "white" or "pro-social" lies), his primary focus – and ours in this volume – is on the category of malicious or self-serving lies.

Even before Donald Trump's inauguration in 2017, the question arose for journalists and their editors on the appropriateness of describing a blatantly false statement by the chief of state as a lie. Some noted that if Trump really believed a proposition, however falsified it had been, he wasn't technically lying. (This is an instance of Costanza's Law, as propounded by Seinfeld's sidekick: "It's not a lie if you believe it".) Others argued that when the content of

---

3 On the standard view, dating back to Kant (see Sorensen's synopsis in Chapter 3), an intentionally false statement can only count as a lie if there is a rational expectation, or warrant, of truth. See also Boogaart, Jansen & van Leeuwen 2021 and Chapter 9 on the irony/sarcasm defense and its abuse.

the proposition is sufficiently egregious and its falsification easily demonstrated, a lie is a lie and should be so identified. Against this background, *Wall Street Journal* editor-in-chief Gerard Baker acknowledged in an interview on Meet the Press on New Year's Day 2017 his discomfort with using the word "lie" in his paper's coverage of false claims (https://tinyurl.com/ykwc25rr, 3:15–4:30 on embedded video).

> "I'd be careful using the word 'lie'," Baker stated. "*Lie* implies much more than just saying something that's false. It implies a deliberate intent to mislead . . . I think if you start ascribing a moral intent, as it were, to someone by saying that they've lied, I think you run the risk that you look like you are – like you're not being objective."

While it is true that *lie* implies more than the assessment of a statement as false and that we cannot always ascertain an intention to deceive, this is precisely what perjury law requires. Is perjury necessarily subjective? The positing of intent is also essential if we are to distinguish murder from manslaughter. How would such legal distinctions be retained if mentalist criteria cannot be invoked?

These are not easy questions, but it is intuitively plausible that the ascription of lies does indeed carry with it "a violent expression of moral reprobation" (in the OED's words) or the ascription of "moral intent" (in Baker's), even if – unlike Augustine in *De Mendacio* or Kant in *On a supposed right to lie from altruistic motives* – we stop short of indicting all lies as intrinsically immoral. This explains the plethora of euphemisms for lying, the dysphemistic extension of the category of lies to practices that technically fall short of the mark, and – more relevant for the purposes of our volume – the tendency for respondents in experimental studies to brand any statement as a lie if it is designed to deceive the recipient unjustifiably, even if the deceptive component is only implicated rather than asserted.

Many philosophers and linguists, including Tom Carson and Marta Dynel in their chapters, distinguish *deceive* from *mislead* on the grounds the former but not the latter is always intentional.[4] It is clear that misleading is unspecified for intention; what is less clear is whether *deceive* and *deception* entail speaker intent. It would be convenient if deception, unlike misleading, were necessarily intentional, and it is certainly reasonable to so stipulate, but it should be recognized that such a move contravenes ordinary usage as displayed in corpora and in standard lexicographic entries. For example, as Carson notes in Chapter 1, the

---

[4] It should be stressed that deception, even presupposing intent, is broader than and arguably also narrower than lying; see Fallis (2009), Carson (2010), and Chapter 1 in this volume for elaboration.

Shorter Oxford Dictionary defines *deceive* as to "cause to believe what is false". But causation is unmarked for intentionality or even animacy.

What of the corresponding nominal forms? Lackey (1983: 241) distinguishes *deception* from *deceit* in terms of the distinction between withholding and concealing information, while treating both as speaker-intended:

> Deception: A is deceptive to B with respect to whether p if A aims to conceal information from B regarding whether p.

> Deceit: A deceives B with respect to whether p if and only if A aims to bring about a false belief in B regarding whether p.

But while deceit may indeed be impossible to carry out inadvertently or non-agentively, deception – contra Lackey, Carson, Dynel, and others – does not require intention, at least outside of artificial stipulation. Sam mishears the weather forecast as predicting a sunny day and relays this misinformation to Joan, with the result that she believes it will be sunny. Joan is deceived in her belief and Sam has deceived her, albeit unintentionally: a case of deception but not deceit on Sam's part. Consider also the case of self-deception (cf. Deweese-Boyd 2016); it is plausible to regard most if not all instances of deceiving oneself as unintentional.

So if we cannot rely on *deceive* to pick out cases of intentional misleading (including but not limited to actual lies), where does that leave us? Consider a modest proposal, which will necessitate a bit of a back story. As Fallis (2006) details in his useful taxonomy, misinformation can take the form of either (i) a false message (or false information, assuming information **can** be false, which is itself a subject of dispute) or (ii) true but misleading information that tends to cause false beliefs. It can arise from an honest mistake, ignorance, unconscious bias, or intentional deception (bad faith). In the last case we are dealing with DISINFORMATION: "the species of misinformation that is intended to mislead" (Fallis 2006: 358).

Purveying disinformation is sometimes conflated with lying, given that in both cases S believes p to be false and S intends that H come to believe p. But an accidentally true statement can be a lie (à la Augustine or Aquinas) – a lie of the mind – without constituting disinformation, while instances of intentional misleading can represent disinformation while falling short of lies, e.g. the misrepresentations of political spinmeisters or Athanasius's celebrated announcement of "He is not far from here" to direct his pursuers away from his own location (Macintyre 1995: 336).

Lying as a subspecies of asserting is an instance of Grice's meaning$_{nn}$, where S counts on H's recognizing S's communicative intention, but this is not necessary to create disinformation, which can just be "out there", yielding false inferences. This is clearest in the case of non-verbal disinformation as in the cases of false displays or deceptive signaling discussed in Chapter 8: scarecrows, frogs puffing up to ward off predators, brochures for swampland in Florida.

Disinformation encompasses cases of intentional misleading as well as lying, while misinformation simpliciter may unintentionally mislead, establishing <*misinformation, disinformation*> as a privative dyad (Horn & Abbott 2012, Deo 2015), i.e. a two-item scale like <*a, the*>, <*finger, thumb*>, or <*rectangle, square*> in which the informationally stronger value is marked or specified for a feature (uniqueness/opposability/equilaterality) for which the weaker value is unmarked. The privative relation of disinformation to misinformation can be projected onto language use via a Google n-gram face-off between the two words as measured through tokens in Google Books (see Figure 1).

Note the spike in *disinformation* around 1987, during the death throes of the Soviet Union. We can begin to track an even higher spike, already underway in 2019 (the last year of included data), presumably related to certain events in the U.S. political sphere.

And now for the modest proposal: It would be convenient to map the *misinformation/disinformation* distinction onto the verbal domain. Alongside *mislead* – unspecified for intentionality – we hereby introduce a new term of art, the more specific *dislead*. We thus obtain a new privative dyad, <*mislead, dislead*> on the model of <*misinformation, disinformation*> or <*kill, murder*>. No need now to characterize speakers as "intentionally misleading" or guilty of "intentional deception": what they're engaged in is *disleading*.

This volume[5] is designed to provide new insights and current research on the distinction between lying and disleading in and outside the courtroom. At least since the biblical commandment against bearing false witness, approaches toward a taxonomy of falsehood have been offered by theologians, ethicists, philosophers of language, linguists, and cognitive scientists, while legal scholars have addressed the relationship between perjury and garden-variety lies, categories that overlap but in important ways remain distinct. Another issue arising in several chapters in the volume is whether a comprehensive semantic analysis might need to admit both strict and loose senses of *lie*. The chapters in this book map the landscape of falsehood by tracing the character and history of lying and its relatives in the light of linguistic, philosophical, and legal argumentation, more recently buttressed by a range of experimental methodologies.

---

5 Thanks here to Janet Giltrow and Dieter Stein for proposing this volume for their Foundations in Language and Law series and for their encouragement along the way; additional thanks to Janet for her careful reading of the chapters. I am also grateful to Kirstin Börgen, Natalie Fecher, Anne Stroka, and their colleagues at De Gruyter for all their help with the manuscript.

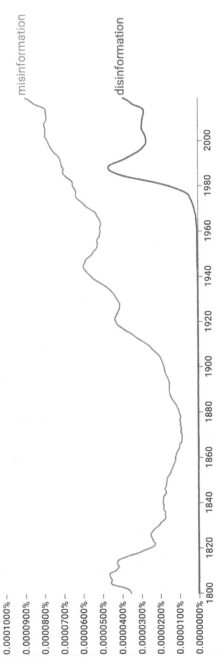

Figure 1: Google n-gram (retrieved Sept. 15, 2021), https://tinyurl.com/y5rufxvo.

One recurring focus in these chapters is the often-cited correlation between the distinction between lying and disleading on the one hand and the pragmatic distinction between what is said and what is implicated on the other (see Saul 2012). At the heart of this correlation is the determination of whether, and when, true assertions with false implicatures should be regarded as lies and whether they in fact are so regarded in naturalistic and experimental settings; these are particularly important questions with direct consequences for the characterization of perjury. But this in turn presupposes the prior establishment of what it is to assert, by no means a straightforward matter. This volume addresses the central issues in the ongoing effort to discern the properties of lying and its relation to disleading – intentional misleading and deception – as well as the relation between lying per se and perjury within the law.

The papers in Part I address foundational issues, the historical and contemporary landscape of the study of mendacity and misinformation. Since Augustine and Thomas Aquinas, questions relating to the definition of and responsibility for the central concepts of lies and truth-telling have been situated at the heart of ethics, speech act theory, and the study of meaning. Tom Carson wrote *the* book on lying and deception (Carson 2010) and has been at the heart of research in this area in the 21st century. He leads off this volume with a comprehensive look in Chapter 1 at the central concepts – falsehood, deception, half-truth, bullshit, withholding of information – and their relation to the traditional characterization of what it is to lie. Bill Lycan, an advocate of philosophical realism, has written extensively on topics ranging from mental states and epistemic justification to conditionals and presuppositions. The ethical dimensions of true but deceptive assertions, one of the questions explored in Carson's essay, constitute the theme of Lycan's contribution in Chapter 2, which focuses on the implications for theories of ethics and moral philosophy of those varieties of deception that fall short of lying. The absolutist position on the ethics of lying has long been associated with Kant, whose makes the case in his 1799 treatise "On a supposed[6] right to lie from altruistic motives" that while deception may sometimes be allowed, lying per se – including paternalistic or "white" lies – is always immoral. This is a point addressed in both Lycan's chapter and Roy Sorensen's, but Sorensen's focus in Chapter 3 is on a different aspect of Kant's thought. Sorensen, the author of seven books on metaphysics, epistemology, and philosophy of language, shows that for Kant, the ordinary run-of-the-mill *a posteriori* lie is indeed the responsibility of the perpetrator, but victims of *a priori* deception have only themselves to

---

6 Note the quantity implicature: a "supposed" (or alleged, or putative, or purported) right, or in the German original a *"vermeintes" Recht*, is likely to be understood as a not quite actual right.

blame. Indeed, since the recipient has no rational expectation of truth in such cases, *a priori* lies aren't really lies at all.

The contributions in Part II, by experimental psycholinguists based in Germany, Poland, and the Netherlands, present empirical evidence bearing on questions of lying, deception, ethics, and speaker commitment. Alex Wiegmann and Neele Engelmann have contributed extensively to the burgeoning literature of experimental studies helping to establish the boundaries of the categories of lies and disleading (intentional deception). The constructed scenarios in the study they present in Chapter 4 are aimed at evaluating hypotheses on whether ordinary (non-lying) deception is morally preferable to lying as well as examining the contextual factors that lead subjects to view a given utterance as a lie. Ronny Boogaart, Henrike Jansen, and Maarten van Leeuwen, part of a team of experimental pragmaticists and argumentation scholars at Leiden that has been investigating a range of defense strategies for denying speaker commitment (cf. Boogaart, Jansen & van Leeuwen 2021), focus in the study in Chapter 5 on the employment (especially by politicians) of one particular strategy, that of rejecting responsibility for troubling remarks on the grounds that the words at issue were not one's own – "I was only quoting . . ." Another form of indirect communication – and deception – employs memes on social media, as explored in Marta Dynel's contribution. Dynel is an international authority on (c)overt untruthfulness, irony, and humor. Her study of multimodal deception in Chapter 6, illustrated with posted memes involving the wearing (or non-wearing) of Covid masks, shows how the practice of reposting can shed light on the complex epistemology of deception, and in particular on the subtle but significant shift from simple "innocent" misleading to true disleading in contexts of humorous communication on social media.

Part III is devoted to a variety of devices for almost but not quite lying, employing the satellites of deception: bullshit, bluffing, puffing, and (purported) sarcasm. In the earlier literature, "non-deceptionists" like Roy Sorensen and Tom Carson have argued that, contra the traditional definition of lying, lies are not necessarily intended to deceive, given the possibility of "bald-faced" lies. In Chapter 7, the philosophers of language Tim Kenyon and Jennifer Saul, who have written extensively on deception and bias, extend the practice of bald-faced lying to the perpetration of "bald-faced bullshit" and tackle the puzzle of why bullshit regularly dispensed by authoritarian politicians like Donald Trump and Boris Johnson remains effective with their base of supporters despite its transparent inconsistency with facts.[7]

---

7 Unsurprisingly, Donald Trump is repeatedly subpoenaed as a star witness throughout our volume – in seven chapters, to be precise, exceeding the six appearances of both his predecessor Bill Clinton and the egregiously crafty Samuel Bronston.

Kenyon and Saul explain this phenomenon and its social implications by invoking the audience-relative character of bullshit. Overlapping with, but not reducible to, bullshit is the practice of bluffing, another form of disleading or intentional misdirection as discussed by Larry Horn in Chapter 8. Horn, a longtime specialist in both pragmatics and poker, surveys bluffing as it extends from social venues – card-playing, business transactions, courtroom ploys – to non-linguistic and even non-human interactions. Bluffing is a high-stakes move that may or may not succeed: bluffers risk the danger of collapse if their bluff is called, while potential bluffees encounter risks of their own, regardless of venue, if they call injudiciously. In Chapter 9, Liz Camp, the dean of insinuation (cf. Camp 2018), investigates the use of indirection, insinuation, sarcasm, and other off-record devices employed to avoiding direct accountability in coercive utterances. She reveals interlocutors' contortions to "avoid catching the conversational hot potato" and draws on an extensive inventory of real-life skullduggery to show that there is no carte blanche for securing plausible deniability. The key standard for judging speech products from threats to epithets and racial jokes, Camp concludes, is determined by our expectations for a reasonable speaker or hearer.

In Part IV we move from lying and deception in ordinary life to the corresponding practices in the legal and forensic domains. When is a lie not just a lie but an actual instance of perjury? When is deception that falls short of lying nevertheless legally actionable as fraud or worse? Roger Shuy, who has testified as an expert linguistic witness in over 50 trials and published over a dozen books on such topics as deceptive ambiguity, language crimes, and the language of perjury, fraud, bribery, and murder, presents an overview of the language of perjury statutes in Chapter 10, before turning to a detailed analysis of the role of the linguist as an expert witness (welcome or unwelcome) in four perjury cases. Saul Kassin is a psychologist widely recognized for his investigations of techniques including bluffing, intimidation, and the use of fake evidence by police and prosecutors leading to false confessions. In Chapter 11, Kassin demonstrates interrogators' techniques in manipulating suspects through the use of implicit promises and threats into confessing to crimes they did not commit. As Kassin explains, juries tend to accept these subtly coerced confessions as clear indicators of guilt, resulting in wrongful convictions. The following two chapters offer complementary empirical studies that examine the factors influencing judgments of lying vs. simple disleading inside and outside the courtroom context. In Chapter 12, the Poland-based linguistic and legal scholars Iza Skoczeń and Aleksander Smywiński-Pohl describe a series of experiments indicating that inferential judgments (lie vs. non-lie) in the courtroom are governed by the strategic context of communication and the presence or absence of objective truth, rather than the speaker's knowledge or assumed intention to deceive. As the authors point out, the threshold for perjury is treated as higher than that for lying per se even within the courtroom setting. Ben Weissman, who works at the intersection of pragmatics, cognitive science, and experimental linguistics, addresses a similar set of questions in Chapter 13: What counts as a lie in the courtroom and out, and what factors influence this judgment? Weissman reconsiders the interpretation of true sentences with false implicatures (investigated earlier by Weissman and Terkourafi 2019) and focuses on the role of discourse genre – casual conversation, formal speeches, testifying under oath – in judgments of lying vs. disleading, as well as the role of implicature type – generalized vs. particularized. This experimental study provides additional evidence to assess earlier proposals (Meibauer 2005, Saul 2012, Horn 2017) on the relation between false implicatures and lies.

Chapter 14 is the unique contribution that is not new to the volume. For the last several decades, Larry Solan – linguist, legal scholar, and the Don Forchelli Professor of Law and Director of Brooklyn Law School's Center for the Study of Law, Language and Cognition and frequent visiting professor at the Yale Law School – has been the leading authority on language, law, and cognition. His trailblazing publications include *The Language of Judges, The Language of Statutes*, and (with the late Peter Tiersma) *Speaking of Crime* and *The Oxford Handbook of Language and Law*. His 2018 paper "Lies, deceit, and bullshit in law" helped inspire the creation of this volume, and I am delighted to be able to include a revised version of that essay here as Chapter 14. Zei gezunt, Larry.

References

Boogaart, Ronny, Henrike Jansen, and Maarten van Leeuwen. 2021. "Those are your words, not mine!" Defense strategies for denying speaker commitment. *Argumentation* 35: 209–235.

Bybee, Joan. 1985. *Morphology: A Study of the Relation between Meaning and Form*. Amsterdam: Benjamins.

Camp, Elisabeth. 2018. Insinuation, common ground, and the conversational record. In Daniel Fogal, Daniel W. Harris, and Matt Moss (eds.), *New Work on Speech Acts*, 40–66. Oxford: Oxford University Press.

Carson, Thomas. 2006. The definition of lying. *Noûs* 40: 284–306.

Carson, Thomas. 2010. *Lying and Deception: Theory and Practice*. Oxford: Oxford U. Press.

Chisholm, Roderick and Thomas Feehan. 1977. The intent to deceive. *Journal of Philosophy* 74: 143–59.

Deo, Ashwini. 2015. The semantic and pragmatic underpinnings of grammaticalization paths: The progressive to imperfective shift. *Semantics & Pragmatics* 8, Art. 14:1–52.

Deweese-Boyd, Ian. 2016. Self-deception. In Ed Zalta (ed.), *Stanford Encyclopedia of Philosophy*. https://plato.stanford.edu/entries/self-deception/.

Fallis, Donald. 2006. Mis- and dis-information. In Luciano Floridi (ed.), *Routledge Handbook of Philosophy of Information*, 357–371. New York: Routledge.

Fallis, Donald. 2009. What is lying? *Journal of Philosophy* 106: 29–56.

Grice, H. P. 1989. *Studies in the Way of Words*. Cambridge: Harvard University Press.

Horn, Laurence. 2017. Telling it slant: Toward a taxonomy of deception. In Dieter Stein and Janet Giltrow (eds.), *The Pragmatic Turn in Law*, 23–55. Berlin: de Gruyter.

Horn, Laurence and Barbara Abbott. 2012. <the, a>: (In)definiteness and implicature. In William Kabasenche, Michael O'Rourke, and Matthew Slater (eds.), *Reference and Referring: Topics in Contemporary Philosophy, vol. 10*, 325–55. Cambridge, MA: MIT Press.

Lackey, Jennifer. 2013. Lies and deception: an unhappy divorce. *Analysis* 73: 236–48.

Lasersohn, Peter. 1999. Pragmatic halos. *Language* 75: 522–551.

MacIntyre, Alasdair. 1995. Truthfulness, lies, and moral philosophers: what can we learn from Mill and Kant? *The Tanner Lectures on Human Values* 16: 307–361.

Mahon, James. 2015. The definition of lying and deception. In Ed Zalta (ed.), *Stanford Encyclopedia of Philosophy*. http://plato.stanford.edu/entries/lying-definition.

Martin, Clancy. 2015. *Love and Lies: An Essay on Truthfulness, Deceit, and the Growth and Care of Erotic Love*. New York: Farrar, Straus, and Giroux.

Meibauer, Jörg. 2005. Lying and falsely implicating. *Journal of Pragmatics* 37: 1373–99.

Meibauer, Jörg. 2014a. *Lying at the Semantics-Pragmatics Interface*. Berlin: De Gruyter.

Meibauer, Jörg. 2014b. A truth that's told with bad intent: Lying and implicit content. *Belgian Journal of Linguistics* 28: 97–118.

Meibauer, Jörg (ed.). 2019. *The Oxford Handbook of Lying*. Oxford: Oxford University Press.

de Montaigne, Michel. 1580. Des menteurs [On liars], *Essais*, Book I, Chapter 9. https://short-edition.com/fr/classique/michel-de-montaigne/des-menteurs

Reboul, Anne. 2021. Truthfully misleading: Truth, informativity, and manipulation in linguistic communication. *Frontiers in Communication*, Article 646820, 6: 1–8.

Saul, Jennifer. 2012. *Lying, Misleading, and What is Said*. Oxford: Oxford University Press.

Shepherd, Susan. 1982. From deontic to epistemic: an analysis of modals in the history of English, creoles, and language acquisition. In Anders Ahlqvist (ed.), *Papers from the 5th International Conference on Historical Linguistics*, 316–323. Amsterdam: Benjamins.

Sorensen, Roy. 2016. *A Cabinet of Philosophical Curiosities*. Oxford: Oxford University Press.

Stokke, Andreas. 2013. Lying and asserting. *Journal of Philosophy* 110: 33–60.

Timmermann, Felix and Emanuel Viebahn. 2021. To lie or to mislead? *Philosophical Studies* 178: 1481–1501.

Traugott, Elisabeth Closs. 1989. On the rise of epistemic meanings in English: an example of subjectification in semantic change. *Language* 65: 31–55.

Twain, Mark. 1880. On the decay of the art of lying. http://www.gutenberg.org/ebooks/2572.

Weissman, Benjamin and Marina Terkourafi. 2019. Are false implicatures lies? An experimental investigation. *Mind and Language* 34: 221–246.

Wiegmann, Alex and Jörg Meibauer. 2019. The folk concept of lying. *Philosophy Compass* 14: e12620.

I Lies and deception: The landscape of falsehood

Thomas L. Carson
Lying, deception, and related concepts: A conceptual map for ethics

Abstract: Roughly, deception is intentionally causing someone to have false beliefs. I argue that there is a limited range of reasonable views about the definition of lying and defend the following claims on which all of the reasonable definitions agree:

> A necessary condition for telling a lie is that one makes a statement or assertion that is insincere, namely, a statement that one believes to be false or doesn't believe to be true. (One can't lie unless one makes an insincere statement.)

> A sufficient condition for telling a lie is that one makes a statement that: a. is false, b. one believes is false, c. one makes with the intention to deceive others, and d. one makes in a context in which one implicitly or explicitly assures others that what one says is true. (Any statement that satisfies conditions a-d is a lie.)

I explain the differences between lying, deception, and withholding information and propose a distinction between misleading and deceiving. Then I explain Frankfurt's concept of bullshit (which involves indifference to the truth of one's statements), the concepts of spin and partial truths, and the distinction between the truth and truthfulness of statements. I conclude by discussing two other important concepts: preventing people from acquiring true beliefs and undermining knowledge and trust in reliable sources of information.

We need to be clear about what we mean whenever we claim that someone performed acts of lying, deception, misleading, withholding information, bullshitting, spinning events, telling half-truths, etc. Any moral assessment of these actions presupposes an account of what it is to do them.

With qualifications, I define deception as intentionally causing someone to have false beliefs. I will not attempt to defend any particular definition of lying. Rather, I will identify a range of reasonable definitions and defend the following two claims on which all of the reasonable definitions agree:

1. A *necessary condition* for telling a lie is that one makes a statement that is insincere, i.e., a statement that one believes to be false (or doesn't believe to be true).

Thomas L. Carson, Loyola University of Chicago, e-mail: tcarson@luc.edu

https://doi.org/10.1515/9783110733730-002

2. A *sufficient condition* for telling a lie is that one makes a statement that: a. is false, b. one believes is false, c. one makes with the intention to deceive others, and d. one makes in a context in which one implicitly or explicitly assures others that what one says is true.

I explain the distinctions between lying and deception, lying and withholding information, lying and failing to correct mistakes, lying and false implicatures, deceiving and misleading, and the distinction between the truth and truthfulness of statements. I also discuss the concepts of bullshit, spin, half-truths, and preventing people from acquiring knowledge by undermining trust in reliable sources of information. This paper proposes a map of the relationships among many different concepts. Each of my definitions should be assessed in terms of its place in the entire account.

1 Lying

Consider the following definition of lying:

> A lie is a statement that is 1. false, 2. insincere (a statement that the speaker believes is false[1]), and 3. is intended to deceive others.

Each of these three conditions is included in some of the definitions of lying that can be found in dictionaries and the work of philosophers and linguists.

Lying requires the use of language. In order to lie, one must make a statement or assertion – it is impossible to lie without making a statement (cf. Fallis 2009: 37–38). A person who nods or shakes her head to answer "yes" or "no" to a question uses language to make a statement. If you deceive another person without using language, for example, if you successfully fake a pass in basketball, you are not lying. Lies must be insincere statements: at a minimum if I believe that what I am saying is true, I can't be lying.[2]

Saint Augustine's definition of lying resembles the foregoing definition, except that it does not include the requirement that lies must be false statements. Augustine defines lying as follows:

---

1 Most people state the insincerity condition in this way, but we might broaden this condition to read "a statement that the speaker doesn't believe is true." In Lying and Deception I am neutral between these two ways of stating the insincerity condition (see Carson 2010: 17–18).
2 On this point see Fallis (2009: 38) and Stokke (2018: 44); for objections to this view see fn. 14.

To lie is to make a statement that one believes to be false (or doesn't believe to be true) with the intention of causing others to have false beliefs.[3]

Most philosophers who write about lying think that true statements can be lies. They favor Augustine's definition over the initial definition stated above, and something resembling Augustine's definition is the closest thing to a standard definition among philosophers.[4] On the other hand, US federal law assumes that lies must be false statements. It defines perjury as lying under oath and holds that statements can't constitute perjury unless they are false.[5]

---

3 In the Enchiridion, he defines lying as follows: "every liar says the opposite of what he thinks in his heart, with purpose to deceive" (1961: 29). He gives essentially the same definition in "On Lying" (1965: 54–56).
4 On this point, see Mahon 2015.
5 Digressive footnote. In the case of Bronston v. The United States, the U.S. Supreme Court ruled that a witness cannot be convicted of perjury if what she says is true (Tiersma 2004: 940; see also Chapter 14). President Bill Clinton appealed to this principle when he defended himself against charges of lying and perjury. During his deposition in the Paula Jones case, he was asked "Did you have an extramarital sexual affair with Monica Lewinsky?" He answered "no." Later he admitted that he had engaged in "inappropriate intimate conduct" with Lewinsky. But he insisted that he hadn't lied earlier (Tiersma 2004: 941). Common usage differs on the question of whether engaging in oral sex constitutes "having sex." In a 1999 study of the ordinary usage of the term "having sex" involving 600 college students, slightly more than 40% of them said that having oral sex constituted having sex (Tiersma 2004: 944). It seems to be debatable whether or not what Clinton said on this occasion was false, therefore, it is debatable whether or not he committed perjury.

Later, when he appeared before the Starr grand jury, Clinton was asked whether he had had "sexual relations" with Lewinsky. The following definition of "sexual relations" was stipulated for purposes of answering the question: "[A] person engages in "sexual relations" when the person knowingly engages in or causes . . . [1] contact with the genitalia, anus, groin, breast, inner thigh, or buttocks of any person with an intent to arouse or gratify the sexual desire of any person . . . "Contact" means intentional touching, either directly or through clothing (Tiersma 2004: 946–947).

Clinton testified that he did not have sexual relations with Monica Lewinsky. Later, during his impeachment trial, he admitted that his testimony was deceptive and misleading, but he claimed that his statement was not a lie and not a case of perjury, because it was true. He claimed that, given the definition of having sexual relations stipulated for his testimony, he did not have sexual relations with Monica Lewinsky. He was not trying to give her sexual pleasure and he did not touch any of the specified parts of her body to sexually arouse her or give her sexual pleasure. However, Clinton to the contrary, his actions with Monica Lewinsky satisfied the conditions of this definition of "sexual relations." He initiated contact between the genitalia of a person (himself) and Monica Lewinsky in order to arouse and gratify the sexual desire of a person (himself) – note the wording of the definition of "sexual relations" – "any person."

On another occasion, Clinton was asked whether he had a sexual relationship with Monica Lewinski. He answered "There is no relationship." His answer was true, but deceptive (there had been a sexual relationship, but it had ended before he said this).

William Lycan (2006: 165) reports a study which found that roughly 40% of college students think that a lie must be a false statement, 40% don't think that a lie must be a false statement, and 20% aren't sure. This survey tested students' reactions to a true statement that the speaker believed was false and made with the intention of deceiving others. Students were asked whether or not this statement was a lie. A recent paper by Turri and Turri involves a similar kind of study. The results of this study support the view that most English-speakers use the word "lie" in a way that presupposes that lies must be false statements. Participants in the study were asked about the following story that featured:

> Jacob, whose friend Mary is being sought by the authorities. Federal agents visit Jacob and ask where Mary is. Mary is at the grocery store but Jacob thinks that Mary is at her brother's house . . . Jacob tells them that Mary is at the grocery store, so that what he says is true despite his intention. (Turri and Turri 2015: 162)

Turri and Turri asked test takers to choose between the following four ways of describing the case (so that answering that "Jacob lied" or "Jacob didn't lie" were not the only options):

> Participants were then asked to choose the option that best described Jacob when he spoke to the agents about Mary's location: (1) he tried to tell the truth and succeeded in telling the truth; (2) he tried to tell the truth but failed to tell the truth; (3) he tried to tell a lie but failed to tell a lie; (4) he tried to lie and succeeded in telling a lie. (Turri and Turri 2015: 164)

88% of the subjects "said that Jacob tried to tell a lie but failed". Presumably, those subjects think that Jacob failed to tell a lie because what he said wasn't false.[6]

1.1 Reasons to think that lying does not require the intention to deceive others

Contrary to what the two forgoing definitions (and most other traditional definitions of lying) contend, I believe that lying does *not* require that the liar intends to deceive others. Consider the following example.

A college Dean is cowed whenever he fears that someone *might* threaten a lawsuit, and has a firm but unofficial policy of never upholding a professor's charge

---

[6] Turri and Turri also asked their subjects about this case by giving them just two possibilities to consider: "Jacob lied about Mary's location" and "Jacob told the truth about Mary's location." Subjects were asked to indicate their level of agreement or non-agreement with these statements. When their options were restricted in this way, most of the subjects agreed that Jacob lied (Turri and Turri 2015: 163).

that a student cheated on an exam unless the student confesses to having cheated. The Dean is very cynical about this and believes that students are guilty *whenever* they are charged. A student is caught in the act of cheating on an exam by copying from a "crib sheet." The professor fails the student for the course and the student appeals the professor's decision to the Dean who has the ultimate authority to assign the grade. The student is privy to information about the Dean's *de facto* policy and, when called before the Dean, he (the student) affirms that he didn't cheat on the exam. He claims that he inadvertently forgot to put his "review sheet" away when the exam began and that he never looked at it during the exam. The student says this on the record in an official proceeding and thereby warrants the truth of statements he knows to be false. He intends to avoid punishment by doing this. Even if he has no intention of deceiving the Dean that he didn't cheat, he is lying. If he is really hard-boiled, the student will take pleasure in thinking that the Dean knows that he is guilty. An objector might say that surely the student intends to deceive *someone* – his parents or future employers. However, this is not the case. The student in my example doesn't care whether or not others know that he cheated – he simply wants to have his grade changed. If it helps, suppose that the will of a deceased relative calls for the student to inherit a great deal of money if he graduates from the college in question with a certain grade-point average. Since the student will receive the money whether or not he deceives anyone, and since he knows that what he says won't cause anyone to have false beliefs, he lies even though he clearly doesn't intend to deceive anyone (from Carson 2010: 21).[7] My

---

7 Setting the record straight. My book Lying and Deception presents three counter-examples to the view that lying requires the intention to deceive: the case of the cheating student, the case of the frightened witness (who lies under oath for fear of being killed by the accused, but who neither hopes nor intends that his testimony will deceive anyone), and the case of a person who lies in order to keep a very solemn promise that he "made on his mother's grave" never to publically acknowledge the fraud committed by a close family member (Carson 2010: 21–23). The case of the cheating student, and the frightened witness are often discussed in the literature, but the last case, which I regard as my best and strongest example, is rarely if ever discussed. Here is that case:

> Suppose that while working in his office, I happen upon evidence that my uncle perpetrated large scale fraud in his capacity as a financial advisor. I ask him about this, and he admits to having committed fraud. He calls in my brother and sister who also work in the office and know about the fraud. My uncle then tells us the whole story but asks us to solemnly swear on our mother's grave that we will never tell anyone else or speak to anyone else (anyone other than the four of us) about this. We all swear to never mention or reveal any of this to anyone else.
>
> After my uncle's death there is a lawsuit against his estate by the victims of his fraud. There is conclusive evidence of his fraud. The evidence includes the testimony of numerous people (including my brother and sister), secret records of the funds he stole, and records of

arguments and similar ones by Roy Sorensen[8] have convinced many people, including Fallis, Saul, and Stokke, that lies needn't be intended to deceive others or require any kind of deceptive intent. But many others, including Lackey, Mahon, Dynel, and Meibauer, are not persuaded. (Figures 1, 2, 4, and 5 illustrate the differences between definitions of lying which require the intent to deceive and those which don't.)

How should we define lying if lying doesn't require the intent to deceive others? It is not enough to remove the intent to deceive condition from the dictionary definition or Augustine's definition. The definitions that result if we do this are much too broad, and count as lies things that are said sarcastically or in jest, e.g., "Hi, I'm George Washington" as said by your friend who comes to your door in a Halloween costume. We need to replace the intent to deceive condition with something else. I think that the best condition to add is this: one's insincere statement (or one's false and insincere statement) must be stated in a context in which one gives others an assurance of its truth and invites others to rely on it. Lies must be statements made in situations in which the speaker implicitly or explicitly gives others an assurance that what she is saying is true. When we lie, we violate an implicit promise or guarantee that what we say is true.[9] This makes sense of the common view that lying

---

secret bank accounts my uncle created to hide the money. There is also conclusive evidence that I knew about the fraud – a handwritten letter from me to my uncle, the testimony of my brother and sister, and wire tapped phone conversations between my siblings and me.

Under oath in court, I am asked if I knew anything about his fraud and whether I ever came across evidence that he committed fraud. Since I believe that I am morally bound by my oath to my uncle (but not by my oath to the court) I deny any knowledge of his fraud and claim that, to the best of my knowledge, no fraud ever occurred. My statement is a lie. What I say is false, I know that it is false, and, since I make the statement under oath, I strongly warrant its truth.

My false testimony is not intended to deceive anyone about matters relating to the lawsuit. I know that my testimony will not cause anyone to believe that my uncle is innocent. Nor do I intend or hope to deceive anyone about what I believe or about anything else. My only intention in this case is to remain faithful to my oath to my uncle. My motives for my actions have been revealed by the testimony of my brother and sister, and I expect everyone to believe their testimony about the time when my uncle told us of his fraud and made us swear not to tell others about it. I am quite happy if everyone knows the whole truth (I think that others will respect my true motives, and I have no fear of being charged with perjury for my testimony). Given all of this, I do not expect or intend my testimony to deceive anyone about anything.   (Carson 2010: 21–23)

8 Many of those who think that there are cases of lying without the intent to deceive follow Sorensen in calling them "bald faced lies."

9 Others who accept this are Saul (2012: 3, 18); Ross (1930: 21); Fried (1978: 67); and (though this is less clear) Hartman (1975, volume II: 286). But Stokke, Fallis, and Sorensen, who agree with me that the intention to deceive is not necessary for lying, do not endorse this condition.

involves a breach of trust. To lie is to invite others to trust and rely on what one says by warranting its truth, while, at the same time, making oneself unworthy of that trust by making a statement (or a false statement) that one believes to be false.

In defense of this condition, I appeal to what linguists call the "transparency thesis," which says that, normally, claiming that what you are stating is true is redundant – the default is that when you make a statement you are saying that it is true and giving others an assurance that it is true. In ordinary contexts, stating "The sky is blue" is equivalent to stating "It's true that the sky is blue." Not all uses of language involve this implicit assurance of the truth of what one says. When telling a joke, or when speaking on April Fools' Day, or when writing a work of fiction, the normal understanding that one is giving others an assurance of the truth of what one says does not hold (for more on this see Carson 2010: 24–25).

1.2 The preferred definition of lying

My preferred definition of lying is the following:

> 1. A person tells a lie provided that: a. she makes a false statement, b. her statement is insincere in that she believes that it is false, and c. she makes her statement in a context in which she implicitly or explicitly assures others that what she says is true.

However, I haven't shown that lies must be false statements[10] and, therefore, can't claim that 1 is preferable to the following:

> 2. A person tells a lie provided that: a. she makes a statement that is insincere in that she believes that it is false, and b. she makes her statement in a context in which she implicitly or explicitly assures others that what she says is true.

Since it is at least debatable whether lies must be intended to deceive others, the following may also be defensible definitions of lying:

> 3. A person tells a lie provided that: a. she makes a false statement, b. her statement is insincere in that she believes that it is false, and c. she intends that her statement will deceive others.

> 4. A person tells a lie provided that: a. she makes a statement that is insincere in that she believes that it is false, and b. she intends that her statement will deceive others.

---

10 The results reported by Turri and Turri provide some support for this view, but their findings have to be weighed against the strong, but not unanimous, consensus among philosophers that falsity is not necessary for lying.

The question of whether lies must be intended to deceive others is distinct from the question of whether lies must involve giving a warranty or assurance of the truth of what one says. There are reasons to think that lying requires giving an assurance or guarantee of the truth of what one says that are independent of the success of my arguments for thinking that the intent to deceive is not necessary for lying (see my discussion of the "transparency thesis" above and in Carson 2010: 24–29). It is possible that lying requires *both* intending to deceive others and warranting the truth of what one says. Therefore, the following two definitions should also be considered defensible:

> 5. A person tells a lie provided that: a. she makes a false statement, b. her statement is insincere in that she believes that it is false, c. she makes her statement in a context in which she implicitly or explicitly assures others that what she says is true, and d. she intends that her statement will deceive others.

> 6. A person tells a lie provided that: a. she makes a statement that is insincere in that she believes that it is false, b. she makes her statement in a context in which she implicitly or explicitly assures others that what she says is true, and c. she intends that her statement will deceive others.

It is a matter of controversy whether or not lies must be false statements. It is also debatable whether or not all lies must be intended to deceive others and whether they must involve giving an assurance of the truth of what one says. However, it is clear that any statement that satisfies all four of the following conditions is a lie: a. the statement is false, b. the person who makes the statement believes it is false, c. the person makes the statement with the intent to deceive others, and d. the person makes the statement in a context in which she implicitly or explicitly assures others that what she says is true.[11] All plausible definitions of lying agree about that – they agree that satisfying conditions a-d is *sufficient* for telling a lie. This is a significant result for purposes of

---

[11] I take the claim that a-d are sufficient for lying to be consistent with Stokke's definition of lying. Stokke defines lying as making an insincere assertion, an assertion one believes to be false, Stokke (2018: 5). According to Stokke, making an assertion "should be understood as saying something and thereby proposing that it become part of the background information that is taken for granted for the purpose of the conversation" (Stokke 2018: 6). Even though my conditions a-d say nothing about proposing that anything be background information or "common ground," I take it that, in any case in which I warrant something as true, then I am proposing that it be taken as true or on the record for the purposes of discussion or inquiry. Stokke would agree. He takes my idea that liars "go on the record" when they warrant the truth of what they say (see my discussion of the cheating student case above) to be equivalent to the view that liars propose that what they say be taken as common ground or background for purposes of conversation. Speaking with reference to my case of the cheating student, he writes:

applied ethics, since a great many cases that are clear examples of lying satisfy all of these conditions. All plausible definitions of lying also agree that a *necessary condition* for telling a lie is that one makes a statement or assertion that is insincere, i.e., a statement that one believes to be false.[12]

According to definitions 1, 2, 5, and 6, the truth of statements is warranted to varying degrees in different situations. Whether or not giving an assurance of the truth of what one says is necessary for lying, giving others a very strong assurance of the truth of what one says tends to make lies worse (because it involves a greater breach of trust). This is important for understanding the case of George Bush and Dick Cheney and the 2003 Iraq War. Bush and Cheney repeatedly said that it was "certain" that Iraq possessed weapons of mass destruction and that it was "certain" that Iraq was actively seeking to acquire nuclear weapons. They asked their country to go to war on the strength of such claims.[13]

1.3 Lying and knowingly failing to correct honest mistakes

In a casual conversation you ask me "have you ever been to Cleveland before?" I answer "no," thinking that what I am saying is true. Then, 30 minutes later during our conversation, I recall that I went to a baseball game in Cleveland with my grandfather in 1965. The conversation has turned to a very different topic and there seems to be no point in correcting my earlier mistake. I don't correct my earlier statement and let my answer "stand on the record." Because my statement was sincere when I made it and because I had no intention of misleading you when I said it, I wasn't lying or trying to deceive you. But suppose that I recall having gone to the baseball game in Cleveland *immediately* as I answer "no" and do not correct my answer. This is arguably a lie. Consider another kind of case. Suppose that one's unintentionally mistaken statement is about a very important matter. For example, a leader says "it is *certain* that Iraq is trying to build an atomic bomb" and urges his country to go to war with Iraq

---

The shared intuition about the case of the cheating student is that the reason the student makes her utterance – despite the fact that both she and the Dean know full well that it is false – is that she wants to "go on the record." This idea lends itself to be explained in terms of the common ground. Namely, to say that the student wants to go on the record is just to say that the student wants it to be common ground that she did not cheat.

(Stokke 2018: 52)

12 But we might want to revise this and say instead that making a statement that is insincere in that one doesn't believe it to be true is necessary for lying, see footnote 1.

13 See Carson (2010: 216–218; 2019: 545–547) for discussions of this.

on that basis. Then, later, the leader discovers evidence that his statement is not true. The leader fails to correct his earlier statement and lets it "stand on the record." This isn't a case of lying or deception at the time he makes the statement, but when he knowingly lets his past false statements "stand on the record" and doesn't correct them, this is tantamount to deception. Claims to the effect that Iraq was actively attempting to acquire nuclear weapons made by Bush and Cheney prior to the 2003 Iraq War fall under this category. Bush and Cheney made a number of false statements about Iraq's alleged aim of acquiring nuclear weapons with the intention of leading the United States into a war. Later, they received clear evidence that some of their earlier statements were untrue. When they failed to correct their earlier mistakes and let them "stand on the record," this was deception and when they continued to repeat some of these statements they were lying (see Carson 2010: 216–218).

1.4 Lying and false implicatures

Suppose that you are gossiping about Mr. Smith and suggesting that he is a philanderer. You truly say "I saw Mr. Smith in an amorous embrace with a very attractive woman in a hotel room in New York last night" but fail to add that you know that the woman in question was his wife. (You know that the person to whom your statement is addressed believes that Smith's wife was in France last night and will take your statement to be evidence of Mr. Smith's marital infidelity.) Clearly, you are trying to deceive the other person. But are you lying? Jörg Meibauer says that this is a lie because it implies something that the speaker does not believe, namely that Mr. Smith had sex with a woman other than his wife last night. In this respect, Meibauer's definition of lying is much broader than standard definitions; he rejects the view that lies must be insincere statements (see Meibauer 2014: 100–103, 125). In a forthcoming coauthored paper, he defines lying as follows:

> A lies to B if and only if there is a proposition p such that 1. A asserts that p to B, and 2. A believes that p is false or there is an implicature q such that
>
> 1. A implicates that q to B, and 2. A believes that q is false.
> (Wiegmann, Willemsen & Meibauer 2021: 18–19)

Given the firmness with which the great majority of those who write on this topic believe that lies must be insincere statements, we should retain the standard way of stating the insincerity condition for the definition of lying but carefully mark the category of statements which the speaker believes to be true but intends to deceive others about something by means of false implicatures.[14] Arguably, they are as bad as most lies and worse than obvious lies that are not intended to deceive others (see Sorensen 2007: 62–63).

2 Deception

I now turn to the concept of deception. A rough definition is as follows: deception is intentionally causing someone to have false beliefs.[15] This definition is mistaken. Intentionally causing someone to believe something that is false isn't always a case of deception. Suppose that Bob is not at home and I intentionally cause you to believe that he is at home. I have intentionally caused you to believe a statement that is false. But it is not a case of deception if I myself believe that Bob is at home. We need to revise and refine the dictionary definition. Here is my proposal:

> A person deceives another provided that she intentionally causes another person to believe something that is false and that she believes is false.[16]

Some people use the word "deception" more broadly, but this definition has the virtue of making deception a morally salient category and making sense of the strong evaluative meaning of the word "deception." It is inconsistent to both 1. use the word "deception" broadly to include cases of unintentionally causing others to have false beliefs and 2. attach strong negative evaluative meaning to deception.

---

14 However, the paper by Wiegmann, Willemsen & Meibauer (2021) discusses examples of true sincere statements that make deliberate false implicatures. They present evidence that most English speakers regard such statements as lies. I can't begin to explain or assess their arguments here. Their claim about false implicatures is consistent with my claim that satisfying conditions a-d (see above) is sufficient for telling a lie. However, if we accept their view then we will want to broaden my claim about sufficient conditions for lying as follows:

> Either of the following is sufficient for telling a lie: 1. satisfying conditions a-d, or 2. making a statement in which one knowingly implicates something that one knows (or believes) to be false.

15 The New Shorter Oxford English Dictionary (New York: Oxford University Press, 1993) defines the verb "deceive" as to "cause to believe what is false."
16 Alternatively, we might want to define deception as "intentionally causing another person to believe something that is false that one doesn't believe is true." In Lying and Deception I am neutral between these broader and narrower definitions of deception (Carson 2010: 48–51).

2.1 Deception vs. misleading

We need to distinguish between deceiving someone and misleading someone. I can't deceive you without misleading you, but I can mislead you without deceiving you. To mislead another person is to cause the other person to have false beliefs (whether intentionally or unintentionally). In order to deceive you, I must intentionally cause you to have false beliefs. So, roughly, to deceive someone is to mislead her intentionally. (See figures 3, 4, and 5.) This account fits well with the ordinary evaluative meaning of the words "mislead" and "deceive." The words "deception" and "deceive" are typically terms of reproach or condemnation. The word "mislead" does not imply the same kind of negative evaluation. The negative evaluative connotations of the term "deception" are often inappropriate in cases in which we unintentionally and blamelessly cause others to have false beliefs. It is possible to unintentionally or inadvertently mislead someone, but it is not possible to unintentionally deceive someone. All this notwithstanding, many people use the word "mislead" interchangeably with "deceive." Further, we sometimes speak about being deceived by natural phenomena, e.g., "don't let the sunshine deceive you, it's very cold outside." So, my distinction between deception and misleading should be taken to be a *proposal* for making our language more precise and perspicuous.

We need to distinguish between unintentionally, but negligently, misleading someone and misleading someone without negligence. Here is a case of misleading someone without negligence. A student dozing in the backrow of a lecture hall is misled by a careful and scrupulously honest lecture. The lecturer's mention of the fact that George Washington Carver attended Iowa State University causes the drowsy student to believe that President George Washington was once a student at Iowa State. In contrast, there are many cases in which people carelessly and negligently mislead others that fall short of deception. An example of this is a glib, careless answer to a question, an answer that one believes is true, but which is false and for which one lacks adequate evidence. Suppose that a patient asks her physician whether a drug she has been prescribed interacts badly with caffeine. On the basis of her recollection of her medical training 30 years ago, the physician quickly answers "no" and then ends the appointment abruptly in order to attend a party. Her answer is incorrect and harms her patient. The physician's statement is not a case of deception or attempted deception, since she believes that it is true. But she has negligently and culpably misled her patient.

In order to be negligent, statements must concern matters of importance. It can't be negligent to carelessly mislead you about something that is extremely unimportant. The standards of care one needs to take in making statements depend on context and the importance of what one says. Careless answers to important questions by overconfident teachers, physicians, lawyers, or politicians are negligent and can be reckless.

2.2 Deception vs. withholding information

There is a clear distinction between deception (or attempted deception) and withholding information. To withhold information is to fail to offer information that would help someone acquire true beliefs and/or correct false beliefs. Not all cases of withholding information constitute deception or attempted deception. A business person who withholds from his clients information about how much he paid for a product that he sells does not thereby deceive (or attempt to deceive) them about his costs. I am not deceiving you if I never reveal to you information about the most embarrassing moments of my personal life. However, withholding information can constitute deception if there is a clear expectation, promise, and/or professional obligation that such information will be provided. For example, a lawyer deceives a client if she fails to inform him that a course of action she advises him to take is illegal and will subject him to severe penalties.[17]

2.3 The difference between deception and lying

Deception differs from lying in two important respects. First, a lie must be an insincere statement. There are many cases of deception that don't involve the use of language to make statements, for example, faking a pass in a football game or wearing a fake beard, wig, and sunglasses to disguise one's identity. Further, true and sincere statements can be used to deceive others. Suppose that I am selling a used car that frequently overheats, and I am aware of the problem. You are a prospective buyer and ask me whether the car overheats. If I answer by making the true statement "I drove the car across the Mojave Desert on a very hot day and had no problems," I am not lying because my statement is true and sincere (I believe that it is true). Even though this statement is true and I believe that it is true, this happened four years ago and I have had considerable trouble with the car overheating since then. I am attempting to deceive you about the condition of the car and its problem with overheating, but I am not lying. (Figures 1 and 2 illustrate the differences between deception and lying.)

---

17 See Dynel 2020 for more on the distinction between withholding, deceiving, and lying.

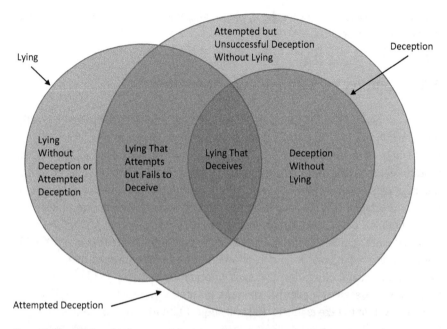

Figure 1: The relationship between lying, deception, and attempted deception on the assumption that lying does not require the intention to deceive others.

3 True vs. truthful statements

We need to distinguish between the truth or falsity of a statement and the truthfulness of the speaker. A truthful statement can be false if the speaker makes an honest mistake about the facts and says what she believes to be true and non-misleading. Similarly, one can make a true statement untruthfully if one believes that one's statement is false and misleading. Roughly, my statements are truthful just in case I believe that they are true and not misleading. My statements are untruthful provided that I take them to be false or misleading. All lies are untruthful statements, but not all untruthful statements are lies. It is often much easier to be sure about the truth or falsity of what someone says than to be sure about the person's truthfulness or untruthfulness, since that requires knowledge of the other person's beliefs and intentions (cf. Bok 1979: 7–13).

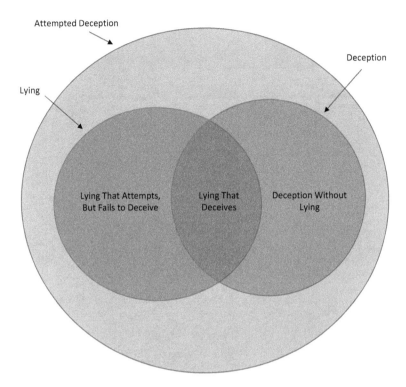

Figure 2: The relationship between lying, deception, and attempted deception on the assumption that lying requires the intention to deceive others.

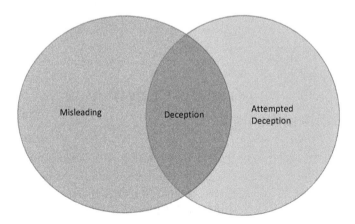

Figure 3: The relationship between misleading, deception, and attempted deception. (Note: All cases of attempted deception are also cases of attempted misleading)

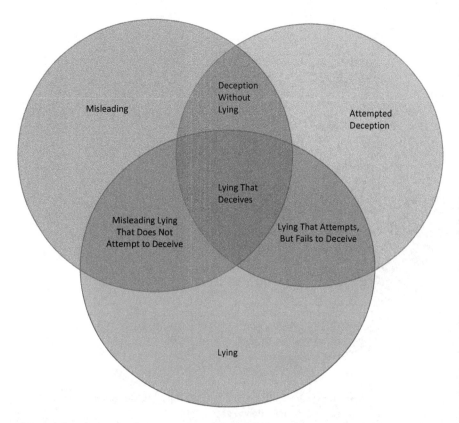

Figure 4: The relationship between misleading, deception, attempted deception, and lying on the assumption that lying does not require the intention to deceive. (The top left circle represents all cases of misleading. The top right circle represents all cases of attempted deception and the lower circle represents all cases of lying.)

4 Deceiving, lying, and Frankfurt's definition of bullshit/bullshitting

Harry Frankfurt says that bullshitting involves attempting to mislead others about oneself in a way that is short of lying. His definition of bullshit is based on Max Black's definition of "humbug," which Frankfurt (2005: 5) takes to be a weaker, more polite, synonym of "bullshit." Black (1983: 143) defines humbug as follows: "deception (deliberate deception) short of lying, especially by pretentious word or deed, of somebody's own thoughts, feelings, or attitudes".

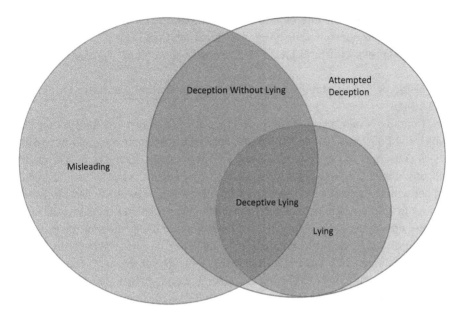

Figure 5: The relationship between misleading, deception, attempted deception, and lying on the assumption that lying requires the intention to deceive. (The left circle represents all cases of misleading. The right circle represents all cases of attempted deception and the smaller circle represents all cases of lying.)

Frankfurt uses this definition as part of his definition of bullshit; he agrees with Black that humbug (bullshit) is "short of lying" and that bullshit (humbug) involves deception, or the intent to deceive others (Frankfurt 2005: 16–19, 54), but he adds that bullshit involves indifference to the truth of what one says. Frankfurt writes:

> It is just this lack of connection to a concern with truth – this indifference to how things are – that I regard as the essence of bullshit. (Frankfurt 2005: 33–34)

> The fact about himself that the bullshitter hides, on the other hand, is that the truth values of his statements are of no central interest to him . . . . the motive guiding and controlling it [his speech] is unconcerned with how the things about which he speaks truly are . . . . his eyes are not on the facts at all . . . He does not care whether the things he says describe reality correctly. (Frankfurt 2005: 55–56)

Frankfurt says that bullshit is "produced without concern for the truth" (Frankfurt 2005: 47). If we combine this with his earlier claims, Frankfurt's definition of bullshit comes to the following: attempting to misrepresent oneself short of lying in a way that exhibits indifference to the truth (of what one says).

Elsewhere (Carson 2010: 58–61; Carson 2016: 56–65) I have argued that Frankfurt is mistaken in the following three claims he makes about the concept of bullshit:

1. Bullshit requires the intention to deceive others.
2. Bullshit does not constitute lying (bullshit is "short of lying").
3. The essence of bullshit is lack of concern with the truth of what one says.

Criticisms by Cohen prompted Frankfurt to concede that 2 is false (Frankfurt 2002: 341).

1 and 3 are false because there are cases of open transparent bullshitting in which the bullshitter has no hope or *intention* of deceiving anyone and in which she is concerned to say only things that are true. Here, is such a case. A student who is writing a long essay on an examination knows that she will get partial credit for her writing something no matter how far off the topic it is. She writes a bullshit answer that is not intended to deceive the grader about anything, including the fact that her answer is pure bullshit. The exam question is: "Briefly describe the facts of the case of Dodge vs. Ford and answer the following question: 'Was Henry Ford morally justified in his actions in this case?' Defend your answer." The student hasn't read the case nor was she in class when it was discussed. Since she doesn't know what Henry Ford did in this case, she can't possibly formulate a coherent argument for thinking that he was or was not morally justified in what he did. So, the student bullshits and produces the following answer:

> In today's increasingly technologically sophisticated, internet-interconnected, multicultural, and multiracial post-modern society, there are many important ethical questions about the role of business in the larger society. These are important questions since business and its actions play such a large role in today's society and have such a great impact on all sectors of society. We have addressed these questions in our class. Milton Friedman is the author of *Capitalism and Freedom*, who taught at the University of Chicago, was born in 1913, won the Nobel Prize in Economics in 1976, and was a major influence on Ronald Reagan and Margaret Thatcher. Friedman holds that the only obligation of business is to make money for the shareholders, provided that it avoids fraud, deception, and unfair competition. Others say that corporations should be run for the benefit of all their "stakeholders" and give back to their communities. Utilitarians hold that corporations should promote the common good in today's society. Henry Ford and the Ford Motor company had many obligations in this case. In this case, it is clear that the obligation to society was the paramount obligation. The company failed to live up to this obligation, to an extent, but this is not a black or white issue. In any case, Henry Ford didn't adequately fulfill his duty to the public. This case has many important implications for today's society and the role of business in today's fast-paced, technologically sophisticated, post-modern, twenty-first century society. (From Carson 2016: 59)

In a slightly different version of this case, the student might be concerned with the truth of what she says. She might know that the instructor will bend over backwards to give her partial credit if he thinks that she may have misunderstood the question, but she also knows that if the things she writes are false she will be marked down. In that case, she will be very careful to write only things that are true and accurate and that she learned in the class, even though she knows that what she writes is not an answer to the question (see Carson 2016: 61). Stokke and Sorensen agree that these examples show that there are cases of bullshitting without the intent to deceive and cases in which bullshitters are concerned to say what's true (Stokke 2018: 152–153; Sorensen 2011). (For more on bullshitting without the intent to deceive, see Kenyon and Saul's chapter on bald-faced bullshit.)

Although I think that I have good criticisms of Frankfurt's definition, I don't have an alternative definition that I am prepared to defend.[18] Lack of concern with the truth of what one says is an extremely important and salient concept, whether or not it is a necessary condition of bullshitting.

Even if we don't accept Frankfurt's definition, it is so well-known that we can refer to examples that count as cases of bullshitting according to his definition as cases of "Frankfurt-bullshitting." Frankfurt's idea that bullshit/bullshitting involves indifference to the truth of what one says is particularly salient. Many people think that this describes many things said by former President Donald Trump.[19] Frankfurt-bullshitting about matters of importance to public policy by people with political authority and/or the power to influence public opinion usually involves a kind of negligence or recklessness. When such people speak on such topics without regard for the truth of what they say they risk greatly harming others.

5 Indifference to the evidence

An important closely related concept is indifference to the evidence and counter evidence that bears on the truth of what one believes or says. Many people are indifferent to, or insufficiently concerned with, the evidence when it conflicts with what they want to believe and/or what they want to say. A clear example of this is Donald Trump's refusal to accept easily verifiable evidence about the World Trade Organization (WTO) that doesn't fit with his view that the WTO treats the US unfairly. While speaking to Gary Cohn, the Chairman of the National Economic Council, Trump said "The World Trade Organization is the worst organization ever

---

18 The best definition that I have seen is Stokke's (a definition he developed in Fallis and Stokke 2017):

> A is bullshitting relative to a QUD [question under discussion] q if and only if A contributes p as an answer to q and A is not concerned that p be an answer to q that she believes to be true or an answer to q that she believes to be false.　　　(Stokke 2018: 147)

19 See Zakaria 2016.

created! . . . We lose more cases than anything." Cohn corrected him and noted that the US won 85.7 % of the cases it brought to the WTO and that the US won cases against China for its duties on US poultry, steel, and autos and for its restrictions on the export of raw materials and rare earths. Trump replied "This is bullshit, this is wrong." Cohn then said "This is not wrong. This is data from the United States trade representative. Call Lighthizer and see if he agrees." Trump said "I am not calling Lighthizer." Cohn concluded this exchange by saying "I'll call Lighthizer. This is factual data. There's no one who is going to disagree with this data" (Woodward 2018: 276–277).

6 Two related notions: "Spin" and "half-truths" or "partial-truths"

"Spinning a story" or "putting spin on a story" involves placing an interpretation on events or facts which, themselves, need not be in dispute. People spin events when they place a particular interpretation on them. Political candidates often spin news stories in such a way as to make themselves and their policies look good and make their opponents look bad. Ideologues of all stripes often spin their interpretations of events so that those events seem consistent with their ideological commitments. Sometimes the interpretations they spin are correct, but sometimes not. If someone spins the interpretation of an event, then his interpretation is likely to be biased and unreliable, but it is not necessarily incorrect.

Interpretations that involve spin can be misleading if they advance unreasonable interpretations of events and incline other people to accept those interpretations. Misleading spin counts as attempted deception if the "spinner" knows or believes that the interpretation he defends is unreasonable or implausible. If spin involves making deliberate false claims about one's state of mind (false statements about how one thinks that events should be interpreted), it almost always involves lying and attempted deception.[20] But spinning the interpretation of facts or events

---

20 Mearsheimer (2011: 16–17) holds that spinning always involves distorting the facts:

> Spinning is when a person telling a story emphasizes certain facts and links them together in ways that play to his advantage, while, at the same time, downplaying or ignoring inconvenient facts. Spinning is all about interpreting the known facts in a way that allows the spinner to tell a favorable story. It is all about emphasizing and deemphasizing particular facts to portray one's position in a favorable light . . . . The basic story being told is distorted, but the facts are not put together so as to tell a false story, which would be a lie.

need not involve any intent to deceive. Some spinners accept their own interpretations of things and sometimes people spin so as to counteract the spin and distorted narratives of other people. A person can spin the interpretation of something without saying anything that is false, but a person can also spin by means of saying things that are false. Interpretations that involve spin can be misleading if they advance unreasonable interpretations of events and incline other people to accept those interpretations.

One common way of spinning events is by stating half-truths. Half-truths or partial-truths are narratives consisting of true statements that selectively emphasize facts that support a particular interpretation or assessment of an issue and selectively ignore or minimize other relevant facts that support contrary interpretations or assessments. For example, a politician spins the interpretation of recent events to support the claim that his policies were successful if he describes the good consequences of those policies in great and vivid detail and omits any mention of the bad consequences of the policies in question. A man's description of his marriage to a friend (or the description he gives to himself) is a half-truth if it includes a long and accurate account of unkind and hurtful things that his wife has said and done to him but mentions only a few of the equal (or greater) number of unkind and hurtful things he has said and done to her.[21]

Even when there is no lying or deception involved, spinning often calls into question one's intellectual honesty. Often, we spin evidence to protect our cherished beliefs when that evidence ought to cause us to question those beliefs.

---

21 Neil Manson believes that spin requires being truthful, but doesn't necessarily involve making true statements; he writes "spin is truthful but need not be true," (Manson 2012: 201). I agree with him that spin can be constituted by false, but truthful, statements. However, Manson is mistaken in criticizing me for denying this. He writes:

> Carson is right to stress that spin is not the same as lying . . . But it is a mistake to frame spin in terms of truth rather than truthfulness.   (Manson 2012: 201)

Manson (2012:201) says that I hold that all spin involves telling half-truths: "He [Carson] identifies spin as the production of half-truths." This is a misinterpretation of my view. In the work he cites, I don't identify spin as the production of half-truths. I only say that "One common way of spinning events involves stating 'half-truths'" (Carson 2010: 57); this is also my present view and I repeat this statement above. For Manson, spinning can't involve making untruthful statements. He is mistaken about this. Consider a long one-sided narrative that counts as a clear case of spin. Imagine that we add a lie or two to the narrative. The result would still be a case of spin. Since all lies are untruthful statements, this would be a case of untruthful spin.

7 Other important concepts: Preventing people from acquiring true beliefs and undermining knowledge and trust in reliable sources of information

Chisholm and Feehan (1977: 143–145) distinguish between what they call positive and negative deception. Roughly, they say that positive deception is causing someone to have false beliefs and negative deception is preventing someone from having true beliefs. Given standard definitions of deception, in terms of causing false beliefs, it doesn't make sense to call preventing someone from gaining knowledge a case of deception. But, in any case, Chisholm and Feehan identify a very important concept: the concept of preventing someone from acquiring true beliefs. There are other important related concepts as well. In addition to acts that prevent people from gaining knowledge of the truth, there are also actions that undermine people's confidence in truths that they know or believe and actions that undermine people's trust in reliable sources of information. The means by which trust and confidence are undermined may or may not involve lying and may or may not involve deceiving others about the truths or sources of information that are the targets of their actions.

American tobacco companies tried to undermine people's confidence in reliable sources of information in order to raise doubts about the clear and overwhelming evidence of the health risks associated with tobacco use. They spent millions of dollars to impugn a large body of careful scientific research. In 1964 the US Surgeon General issued a report warning smokers about the connection between cigarette smoking and lung cancer. Smokers are 10 to 20 times more likely to get lung cancer than non-smokers and are also "more likely to suffer from emphysema, bronchitis, and heart disease. The more a person smoked, the worse the effects," (Oreskes and Conway 2010: 22). In 1967, the Brown and Williamson tobacco company said "There is no scientific evidence that cigarette smoking causes lung cancer and other disease" (Oreskes and Conway 2010: 23). An internal memo by a tobacco industry executive from 1969 reads "Doubt is our product, since it is the best means of competing with the 'body of fact' that exists in the minds of the American public" (Oreskes and Conway 2010: 34). The industry succeeded in creating doubts about the harmfulness of tobacco. As of 1992, 25% of Americans "still doubt that smoking is dangerous at all" (Oreskes and Conway 2010: 33) and as of 2006 almost 25% of Americans believed "there's no solid evidence that smoking kills" (Oreskes and Conway 2010: 241, 335 n4).

Another example is the very well-funded campaign by petroleum industry to create doubts about the very strong evidence that greenhouse gases are causing climate change. In July 1977, James Black, a senior Exxon scientist, told some of Exxon's leaders: "There is general scientific agreement that the most likely manner in which mankind is influencing global climate is through carbon dioxide release from the burning of fossil fuels" (McKibben 2019: 72). As McKibben relates:

> A year later, he spoke to a larger pool of the company's executives. Independent researchers, he said, estimated that a doubling of carbon dioxide concentration in the atmosphere would increase average global temperatures by 2 to 3 degrees Celsius (3.6 to 5.4 Fahrenheit). Rainfall might get heavier in some regions, and other places might turn into a desert.

Exxon and other oil companies acted on this information in their drilling operations: they "built their new oil drilling platforms with higher decks to compensate for the sea-level rise that they now knew was coming" (McKibben 2019: 74). In the late 1980s, Shell Oil Company scientists "predicted that . . . carbon dioxide levels could double as early as 2030" and that this would lead to "runoff, destructive floods, and the inundation of low-lying farmland" (McKibben 2019: 74). Instead of publicly acknowledging these very alarming findings, the oil companies kept them hidden from the public and engaged in a very well-funded campaign of misinformation. In 1988, Exxon's public affairs manager wrote an internal memo in which he recommended that the company "emphasize the uncertainty" in the scientific evidence about climate change (McKibben 2019: 75–76). Oil companies helped to create "the Global Climate Coalition", an organization that opposed fossil fuel taxes and the 1997 Kyoto Protocol on climate.

> Two months before the Kyoto meeting, Lee Raymond (Exxon's president and CEO, and the man who had had oversight responsibility for the science department that in the 1980s produced the unambiguous findings about climate change) . . . insisted that the Earth was cooling, said that the idea that cutting fossil fuel emissions could have an effect on the climate "defied common sense," and declared that, in any event, it was "highly unlikely that the temperature in the middle of the next century will be affected whether policies are enacted now, or twenty years from now."[22]　　(McKibben 2019: 76)

---

22 When he retired in 2006, Raymond took a $400,000,000 retirement package (McKibben 2019: 77).

References

Augustine. 1965. *Lying* and *Against Lying*, in *Treatises on Various* Subjects, *Fathers of the Church, Vol. 16*, Ray J. Deferrari (ed). Washington, DC: Catholic University Press.

Augustine. 1961. *Enchiridion*, Henry Paolucci (ed.). Chicago: Regnery.

Black, Max. 1983. *The Prevalence of Humbug and Other Essays*. Ithaca: Cornell University Press.

Bok, Sissela. 1979. *Lying*. New York: Vintage Books.

Carson, Thomas. 2010. *Lying and Deception: Theory and Practice*. Oxford: Oxford University Press.

Carson, Thomas. 2016. Frankfurt and Cohen on bullshit, bullshitting, deception, lying, and concern with the truth of what one says. *Pragmatics & Cognition*, 23 (1): 54–68.

Carson, Thomas. 2019. Lying and history. In Jörg Meibauer (ed.), *The Oxford Handbook of Lying*, 541–552. Oxford: Oxford University Press.

Chisholm, Roderick and Thomas Feehan. 1977. The intent to deceive. *The Journal of Philosophy* 74 (3): 143–159.

Cohen, G. A. 2002. Deeper into Bullshit. In Sarah Buss and Lee Overton (eds.), *Contours of agency: essays on themes from Harry Frankfurt*, 321–339. Cambridge, MA: MIT Press.

Dynel, Marta. 2011. A web of deceit: A neo-Gricean view on types of verbal deception. *International Review of Pragmatics* 3 (2): 137–165.

Dynel, Marta. 2020. To say the least: Where deceptively withholding information ends and lying begins. *Topics in Cognitive Science* 12: 555–582.

Fallis, Don. 2009. What is lying? *Journal of Philosophy* 106 (1): 29–56.

Fallis, Don and Andreas Stokke. 2017. Bullshitting, lying, and indifference toward truth. *Ergo* 4 (10): 277–309.

Frankfurt, Harry. 2002. Reply to G. A. Cohen. In Sarah Buss and Lee Overton (eds.), *Contours of agency: essays on themes from Harry Frankfurt*, 340–344. Cambridge, MA: MIT Press.

Frankfurt, Harry. 2005 [1986]. *On Bullshit*. Princeton: Princeton University Press.

Fried, Charles. 1978. *Right and Wrong*. Cambridge: Harvard University Press.

Hartman, Nicolai. 1975 [1932]. *Ethics*. Translated by Stanton Coit. Atlantic Highlands, NJ: Humanities Press.

Lycan, William. 2006. On the Gettier problem problem. In Stephen Heatherington (ed.), *Epistemology Futures*, 148–168. Oxford: Oxford University Press.

Mahon, James. 2015. The definitions of lying and deception. *Stanford Online Encyclopedia of Philosophy*.

Manson, Neil C. 2012. Making sense of spin. *Journal of Applied Philosophy* 29 (3): 200–213.

McKibben, Bill. 2019. *Falter: Has the Human Game Begun to Play Itself Out?* New York: Henry Holt.

Mearsheimer, John. 2011. *Why Leaders Lie: The Truth About Lying in International Politics*. New York: Oxford University Press.

Meibauer, Jörg. 2014. *Lying at the Semantics Pragmatics Interface*. Berlin: De Gruyter.

Oreskes, Naomi and Eric Conway. 2010. *Merchants of Doubt*. New York: Bloomsbury Press.

Ross, W. D. 1930. *The Right and the Good*. Oxford: Oxford University Press.

Saul, Jennifer. 2012. *Lying, Misleading and What Is Said*. Oxford: Oxford University Press.

Sorensen, Roy. 2007. Bald-faced lies! Lying without the intent to deceive. *Pacific Philosophical Quarterly* 88:251–264.

Sorensen, Roy. 2011. Review of *Deception* and *The Philosophy of Deception*. *Times Literary Supplement*. 4 March 2011, 22–24.

Stokke, Andreas. 2018. *Lying and Insincerity*. Oxford: Oxford University Press.

Tiersma, Peter. 2004. Did Clinton lie?: defining 'sexual relations.' *Chicago-Kent Law Review* 79 (3): 927–958.

Turri, Angelo and John Turri. 2015. The truth about lying. *Cognition* 38 (1): 161–188.

Wiegmann, Alex, Pascale Willemsen, and Jörg Meibauer. 2021. Lying, deceptive implicatures, and commitment. To appear in *Ergo*. Preprint, PsyArXiv. https://doi.org/10.31234/osf.io/n96eb. https://osf.io/n96eb.

Woodward, Bob. 2018. *Fear: Trump in the White House*. New York: Simon and Schuster.

Zakaria, Fareed. 2016. The unbearable stench of Trump's B.S. *Washington Post*, 8.4.2021. https://fareedzakaria.com/columns/2016/08/04/the-unbearable-stench-of-trumps-b-s.

William G. Lycan

The morality of deception

Abstract: Outright lying is one end of a scale or spectrum, with deliberately misleading very close to it. Then:misleading *somewhat*; concealing facts; "lies of omission"; withholding information; not disclosing; lack of transparency; . . . ?. Many people talk as if these items are listed by degree of deceptiveness, and as if that order presumptively corresponds to degree of moral objectionability.

This paper defends two claims:

(1) No good arguments support the intuition that plain flat lying is, other things being equal, morally worse than deceiving merely by implicature.

(2) Although in some contexts one can deceive by remaining silent, and there are other cases in which silence is morally wrong, there is a large and principled break in the aforementioned scale, between positive lying *or* misleading and merely not revealing something that one knows.

Special attention is paid to the semantic and pragmatic relations between questions and answers.

In this chapter I shall consider some types of deliberate verbal deception that are not cases of lying. In writing that opening sentence I could naturally have said, types that "fall short of" lying, or "not *actually* lying," or "not cases of *outright* lying." How so? The implication would be that such cases are *on the way to* lying, perhaps that they lie on a scale at one end of which is lying. But if there is such a scale, what would the other end be? Just never knowingly deceiving anyone? An even purer shining honesty in all actions verbal and nonverbal? Utter informational transparency? Posting everything you know on Facebook? In any case, there is a general idea that, morally, lying is worse than other forms of deception.[1] One of my purposes here is to probe that idea.

A second and more practically significant purpose is to argue that there is a natural and morally important break in the foregoing scale, roughly between speaking deceptively and merely not revealing something one knows.

In aid of the first purpose, I shall assume a distinction, based on characterizations of lying by Carson (2006, 2010), Fallis (2009), Saul (2012) and others,

---

[1] My discussion in this chapter concerns commonsense morality only. Some of it will bear on legal matters such as that of perjury, but I am not qualified to address those.

William G. Lycan, University of Connecticut, e-mail: william.lycan@uconn.edu

https://doi.org/10.1515/9783110733730-003

between actual lying and merely misleading. I shall use "mislead" a bit stipulatively to mean purposeful verbal deception that is not lying, and "deception" to mean deliberately causing or encouraging a false belief.[2]

1 On the concept of lying

I shall not enter into the fine details of defining "lie," because some of the disputed and/or borderline cases will not concern us in this chapter.[3] I will only offer four remarks on the concept that will be pertinent.

First, "lie" has a feature that is shared by several other words, such as "desire" and "know": In the language, it has both a broad or loose use and a much stricter use (where this is not merely a matter of vagueness). Each use is acceptable to competent speakers; the broad use is not incorrect. But contextual purposes, especially when theorizing, may require the stricter use.[4] The broad use of "lie" does not even require dishonesty or deception, but covers mistakes and slips of the tongue as well as "lies of omission" and any sort of misleading; but we all also recognize one or more stricter senses, including of course the one that excludes speaking truly but deliberately misleading by strong implicature.[5]

Second, for purposes of this chapter I shall assume that even "lie" in the strictest sense is a prototype concept.[6] There is a paradigm case, defined by a list of criteria, but no one of the criteria is strictly necessary (at least not uncontroversially

---

2 Of course both "mislead" and "deceive" have nonmoral and nonpurposive uses, including the obvious ones predicated of facts in the world that involve no agents at all. And in this chapter I shall not consider completely nonverbal deception.

3 In particular, I shall not address "bald-faced" lies; bullshit; or the Sartrean (and Thomistic) question of whether a lie can be accidentally true.

4 Compare "desire": It has a quite general use in English which includes nearly all conative or motivational mental states, "pro-attitudes" as philosophers call them. But even in English, we often use "want" much more strictly – what has been called "desire proper" – that excludes hopes, wishes, intentions, normative beliefs, urges, . . . (To see that there is a large difference, note that if I do voluntarily and deliberately do something, I must have wanted or desired, broadly speaking, to do it; but also, I very often do things I have *no desire at all* to do, for other sorts of motivating reason.) The phenomenon is related to, but not quite the same as, Lasersohn's (1999) "pragmatic slack."

5 For some of the relevant data, see Horn (2017a, 2017b). Of particular interest is the "lexical clone" phenomenon, as in "That wasn't a LIE lie" (2017b).

6 See Coleman and Kay (1981). It is a particular type of what philosophers following Wittgenstein call a "family resemblance" concept. For the record, I do not agree that "lie" perfectly fits this pattern, but the differences do not matter here.

so), and they may be weighted differently in different contexts.[7] As I formulate them, the criteria are: (1) The speaker asserts a proposition P; (2) the speaker believes P to be false; (3) P is false; and (4) in asserting P, the speaker intends to deceive the hearer into believing P. I believe (1)–(4) are conceptually sufficient for lying. Typical cases of misleading are ones in which the speaker asserts, not the falsehood P, but a truth which merely implicates or otherwise conveys P.

Third, even the strict concept admits of many genuinely borderline cases. To catalogue and explain them would require at least a whole chapter in itself, but I shall just mention a few.[8]

Borderline cases of asserting (at all): Utterance while mentally impaired. Nonverbal signals such as nods, fist pumps, or the "OK" gesture. Questionable tone (especially suspected ironic tone). Utterances having only indirect assertive force ("Did you know that . . . ?"; "Now, why do you think I . . . ?"; "I bet you $1,000 that . . ."). *Hedged* assertions. Doubt as to whether an utterance is audible. Assertion that is not *to* the hearer in question (eavesdropper or other third-party cases).

Asserting P in particular: What may be only a conventional phrase ("Pleased to meet you"). Sentences having a small ambiguity when speaker-meaning does not decide between the interpretations. Meaning relations that are disputed as between entailment and implicature (e.g., temporal and causal uses of "and"). "Explicatures" and/or "free enrichments" not determined by sentence meaning but allegedly still part of what is said (Carston, 1988). Parts of sentence meaning that are not "at-issue" content, such as are carried by adjoined adjectives ("Fat Tommy got up and . . ."). Everyday exaggeration. Mild figurativeness. Other allowable disparity between sentence meaning and speaker-meaning.[9]

Believing: There are a number of very belief-like attitudes that can loosely be called "belief" in English but are not strictly factual belief.[10]

Truth: Vagueness of predicates and other lexical items, and purpose-relative standards of strictness.

Fourth remark: I shall go along with the majority assumption (Adler (1997), Horn (2009, 2017b), Saul (2012), Mahon (2016)) that the distinction between lying and misleading tracks that between asserting and conveying in some less direct way such as, paradigmatically, implicating.[11]

---

7 Horn (2017a, 2017b) develops this model in detail.

8 Some will be of interest for legal purposes, others not.

9 "Allowable" is in there to exclude outrageous "mental reservations," on which see Saul (2012), pp. 101–09; Horn (2017a), pp. 40–44.

10 See, e.g., Matthews (2013); Van Leeuven (2014).

11 Meibauer (2014a) resists that assimilation and proposes a considerably more inclusive definition of "lie" (p 125). (See also Falkenberg (1982) on "indirect lying.") That definition simply

Before we move on to our squarely moral concerns, I want to devote a section to a topic that is directly relevant but has never to my knowledge been discussed in this connection by philosophers. It is closely related to matters of implicature but is not merely a sub-area of that phenomenon.

2 Questions and answers

The relations between questions and answers raise special issues about lying, misleading and truth.

A question has a range of possible answers, indeed a range of possible true answers. A yes-no question may be answered "Yes" or "No," but that is not the end of it; the question could be answered echoically, or by making a statement that semantically entails the positive or the negative. Conversationally, it can be answered by strong implicature. Even less formally, it can be answered in context by quite a weak implicature that exploits both background information and the assumption that the responder *is* answering the question. Thus: "Are you going to the symphony this evening?" (a) "No." (b) "I am not going." (c) "I am not going anywhere at all this evening." (d) "The only concerts I go to are choral concerts." (e) "Twentieth-century music bores me to death" / "Our godson is visiting from Oregon." Naturally, answers by implicature are cancellable ("But I'm going anyway"). In fact, they are also cancellable in a notably deeper way, for their status *as answers* may be retracted (see below).

Notice, in any case, that replies by implicature, such as (d) and (e) – indeed any replies that are not *semantically* answers to the question – depend on coöperation and a degree of trust.[12] Strictly speaking, they are not answers, and are not counted as such in forensic or other formal contexts. Even in an everyday conversation such as a not-entirely-friendly domestic exchange, a spouse may rejoin "That's not what I asked."[13] In fact, somewhat to my surprise, answers by semantic entailment (c) are sometimes not counted as answers. From a biography of Henry Kissinger (Kalb and Kalb, 1974, p. 400): "Later, Kissinger would wonder about the

---

rules in deception by implicature (though, as Meibauer points out, it does not itself contain the notion of deception). But if we are using "lie" in its acknowledged strict sense, that is just too broad. Horn (2017b, pp. 163–64) cites data indicating that survey respondents agree.

12 Nowhere to my knowledge does Grice discuss questions, but the Maxim of Relation in spirit affords the notion of answering by implicature.

13 Saul (2012, pp. viii, 70) gives an interesting example of a would-be-friendly answer by implicature; I shall discuss it in sec. 4 below.

interview [that he had given to Oriana Fallaci on November 4, 1972] and sigh, 'I couldn't have said those things, it's impossible.' But he never denied it."[14]

Wh-questions admit more variety because they force the responder to choose a description: "Who is that man?" "When will you be leaving on your vacation?" "Where did you live before you moved here?" And the choice of description will depend on various factors, one of which, ideally, will be the responder's belief about the questioner's interest in asking. Some descriptions, though correct, would be so unhelpful as to be hardly answers at all: "That man? He's the one you're pointing at." "We'll leave when we decide to." "We moved here from our previous home." *Who*-questions and their answers are especially interest-relative: "Who is that man?" might elicit just a name, which might or might not be helpful, but, whatever the questioner's purpose is, very few descriptions even among the true ones would serve it. "He's the only man in the room who ate any Rice Krispies on January 4, 2002"; "He's my daughter's best friend's grandfather"; "He's my boss"; "He's your boss"; "He's your biological father." (See Boër and Lycan (1986).)

I mentioned the hearer's assumption that the responder is answering the question. That is a crucial and sometimes tricky matter. Normally, when someone asks a question and the hearer immediately says something to the speaker and without discourse comment, the hearer is answering the question. (Which is why (e) above qualifies.) And we quite automatically make that assumption, which we may evocatively call the Prompt Answer Principle. "Why were you late this morning?" – "We had spaghetti at our house last night." Even charitably assuming the implicature that the speaker *had eaten* some spaghetti, the questioner would think, "What? How does eating spaghetti impede getting to class on the following day? Was it tainted?"

At this point, a special feature of *why*-questions shows itself. It may seem to be a version of the purpose-relativity of *who*-questions, but it runs deeper than that: A sentence asserted in answer to a *why*-question may be a perfectly true sentence and with no misleading implicatures of its own, and yet be a false answer to the question. If that sounds paradoxical, consider this exchange: "Why does a toy boat float on still water in the bathtub?" – "Water is a clear and transparent liquid." The reply does answer the question, or so I stipulate regarding the envisaged context. The sentence uttered is true. The sentence itself is not misleading, or at

---

14 Though it may be responded that Kissinger's reply was indirect and did not unequivocally entail the negative answer, since modal auxiliaries are evaluated relative to classes of "real" and/or "relevant" possibilities (Lycan, 1994; see this work on what I there called the "lunatic relativity" of modal expressions more generally). Compare, from a Lawrence Sanders novel: "'And the insurance?' Callaway asked. 'When may the beneficiaries expect to have the claim approved?' Dora smiled sweetly. 'As soon as possible,' she said, and shook his hand." (Sanders, 1992, p. 38).

least has no false implicatures. Yet it is a false answer to the question; that water is clear and transparent is not the reason and does not explain why a toy boat will float on it. (The true answer would have to cite Archimedes' Principle and the boat's small weight compared to that of the water displaced.)

What has gone linguistically wrong? The true or correct answer to a *why*-question must give the relevant reason or explanation, that being what "why" means. To be more explicit, the responder could and perhaps should have said, "*The boat floats because* water is a clear . . .," in which case the sentence s/he uttered would itself have been false. It is plausible that every answer to a *why*-question in some sense contains a tacit "because," and so is not a true answer unless the corresponding ". . . because . . ." sentence is true. It might even be maintained that a sentence used to answer a *why*-question actually has the question's propositional content plus "because" elliptically in its underlying logical structure, though to say that would require specifically syntactic evidence of ellipsis and I know of none.

A type of *why*-question that commonly figures in moral evaluations is that which seeks a person's reason for what s/he has done. And here, the person in question may have a motive for deceiving. "Why did you sell your yacht?" – "I didn't have enough money to pay my large telephone pledge to Doctors Without Borders." – Which sentence was true, but in fact the responder had no intention of making any charitable donation; she needed the money to buy a year's supply of Scotch for herself (and she did use the proceeds for that purpose).[15]

An answer regarding motives can be both true *and* a partly true answer but viciously misleading, as when the responder cites a reason that was, indeed, a reason, but a minor one and far from the responder's main reason, which s/he means to keep well hidden.

As noted above, the Prompt Answer assumption is itself cancellable: " – Oh, sorry, I wasn't answering your question, I was thinking about something else." The responder should then give the answer, but of course might instead evade the question, decline to answer it, or remain mutinously silent.

I have been assuming that the Prompt Answer assumption is motivated and justified by a background conversational principle, a norm of the same sort

---

15 An example from a (fictional) 19th-century legal proceeding: The U.S. Navy has impounded J.C. Spring's ship on a charge of slave-trading. "'Captain Spring, you say you brought palm oil from Dahomey to Roatan – an unusual cargo. Why then was your ship rigged with slave shelves?' 'Slave shelves, as you call them, are a convenient way of stowing palm oil panniers,' says Spring. 'Ask any merchant skipper.'" (Spring's ship had in fact carried slaves from Africa to the West Indies, and the shelves were used entirely for confining hundreds of slaves in appalling conditions.) Fraser (1971, pp. 265–66).

as generates implicatures, such as Grice's Coöperative Principle or Sperber and Wilson's (1986) Communicative Principle of Relevance. A finicky philosopher might balk at the normative idea, and say that we as audience simply *assume* that the responder is answering the question, and that is the questioner's problem and ours. I do not recall Grice's expressing any view on this matter; but a Relevance theorist would have plenty to say on behalf of the normative understanding: However the details may be spelt out, the "optimal relevance assumption" for the questioner and for the audience *is* that the responder is answering the question, and all parties know it.

In fact, it would seem that some question-answer relations are linguistically mandated: plain "Yes" or "No" in direct response to a yes-no question, and especially echoic responses even when subsentential: "Yes, I am," or "No, she didn't," or just "I am" / "She didn't." Those, I concede to the Prompt Answer principle, are unequivocally answers and the speaker is responsible not only for their truth but for their correctness as answers. Moreover, they are assertions, and are squarely lies if deliberately false.[16]

(And that raises the question of whether Prompt Answer is actually a brute discourse *convention*, though naturally with a defeasibility clause to cover the cancellable cases. If so, what would show that? I do not know how to decide whether some regularity is a brute convention with no further rationale.)

In light of this discussion let us revisit a now famous legal example, the exchange in *Bronston v. United States*:[17] "Have you ever [had a bank account in a Swiss bank]?" – "The company had an account there for about six months, in Zürich." The fact was that Bronston himself had had one, and he was charged with perjury. There was much legal discussion over whether Bronston's deliberate misleading merited the perjury charge, as well as continuing philosophical discussion (Saul (2012), Horn (2017b)); but what strikes me, indeed stands out a mile, is that Bronston *simply did not answer the question*. Strangely, the lawyer did not call him on that, though it screams for "That's not what I asked you."[18]

---

16 And this is a case in which an equivocation may be a flat lie rather than true-but-misleading. If I know perfectly well what you meant by the words you used in framing your question, but I (pseudo-)answer it using the same words with different meanings and the sentence I utter translated into what you meant is false, I have lied.

17 See Solan and Tiersma (2005) and Chapter 14. The case figured as a precedent in evaluating U.S. President Bill Clinton's testimony during his impeachment hearings.

18 In his majority opinion overturning the perjury conviction, Chief Justice Burger mentioned in passing that Bronston's putative answer was "not responsive," but all ensuing discussion focused on its literal truth as against its false implicature. As Solan and Tiersma say, "If a witness gives a literally true but unresponsive answer, the solution is for the lawyer to follow up with more precise questions, not to instigate a federal perjury prosecution" (p. 215).

Another interesting example is mentioned by Horn (2017b, pp. 162–63), regarding the use of cardinal numerals. Green (2006) and Meibauer (2014b) raised the question of whether a numeral *n* must mean exactly n, or can mean only at least n. (The forensic issue would be whether a deceptive statement containing *n* would be a flat lie or only a misleading quantity implicature.) Horn reports that there is "current consensus [among linguists] that cardinals – *in particular when occurring as fragment or focused answers to questions* – express 'exactly n' and do not merely implicate the upper bound" [italics mine]; Horn (2009, p. 8) had argued that cardinals differ from other scalars in question contexts.

3 Morality

For purposes of this chapter I shall (except where noted) suspend my own moral theory and take an ecumenical view of moral assessment. There are quite a number of factors and considerations that are morally significant and agreed to be so by all but philosophical ideologues. (Little of the debate among present-day moral philosophers takes the form of insisting that one or more of the considerations should be written off as simply irrelevant, much less arguing that just one of the factors by itself determines all moral value. Rather, the main disagreements – even quite sharp ones – are over which considerations are more fundamental and which might be explained in terms of others.)

Some of them, in no particular order: harm; the Golden Rule and other forms of universalizability; happiness, welfare and/or well-being; autonomy; respect for persons; beneficial habits and policies; duties of office; virtues such as courage, honesty and kindness; strengthening or weakening of personal relationships; integrity; concern for the public good. I shall continue to assume that each of those is morally relevant, though along the way I shall confess one or more of my own substantive and controversial positions in moral philosophy.

4 Moral generalizations about lying vs. misleading

It should be uncontroversial (bar the most maniacal Kantians) that some cases of misleading are morally worse than some cases of lying.But would it be true to say that, *other things being equal*, outright lying is morally worse than merely misleading?Call that the Defeasibility view, the idea being that if a particular case of misleading is morally worse than or even as bad as the corresponding lie would have been, there must be some special circumstantial explanation of how that is so.

Many philosophers other than Kant have defended the Defeasibility view. And ordinary people talk as if they accept it; recall phrases such as "didn't actually lie," and defenses such as "Every word I said was true!" Moreover, people who need or want to deceive regularly make efforts to mislead rather than just to lie. As has often been pointed out, misleading usually costs a speaker more thought and ingenuity, and so the nonlying deceivers must consider the difference worth the trouble in leaving them less open to moral condemnation. Still further, in some legal or otherwise forensic contexts, lying is heavily penalized while mere misleading – even deliberate, crafted and craven misleading – is not.

Jennifer Saul (2012) vehemently opposes the Defeasibility view.[19] She rebuts the main arguments in its favor, and works at explaining away the Defeasibility intuition. I shall base my discussion largely on hers.

As counterexamples to Defeasibility, she offers three cases in which she sees no moral difference whatever between lying and misleading (pp. 72–73). They are ones in which the consequences of the deception are not just negative but likely fatal: an HIV-infected person tells a sex partner (truly) that she "do[es]n't have AIDS"; a cook tells a dinner guest who has a violent peanut allergy that he "didn't put any peanuts in," when he is deliberately poisoning her with peanut *oil*; and the Murderer At The Door, to whom (contra Kant himself) if anything one should lie outright as convincingly as possible rather than merely implicating or suggesting the needed falsehood.

I myself tend to accept Saul's judgments on cases of that kind. But – full disclosure – I personally am an act-consequentialist. Nor need one be a complete consequentialist in order to grant that very grave consequences are morally relevant, sometimes overwhelmingly so; even some nondoctrinaire Kantians concede that (e.g., Hill, 2016). It is no surprise to learn that the Defeasibilty view is incompatible with strong versions of consequentialism. Obviously, misleading someone often has at least as bad consequences as lying would have, and it is an entirely empirical matter how often that is. Moreover, in some circumstances lying may have better consequences.[20] And, Defeasibility advocates will argue, so much the worse for

---

19 Her statement of it: "Holding all else fixed, lying is morally worse than merely deliberately attempting to mislead, and successfully lying is morally worse than merely deliberately misleading" (p. 72; on p. 74 she weakens it to allow for "certain special cases").

20 For the same reason, a Golden Rule comparison would not help: If you are to be deceived, you probably do not care whether it is lying or misleading done unto you, unless there are special circumstances.

consequentialism; indeed, less dramatic examples of this type have themselves been put forward as plain objections to consequentialist theories.

Saul mentions two other kinds of case in which she can see no moral advantage of misleading over lying: a deathbed reassurance that protects the dying woman from being distressed by an awful truth (p. 70), and a case of non-self-serving tact or politeness (p. 127). In each, she maintains that morally, the agent might as well simply lie. She acknowledges that "if their motivation for choosing [to mislead rather than lie] . . . is a good one, we are likely to approve (or at least to not judge it so harshly" (p. 128), but she explicitly insists that matters of motivation merely "reveal . . . [something admirable] about a person" (p. 127) and do not affect the moral status of the action itself. Well; so much for Kant on the uniquely valuable good will!

Thus, Saul's verdicts on cases are not likely to convince non-consequentialists or even pluralists who hold that motivations and other factors should and do affect the evaluation of actions. But let us turn to the main philosophical defenses of the Defeasibility view.

(1) She neatly disposes of my appeal (above) to legal contexts. Although no one thinks that legal prohibitions/permissions universally coincide with moral ones, it is easy to assume that the strong legal differences between lying and mere misleading are based on more fundamental moral ones that are generally to be respected. But Saul (pp. 95ff., following Solan and Tiersma (2005), pp. 215–16) offers plausible independent explanations of the legal focus on what has been strictly asserted as opposed to implicated: Courtroom proceedings are by their nature adversarial and very formally so; for any witness' testimony, there is a highly trained official opponent whose job it is to see that the relevant question has been answered literally and to guard against evasion and weaseling by way of implicature. That is why the process includes cross-examination.

Similar considerations pertain to some less formal but still recognizably adversarial or contentious contexts and to written contracts and formal agreements of various kinds. It hardly follows that everyday morality makes much of or even should respect the difference between lying and deliberately misleading.

A generalization argument or a rule-consequentialist argument might be made. Obviously the general good requires that people not go around deceiving others whenever they feel like it, but it also requires that there be exception clauses that allow deceiving under circumstances of this type or that. Would it be beneficial to allow more exceptions for misleading than ones for lying? Perhaps if misleading were allowed more readily than lying, a greater general respect for truth would be preserved, and such respect would be a good thing. But that is empirical speculation, and lends little support to Defeasibility as a normative principle.

(2) On one interpretation of Kant (Adler, 1997), the moral difference goes back to the allocation of responsibility for a hearer's being deceived. A liar bears full responsibility, but when the deception is only by implicature, the hearer is at least partly responsible, by having voluntarily drawn an inference (however natural) in order to arrive at the false belief. True, the speaker mendaciously invited the inference, but the hearer had to collaborate by drawing it.[21]

This argument is a special case of the more general Kantian concern for autonomy. A liar simply manipulates the victim's information store, while the misleader only issues an attractive invitation and the hearer accepts the invitation or not. (The deceiver is still "using the hearer as a means" and violating Kant's principle of respect for persons, but not in a way that simply bypasses the hearer's autonomous will.)

To Adler's responsibility argument Saul makes two main replies. First (p. 82), at least for strong implicatures, the hearer *must* make the relevant conversational inference if s/he is to understand the speaker as being coöperative at all; to that extent, accepting the implicature is no more voluntary than is accepting an outright assertion as true. In normal conversation we do not even notice the difference between what is strictly asserted and what is only strongly implicated. (As regards autonomy, notice also that when an assertion is made to us, we do always have the choice as to whether to believe it; accepting it is up to us. A liar does not simply manipulate our information store while we are asleep or helpless, with no action whatever on our part.)

Second (pp. 83–84), even when the victim of a wrong has knowingly increased the likelihood of such a wrong being done, that does not diminish the magnitude of the wrong. (Saul gives the example of the mugging victim who has stupidly and/or rashly strolled through a dangerous part of town "with his bulging wallet clearly visible in his back pocket.") In such a case someone will say that the victim "had only himself to blame," but (a) that is obviously an exaggeration, and (b) even if the victim does bear some degree of "blame," it does not make the mugger's act any less morally wrong.

The second argument is too quick, because Saul assumes that the victim's responsibility is "causal" only, and that he did not actively contribute to the wrongful act itself. Suppose the mugging victim had worn a sign that said "Wallet here ↓ with over $200 inside!," and maybe even helped the mugger free the wallet from his back pocket. Then it is not so obvious that the mugger's act is quite as

---

21 A strong version of this view is Green's (2006) "Caveat Auditor" principle: "A listener is responsible for ascertaining that a statement is true before believing it" (p. 165). Saul points out (p. 77) that even if that very occasionally applies in everyday cases, it could hardly be in force as a general norm.

wrong as it would have been without the victim's coöperation. – Yet, deliberate misleading is still deliberate. Wilson and Sperber (1986) point out that "By encouraging you to supply [a certain assumption and conclusion], I take as much [normative] responsibility for the . . . truth [of a clearly implicated proposition] as for the truth of the proposition I have explicitly expressed . . . . The speaker is committed to the truth of all determinate implicatures conveyed by her utterance, just as much as if she had expressed them directly" (pp. 61–62).[22] That may not be true of very weak implicatures, but we cannot assume that there is always a division of responsibility big enough to warrant Defeasibility as a general rule.

(3) Chisholm and Feehan (1977), Simpson (1992), Faulkner (2014) and Strudler (2010) appeal to the notion of *trust* in distinguishing lies from other deceptions. A lie is a "breach of faith" (Chisholm and Feehan, p. 153[23]) – a betrayal of trust, and betrayal is a serious matter. Strudler (2009) sharpens this argument by distinguishing trustworthiness in the moral sense from extraneously guaranteed reliability (as when we know that our interlocutor has a good reason to be truthful this time),[24] and by emphasizing that trustworthiness is not just a general moral mandate but is an important feature of the relationship between two people.

However, detailed explanation is needed as to why deliberately deceptive strong implicature is not as bad a breach of trust as is an outright lie (never mind whether lying has *further* properties that make it morally worse). Bernard Williams (2002) explicitly maintains that it is just as bad, and deplores the "fetishizing of assertion" (p. 100). Heffer (2020), whose entire analysis of untruthfulness in a very general sense is based on a notion of trust, adds that "any intent to deceive in whatever linguistic form betrays our trust that the speaker will be truthful" (p. 33).

Strudler (2010) does offer additional explanation, arguing that although when I learn you have misled me there is a reduction of trust, in that I start to scrutinize your assertions more warily, if I find out that you have simply lied, trust just evaporates and we can no longer communicate productively. But Saul (p. 79) points out that that is an empirical generalization and a dubious one.

---

22 Contrary to early interpretations of Grice, not all of an utterance's implicatures are speaker-meant. But in the cases of deliberate deception that concern us here, the implicatures plainly are speaker-meant.

23 Though they complicate the matter by also invoking rights.

24 It is an interesting fact that in practice, *reassurances* have some trust- or confidence-boosting effect: "Look, I know I've lied to you in the past, but I'm not lying this time." (The speaker may or may not go on to explain further.) I am inclined to say that such a reassurance even gives the hearer some reason, however slight, for trusting the speaker this time; but I do not know why.

Too much depends on the details of particular cases and on surrounding circumstances. And, I add, even if the generalization is statistically true, it does not support the normative Defeasibility thesis. Finally, Clea Rees (2014) argues in ingenious detail that (other things being equal) misleading by implicature involves a *greater* breach of trust than does simply lying.

(4) Asserting is a speech act in Austin's illocutionary sense, and so has both a conventionally determined social role and speaker-based felicity conditions. On each of those types of ground, an assertion is normatively supposed to be true: It is a constitutive feature of asserting that the speaker is both putting forward the asserted content as being true and warranting its truth to the hearer, inviting the hearer to rely on it.[25] Should the assertion prove false, then absent special mitigating conditions, the hearer has at least slight grounds for complaint. Moreover, an assertion is *eo ipso* illocutionarily defective (via the "sincerity condition") if the speaker does not believe it to be true.[26] None of those things holds for implicatures or merely implicated content. There is no such speech act as "implicating." (In fact, it would be nearly contradictory to say "I hereby implicate that . . . .") That is a *categorial* difference between lying and misleading.

Timmermann and Viebahn (2021) make a related point, à la Brandom (1983, 1994), that a speaker who asserts P incurs a social-epistemological responsibility to defend P and to respond to any challengers who question P, by giving reasons. (One who merely implicates P also faces some potential challenges, but "weaker" ones, of a different kind (p. 1490).) Asserting makes a commitment that implicating does not, and that too is a categorial difference.

Although Saul makes use of the notion of warranting in defining "lie" (pp. 10–12), she does not address these arguments for Defeasibility. And the categorial differences provide potentially solid conceptual ground for a systematic moral difference. But in fact, no moral difference strictly follows. The norms that figure in characterizing assertion are linguistic norms. An assertion is *illocutionarily* defective if it is not both true and believed to be true; when either condition fails, the speaker who made the assertion has violated a linguistic convention. Likewise,

---

25 As Carson (2006) notes, W.D. Ross went so far as to say that a liar "break[s] an implicit promise to tell the truth that one makes whenever one uses language to make statements" (p. 292), and Carson develops this sympathetically (pp. 294–96). Charles Fried (1978) went much further: "Every lie is [n.b.,*itself*!] a broken promise, and the only reason that this seems strained is that in lying the promise is made and broken at the same moment. Every lie necessarily implies – as does every assertion – an assurance, a warranty of its truth" (p. 67). That is surely too strong, especially since the idea of an "implicit promise" is puzzling, seemingly more specific than just everyday tacit agreement. On tacit agreement, see below.
26 But see Pruss (2012).

Timmermann and Viebahn's point about social epistemology is a good one, and the word "commitment" naturally alerts the moral sense, but the commitment is not itself moral in kind. (They acknowledge that it is not per se a moral commitment, but insist that it will be morally relevant in one way or another.)

In the context, there may well be a further argument to show that a speaker who violates an illocutionary or a social-epistemological norm of assertion has also broken a moral rule, especially if the belief condition fails. But to support Defeasibility, either that argument would not also apply to deliberate misleading, or there would have to be a further moral argument that prohibited lying but not misleading. And it would have to be a stronger argument than any of the three we have considered so far.

The best candidate that occurs to me would be an appeal to tacit agreement. It is plausible to say that successful communication depends on tacit agreements and in particular that the warranting feature of assertions does include at least a tacit agreement among the speech community to assert only propositions that one believes to be true. Lying would *eo ipso* violate that agreement.

But even that would not prove Defeasibility. For implicature too is governed by, indeed generated by, tacit agreements: the Coöperative Principle, or the Communicative Principle of Relevance, or whichever such general norm does drive the derivation of implicatures and other unasserted but communicated contents. Deliberately deceptive implicating violates an agreement too.

We have not found a strong argument for the Defeasibility view. But as Saul concedes, there is still a widely shared Defeasibility "intuition." It needs at least explaining away, and she offers to explain it away by distinguishing sharply between evaluating actions and evaluating motives and/or character. We have already seen one instance of that strategy in her treatment of the deathbed reassurance case; let us now turn to character and character traits.

Returning to the deathbed case (pp. 87–88), Saul now takes a slightly different line. She contrasts it with a case in which you, the visitor, choose to mislead rather than lie, but not because you care at all about the dying woman's feelings or about truth – only because you think that a legal action is going to ensue regarding her estate and if you are found to have lied to the woman it will cost you a great deal of inheritance money. Here we do not admire your kindness or your virtuous respect for truth, because there is none of either, and, Saul maintains, we are also less inclined to judge that the misleading was morally any better than lying. She further cites cases in which a dishonest politician chooses to mislead rather than to lie only to preserve public deniability and/or some undeserved self-respect (pp. 90, 92).

Saul quotes an early version of Adler (2018), who had written that "the misleading implicature is an effortful compromise between sincerity, resistance to gratuitous hurt, and encouraging good feeling in another" (Saul, p. 78) and she agrees that the compromise speaks well of the agent as a person, but she insists that just to recognize the virtues standing behind the choice does not at all support the view that the misleading itself was a better action.

Here again I am inclined to agree and to applaud Saul's enforcing of a distinction that is easily and often trampled. But I am not a virtue theorist; and it is clear that Saul has no more sympathy for Aristotle and virtue ethics than she has for Kantian views. Virtue ethics (see, e.g., Hursthouse and Pettigrove (2018)) judges actions precisely by the virtues and/or vices they manifest and/or from which they proceed. No virtue sympathizer will accept Saul's simple write-off of virtue considerations as affecting the moral status of an action.

It should be noted that virtues and vices, independently of consequences, can seem to make misleading actually worse than the corresponding lie. Of President Bill Clinton's deliberate misleading in regard to his relationship with Monica Lewinsky, Saul remarks that "[s]ome, in fact, would think that Clinton's careful misleading is somehow sneakier and more reprehensible than an outright lie" (p. 2). And it is easy to think of cases in which the choice to mislead rather than to lie is based entirely on a particular vice: self-righteousness, pride, or simple vanity. A good friend of mine prides himself on never lying to anyone: "A person who has to *lie* in order to dupe, hoodwink, and bamboozle is an amateur." I would agree with Saul that there is nothing redemptive about misleading rather than lying for that sort of reason.

Moreover, personal relationships sometimes require an honesty that favors lying over misleading. Jenny in Kingsley Amis' *Difficulties with Girls* does not want to know about her husband Patrick's frequent infidelities, and she positively resents one of his confessions: "Here I was, I was just starting to get used to the idea, or trying to, . . . and here you go like a bull in a china shop, pushing and shoving things out in the open . . . . Of course you're the one that's better off, showing me how much you care, and forget about how I might feel about it, being told, that's not the same as being sure in your own mind. You selfish pig" (p. 210). Jenny had been happier with Patrick's frequent lies; and we may imagine that she would also have preferred the straightforward lies to smarmy evasions and too carefully crafted implicatures.[27]

Where does all this leave the Defeasibility view? We have seen no very good argument for it. But it would be hard to make a direct argument against it,

---

[27] Having given an example of "bald faced" lying, Horn (2017b) cites a similar sentiment on the part of the character Emily Tallis in Ian McEwan's (2001) *Atonement*: "Even being lied to constantly, though hardly like love, was sustained attention; he must care about her to fabricate so elaborately and over such a long stretch of time" (p. 139).

without appeal to a particular moral theory, such as Act-Utilitarianism as opposed to some particular Kantian view or a fairly detailed virtue theory. I think our discussion has also shown that much depends on the details of particular cases.[28] I conclude, disappointingly, that although Defeasibility may be true,[29] direct argument cannot decide either way, and Defeasibility must stand or fall with more general moral principles. By itself, the Defeasibility intuition carries little if any weight and in the end is not very interesting. Certainly it should not be taken as a guide to moral decision.

Until now this chapter has addressed acts of verbal deception. I now turn to differences between acts of deception, deceiving by omission, and omissions that may be culpable but should not be called deceptive.

5 The morality of silence

People are sometimes condemned for not telling the "whole" truth. Likewise for "concealing" facts, and for "withholding" information. The moral objection is not to what they said but to what they did not say. But as I noted in my opening description of the "scale" of honesty or transparency, that distinction is not highly respected either by ordinary people or in the literature. I want to make a modest plea for it here, and call attention to an important moral difference.

Famously, Mark Twain spoke of a "silent lie."[30] Really? Strictly speaking? The first question to be asked is, can deliberate silence constitute an assertion? Yes, it can, if there is significant stage-setting along the general lines of "Speak

---

28 Here I agree with Timmermann and Viebahn (2021). They document that claim at length and convincingly. One factor that varies greatly with circumstances is the reliability of one's method of deceiving: e.g., sometimes "[a][misleading response might,. . .precisely in virtue of its non-committal nature, give rise to suspicion that would be avoided by a plain lie" (p. 1493). But Timmermann and Viebahn also emphatically appeal to some people's in some circumstances having a "fundamental claim to truthfulness" (pp. 1496ff); I do not buy that, unless they mean just what we ordinarily call the "right to know," on which see my sec. 5 below. However, my position on Defeasibility is on the whole closer to Timmermann and Viebahn's than to anyone else's I have read.

29 For the record, I find myself still having the intuition, despite the debunking arguments we have seen. That in itself is of no interest, but as I have mentioned, I am an act-consequentialist and I have no tendency toward virtue theory. According to my own moral theory, which I not only defend in print but mostly believe, I *should* be firmly and unreservedly with Saul.

30 "On the Decay of the Art of Lying," 1880 (http://www. gutenberg.org/ebooks/2572); thanks to Horn (2017a) for the reference. But Twain was deliberately using "lie" very broadly, to include even facial expressions and bodily postures.

now or forever hold your peace." A deliberately extreme example: You are a member of a club of some kind, and the club has meetings that are quite formal and heavily rule-governed. You arrive for a meeting and when you check in at the door and show your ID, the porter or bailiff says, "Today we're practicing *silent speech*." What is that? "From time to time during the meeting we'll hold up a big purple placard with a curlicued border that has a sentence written on it. If you endorse and accept the sentence as true, do nothing; if you disagree with it or just don't have an opinion, say so. OK? Are you prepared to observe this convention?" You agree to observe it – if necessary, signing a paper – and go on into the meeting. At one point, the chair holds up one of the purple placards that reads, "W.V. Quine was a Harvard philosopher." You remain silent. Later the chair holds up a placard that says "W.V. Quine was born in Kenya." You remain silent; moreover you do so voluntarily – nothing prevents you or inhibits you or even slightly distracts you. It seems to me you have asserted, or at least endorsed, the claim that Quine was born in Kenya.

The question, though, is whether weaker versions of his sort of thing happen in real life. No doubt they do, especially in contexts that are in one or another way formal, but for a silence to be an actual assertion, there must be some conventional element in the context in virtue of which the silence qualifies as one. (Cf. the perhaps conventional element I mentioned in sec. 2 above, regarding questions and answers.) As before, asserting is an in part conventionally defined act. Granted, the convention need not be explicitly specified; there are unspoken conventions. But here I would not be too liberal and speak merely of "tacit agreements" either, for there are plenty of tacit agreements that could not be called conventions.[31]

Here is a nice example of a borderline borderline case, from George Macdonald Fraser's *Flashman and the Dragon* (1985). (My repetition of "borderline" is intentional.) Flashman is conversing with the adventurer Fred T. Ward, who gets the (false) impression that Flashman has been demoted in rank owing to an earlier quasi-military scrape on the Pearl River: "You don't mean they broke you? . . . Gee, I'm sorry about that! I sure am . . . . Over a passel o` guns. Well, I'll be!" (p. 62). Weeks later, after the battle of Sungkiang, Flashman and Ward meet again. Ward asks, "You back in British service, or what? I thought you said they busted you over that Pearl River business." Flashman: "No-o, you said that, and I didn't contradict you" (p. 139). – Here *I* believe Flashman is clearly in the right; he never told Ward that he had been demoted.

---

31 The *Ur*-work on unspoken convention was Lewis (1969); for what I consider a neologistically weak notion of convention, see Millikan (2017).

But my wife Mary Lycan disagrees. She would say that the original conversation was a friendly one and Ward was being if anything very sympathetic; he raised the hypothesis of demotion with question intonation, though he did not *quite* invite Flashman to comment, and Flashman did not correct him. If it is too strong to hold that Flashman *asserted* that he had been demoted, M. Lycan maintains that by his silence he committed himself to that statement. We can imagine a slight variant in which Ward had added "Is that right?" and Flashman still said nothing. I myself do not consider the original (as written) even a borderline case, and I certainly would never judge that Flashman lied; but I am not sure how many nonphilosophers would agree, and so I elevate the case to borderline borderline.[32]

There are degrees of silent commitment. The next weaker grade is a case wherein A knows something, that P, and B has a strong and legitimate interest in whether P, and the topic comes up at least tangentially, and A says nothing. But this, I maintain, is different in kind. I agree right away that B has a complaint against A, in that A has withheld information that A knows B needed. But A has in no sense said that not-P, much less lied. We shall return to this kind of case below.[33]

(I have encountered an odd obverse phenomenon. Sometime in the 1980s, a Gay Pride group in North Carolina declared "statewide Blue Jeans Day," on which day one was urged to wear jeans as a strong public declaration of support for gay rights. This was, of course, mostly a gag, since obviously the Gay Pride group and everyone else knew that a very large number of North Carolinians wear jeans every day and may not even own any alternate clothing. Is there a linguistic equivalent? That would require there to be something that people say all the time, such as "Hello" or "Thank you," and an alleged local temporary convention that that expression is to mean that the speaker strongly supports gay rights or whatever.)

Let us now leave the issue of silent lying and move on to a clearer case of verbal deception by omission: false implicatures that are generated precisely by

---

32 I also wonder if there is a gender difference here; a sociolinguist could tell us.

33 Once a guest on Oprah Winfrey's talk show (I do not remember the guest's name or the date of the episode) recounted that a new neighbor had bought the house next door to hers. The neighbor did not formally introduce himself or ever converse at any length with the Oprah guest, but would wave and say hello over the fence if the two happened to see each other outdoors. At some point the guest somehow learned that the neighbor was an ex-convict who had recently been released from prison. She was indignant at not having been informed of this, and expressed her indignation on national television by saying, "He *lied* to me!" I *trust* no one would defend the guest's charge taken literally.

withholding hyper-relevant information. Some of the implicatures thus generated are very strong ones.

The best examples I know are cases in which what is said is pointlessly specific. Suppose I wish to make the Mayberry Town Council look silly, and I say, "Did you know that in Mayberry, it is illegal for a grocery store to sell canned tomato sauce between 3:30 and 4 p.m. on Memorial Day?" Dumbfounding! But I have withheld the information that Mayberry simply does not allow retail establishments to open on national holidays. Or suppose I tell you that Smedley College has never in its history promoted a woman to tenure. Sounds very bad! But the fact is that as a matter of policy, Smedley never hires any junior person to tenure-track in the first place, and a fortiori never promotes any faculty member from the ranks. (All tenured faculty including women are hired from the outside, and as a matter of fact 65% of Smedley's tenured faculty are women.)

Though they are not lies, the statements in these cases are viciously misleading and deliberately deceptive by implicature, so they are clear cases of linguistic deception. And they work precisely by withholding information. Therefore, remaining silent about something one knows can be the key component of a wrongful linguistic deception.[34] As yet there is no break in the honesty/transparency scale.

Rather, my case for the break begins with the following: cases of silence that do *not* contribute to the linguistic conveying of false information. Of course, such a silence may be morally wrong, for any of a number of reasons, and, further, it may constitute an action that is dishonest or untransparent in a very broad sense; but it will not be wrong in the same way or for the same reasons that verbal lying and deliberate misleading are wrong (when they are). In particular, the reasons will not be specific to the ethics of communication,[35] but will arise in the ordinary way from one's general moral theory – Intro Ethics business as usual. And there is here a significant difference between commission and omission, between negative and positive duties if you like. Pure

---

34 Marta Dynel (2018) makes the same point, on the basis of several other types of example, but she is a little readier than I to see implicatures and ditto deception.

35 Of course, not all moral objections to plain lying, such as a Benthamite Act-Utilitarian's, are specific to the ethics of communication either, but some theorists' are. For example, Strudler (2009); and especially Shiffrin (2014; note her title, "*Speech Matters*"), who argues that lying is distinctively wrong, even when it does not deceive: "Speech. . .plays a special role in allowing us access to. . .necessary information [about the speaker – the information that affords accurate, discretionary self-revelation], helping us to overcome the opacity of one another's minds" (p. 10). "[T]o lie is to sully the one road of authoritative access to oneself and thereby cut oneself off from community with others" (p. 23); lying conceptually subverts the processes that uniquely enable our direct communication with each other.

consequentialists will have no truck with differences of that kind, but then neither will pure consequentialists see any point to the honesty/transparency
scale, or grant that there is any particular "ethics of communication," per se.

Now, as to words like "conceal" and "withhold": To conceal information is
to hide it, to take steps to prevent someone from learning it when in the ordinary course of events they would or might have learnt it. Concealing is a positive action and obviously subject to moral evaluation. But it is not an act of
conveying information. "Withholding" is weaker, and not (ordinarily) a positive
action. Yet the word at least connotes something done by choice, and it connotes something of potential moral significance; we would not say that in publishing this chapter I, WGL, have *withheld* from you the information that I was
born in Milwaukee, USA, or for that matter that Bach's fourth Two-Part Invention is in the key of D minor. Carson (2010, p. 54) speaks of "keeping someone
in the dark," and defines it in terms of knowing that the victim wants the information in question and/or "occupying a role or position" in which one is expected to provide such information. Though less strongly, much the same goes
for negative terms such as "not divulging," "not disclosing" and "not revealing." Actually I cannot think of a term that means strictly just *not telling*, with
no further connotation whatever – not even "not telling."

It should be noted that not all withholding, indeed not all concealing, even
begins to raise a moral issue. I am thinking of cases in which the information
(a) for privacy reasons is no one else's business and/or (b) it does not matter to
anyone concerned. E.g., in philosophy papers, including this one, I sometimes
use examples involving friends or relatives of mine, and I may disguise those
people a bit, though without either stating or strongly implicating any falsehood. (For my expository purposes it does not matter even whether the example
involves a real person. Why, then, do I bother? I am not sure.)

In what sorts of cases, then, is an agent open to moral criticism for silence
per se? I shall not attempt a survey, but merely note a few. (1) Obviously there
may be legal or formal contractual requirements, or I otherwise "occupy a role
or position" in which I am expected to provide certain information. (2) There
are other times when we say, for whatever reason, that such-and-such a person
"has a right to know." Even then it would not follow that any particular other
person has a correlative duty to tell them, but suppose that I am the only one
who is in a position to do so. (3) Not telling someone something would have
very bad consequences, possibly for many. (4) Due to the nature of a personal
relationship, I might naturally be expected to share certain information, background or circumstantial, about myself. Marriage is the obvious such relationship, and clear examples involve health-related or biographical information,
even when it is no business of anyone's other than the spouse's. (5) Due to the

nature of a personal relationship, I might be expected to share information that I know will be helpful to the other person given her/his projects or interests.

But in few if any of these cases is it accurate to speak of *deception*, even when my silence has been very wrong and richly deserving of moral blame. In few is it accurate, except very broadly, even to speak of dishonesty (assuming as before that there is no communicative convention in play). That is one reason why I see a substantial break in the honesty/transparency scale.

Now, here is a questionable case: What if I said something, wholeheartedly believing and meaning it and with no smidgen of deception or dishonesty, but later came to believe otherwise, yet did not then inform the original hearer of that change? (Notice again that there had been no false implicature or other misleading.) Carson (2010) cites examples of this kind that do qualify as deceptive. They are ones in which I had been trying to sell you something or at least persuade you to do something. "My failure to report my earlier mistake to you counts as deception. When I realize my mistake and fail to take advantage of my opportunity to correct it, I am intentionally causing you to persist in believing something that is false" (p. 53).

I do not buy "*causing you to* persist," but I agree there is deception going on (assuming you have not yet done what I wanted you to do, and that I still want you to do it, and that you will do it only because you believe what I told you). I explicitly tried to persuade you to do X, offering a set of inducements or other reasons, and I now know that one or more of the reasons was false. Not to retract would be dishonest and wrongful manipulation. (We might construe that as a case of misleading by *extended* implicature, if there were such a thing as extended implicature.)

But now subtract the elements of persuasion and self-interest: As in Carson's examples, I say something, believing it and meaning it and with no false implicatures, but then come to learn that it is false. In this spare version, the hearer is unaffected. Nothing at all hangs on it; the hearer may continue to believe what I said, but does not care one way or the other, and I know that. Assuming there are no further, extraneous considerations, I have no moral obligation to seek out the hearer and correct what I said. (Being a philosopher or otherwise truth-obsessed, I *might* do that, but only for my own peace of mind and not because it is a moral duty.) What this shows is that the deception in Carson's original examples is wrong only because of general moral principles governing interpersonal transactions and balancing of interests, not solely and *eo ipso* because I had caused the hearer to believe something false. So this sort of case does not after all threaten my claim that when silence is morally wrong (and not because it generates a false implicature), that must be for circumstantial reasons proceeding from general moral theory, rather than because of anything specific to the ethics of communication.

There are intermediate cases. Quite a common one is an everyday statement of intention, as in "Tomorrow I'm going to go to Walmart and buy a new bathroom scale," after which the speaker forgets about doing that or changes her/his mind.[36] I am speaking of a statement which merely expresses the intention, not of a promise or other declaration on which the hearer is authorized to rely. If the speaker has some reason to suppose that the hearer *will* in some way rely on the prediction, there is at least a slight obligation to notify the hearer by retracting But, here too, silence will be wrong only when there is such reason.

Returning for a moment to the matter of trust: Can silence betray a trust? Of course it can. But we must ask, a trust to do (or not to do) what? And all my foregoing arguments will apply.

6 Summary

I have defended two main claims:

First, there is no strong argument for the thesis that, other things being equal, outright lying is morally more objectionable than deliberate misleading by implicature; the Defeasibility view must stand or fall with one's more general moral principles. And even if the thesis is true, it should not per se be adopted as a guide to moral decision.

Second, there is a significant moral difference between positive lying *or* misleading and merely not revealing something that one knows.[37]

References

Adler, Jonathan. 1997. Lying, Deceiving, or Falsely Implicating. *Journal of Philosophy* 94: 435–52.

Adler, Jonathan. 2018. Lying and Misleading: A Moral Difference. In Eliot Michaelson and Andreas Stokke (eds.), *Lying: Language, Knowledge, Ethics, and Politics*. Oxford: Oxford University Press.

Boër, Steven and William Lycan. 1986. *Knowing Who*. Cambridge, MA: Bradford / MIT Press.

Brandom, Robert. 1983. Assertion, *Noûs* 17: 637–50.

Brandom, Robert. 1994. *Making It Explicit*. Cambridge, MA: Harvard University Press.

Carson, Thomas. 2006. The Definition of Lying, *Noûs* 40: 284–306.

---

36 A person who shares my household is astonishingly *anti*-reliable in this regard.

37 Thanks to Tom Carson, Jenny Saul and especially Larry Horn for very helpful comments on a previous draft.

Carson, Thomas. 2010. *Lying and Deception: Theory and Practice*. Oxford: Oxford University Press.

Carston, Robyn. 1988. Implicatures, Explicatures, and Truth-Theoretic Semantics. In Ruth Kempson (ed.), *Mental Representation*. Cambridge: Cambridge University Press.

Chisholm, Roderick and Thomas Feehan. 1977. The Intent to Deceive. *Journal of Philosophy* 74: 143–59.

Coleman, Linda and Paul Kay. 1981. Prototype Semantics: The English Word *Lie*. *Language* 57: 26–44.

Dynel, Marta. 2018. To Say the Least: Where Deceptively Withholding Information Ends and Lying Begins. *Topics in Cognitive Science* 12: 555–82.

Falkenberg, Gabriel. 1982. *Lügen: Grundzüge einer Theorie Sprachlicher Täuschung*. Tübingen: Niemeyer.

Fallis, Don. 2009. What is Lying?. *Journal of Philosophy* 106: 29–56.

Faulkner, Paul. 2014. The Moral Obligations of Trust. *Philosophical Explorations* 17: 332–45.

Fraser, George Macdonald. 1971. *Flash for Freedom*. New York: New American Library.

Fraser, George Macdonald. 1985. *Flashman and the Dragon*. New York: Penguin Books.

Fried, Charles. 1978. *Right and Wrong*. Cambridge, MA: Harvard University Press.

Green, Stuart. 2006. *Lying, Cheating, and Stealing: A Moral Theory of White-Collar Crime*. Oxford: Oxford University Press.

Heffer, Chris. 2020. *All Bullshit and Lies?* Oxford: Oxford University Press.

Hill, Thomas. Jr. 2016. Human Dignity and Tragic Choices. *Proceedings and Addresses of the American Philosophical Association*, Presidential Address (Eastern Division): 74–97.

Horn, Laurence. 2009. Implicature, Truth, and Meaning. *International Review of Pragmatics* 1: 3–34.

Horn, Laurence. 2017a. Telling It Slant: Toward a Taxonomy of Deception. In Janet Giltrow and Dieter Stein (eds.), *The Pragmatic Turn in Law*, 23–55. Berlin: De Gruyter.

Horn, Laurence. 2017b. What Lies Beyond: Untangling the Web. In Rachel Giora and Michael Haugh (eds.), *Doing Pragmatics Interculturally*, 151–174. Berlin: De Gruyter.

Hursthouse, Rosalind, and Glen Pettigrove. 2018. Virtue Ethics, *Stanford Encyclopedia of Philosophy* (Winter 2018 Edition), E.N. Zalta (ed.), URL = https://plato.stanford.edu/archives/win2018/entries/ethics-virtue/.

Kalb, Marvin and Bernard Kalb. 1974. *Kissinger*. Boston: Little, Brown & Co.

Lasersohn, Peter. 1999. Pragmatic Halos. *Language* 75: 522–51.

Lewis, David. 1969. *Convention*. Cambridge, MA: Harvard University Press.

Lycan, William. 1994. Relative Modalities. In *Modality and Meaning*. Dordrecht: Kluwer Academic Publishing.

Mahon, James. 2016. The Definition of Lying and Deception. In Ed Zalta (ed.), *Stanford Encyclopedia of Philosophy* (Winter 2016 Edition), URL = https://plato.stanford.edu/archives/win2016/entries/lying-definition/.

Matthews, Robert. 2013. Belief and Belief's Penumbra. In Nikolai Nottelmann (ed.), *New Essays on Belief*. London: Palgrave Macmillan.

McEwan, Ian. 2001. *Atonement*. New York: Doubleday.

Meibauer, Jörg. 2014a. *Lying at the Semantics-Pragmatics Interface*. Berlin: De Gruyter.

Meibauer, Jörg. 2014b. A Truth That's Told with Bad Intent: Lying and Implicit Content. *Belgian Journal of Linguistics* 28: 97–118.

Millikan, Ruth. 2017. *Beyond Concepts: Unicepts, Language, and Natural Information*. Oxford: Oxford University Press.

Pruss, Alexander. 2012. Sincerely Asserting What You Do Not Believe. *Australasian Journal of Philosophy* 90: 541–46.

Rees, Clea. 2014. Better Lie!. *Analysis* 74: 59–64.

Sanders, Lawrence. 1992. *The Seventh Commandment*. New York: Berkley Books.

Saul, Jennifer. 2012. *Lying, Misleading, and What is Said*. Oxford: Oxford University Press.

Shiffrin, Seana Valentine. 2014. *Speech Matters*. Princeton: Princeton University Press.

Simpson, David. 1992. Lying, Liars and Language, *Philosophy and Phenomenological Research* 52: 623–39.

Solan, Lawrence and Peter Tiersma. 2005. *Speaking of Crime: The Language of Criminal Justice*. Chicago: University of Chicago Press.

Sperber, Dan and Deirdre Wilson. 1986. *Relevance: Communication and Cognition*. Cambridge, MA: Harvard University Press.

Strudler, Alan. 2009. Deception and Trust, in Clancy Martin (ed.). *The Philosophy of Deception*. Oxford: Oxford University Press.

Strudler, Alan. 2010. The Distinctive Wrong in Lying. *Ethical Theory and Moral Practice* 13: 171–79.

Timmermann, Felix and Emanuel Viebahn. 2021. To Lie or To Mislead?. *Philosophical Studies* 178: 1481–1501.

Van Leeuwen, Neil. 2014. Religious Credence is Not Factual Belief. *Cognition* 133: 698–715.

Williams, Bernard. 2002. *Truth and Truthfulness: An Essay in Genealogy*. Princeton: Princeton University Press.

Wilson, Deirdre and Dan Sperber. 1986. Inference and Implicature. In Charles Travis (ed.), *Meaning and Interpretation*, 45–75. Oxford: Basil Blackwell.

Roy Sorensen
Kant tell an *a priori* lie

Abstract: An *a priori* lie is a lie that contradicts an *a priori* truth. Rather sportingly, the *a priori* liar leaves himself open to refutation by armchair methods such as calculation. My thesis is that Immanuel Kant precludes the existence of *a priori* lies. For asserting a proposition requires raising a rational expectation of its truth. If the hearer believes the negation of an *a priori* proposition, Kant blames the deceived, not the deceiver. We are sometimes permitted to believe *beyond* the evidence but never *against* the evidence. The impossibility of *a priori* lies vindicates the advocate's assumption that one cannot lie with deliberately invalid deductions or by engaging in insincere legal semantics.

A cartoon by Dan Piraro features a bookstore cashier: "Two *Math for Dummies* at $16.99 each. That'll be $50." An *a priori* lie is a lie whose content is the negation of an *a priori* proposition. To believe an *a priori* lie is to believe a proposition *contrary* to what is non-experientially warranted. Neglecting this evidence makes the *deceived* culpable for the deception. Does the deceiver share responsibility?

Plato depicts the sophists as never accepting blame for *a priori* deception. After all, the sophists openly advertise their ability to make the weaker argument appear the stronger. Insincere inferences are inevitable given an adversarial legal system. Defendants pay lawyers to persuade rather than discover the truth. That is why Cicero boasts of having thrown dust in the eyes of jurors (Smith 1995). Advocates who restricted persuasion to sincere inferences would be outcompeted by less inhibited arguers.

The Enlightenment emphasized an individualistic rationale for permitting *a priori* deception. Although you must rely on others for knowledge of premises, you are on your own when assessing the validity of an argument. Samuel Johnson concludes that a lawyer bears no responsibility for the deceptive consequences of his arguments. The advocate's role is to construct arguments. Responsibility for appraising the arguments is borne by judges and juries. Section 1 summarizes Johnson's cognitive division of labor.

Acknowledgement: This essay has been improved *a priori* by Tom Carson and *a posteriori* by Laurence Horn and Ian Proops. More help came from audiences at the EST group at St. Andrews University and ERGo group at the University of Texas at Austin.

Roy Sorensen, University of Texas – Austin, e-mail: roy.sorensen@austin.utexas.edu

https://doi.org/10.1515/9783110733730-004

Section 2 turns to Immanuel Kant's refined application of developmental individualism to *a priori* lying. Kant defines "enlightenment" as humanity's emergence from self-imposed immaturity. Each human being is designed to think for himself. But most are timid and lazy. Coddling mothers and domineering fathers exacerbate this reluctance to suffer the falls necessary to walk on one's own. Instead of attending to their own finances, these Peter Pans rely on guardians such as cashiers. For Kant, the epistemological root of this immaturity is the following myth of *a priori* testimony: Just as one can learn from *a posteriori* testimony, one can learn from *a priori* testimony. My main thesis is that Kant deracinates this myth by precluding the possibility of *a priori* lies.[1] His eliminativism about *a priori* lies is a surprising resource for lawyers who endorse high standards of honesty (as measured by their willingness to be disbarred for legally significant *a posteriori* lies) and yet defend courtroom deception by means of deliberately fallacious arguments and insincere legal semantics.

Section 3 distinguishes prudential reasons for *a priori* truth-telling from moral reasons. This leaves *a priori* lying a tactical option for advocates who must cope with bans on *a posteriori* lying. Section 4 reviews how the mercenary sophists set the stage for a moralistic second act. Kant spotlights autonomy as a rationale for ranking misleading truth-telling over lying. Section 5 continues to my punchline: the autonomy rational commits Kant to rank *a priori* false-telling above misleading truth-telling. In section 6, this unexpected high score for *a priori* deception is buttressed with Kant's whipsawed account of testimony: we are obliged to heed *a posteriori* testimony while forbidden to heed *a priori* testimony.

This asymmetry, the focus of section 7, is enticing for fields where there are poor prospects for empirical warrant: mathematics, metaphysics, ethics. Section 6 shows how it opens a workspace for lawyers. As long as their deception is *a priori*, sophistical lawyers can meet Kant's lofty moral standards. For Kant treats *a priori* lies as non-lies. He is committed to bold toleration of deceptive falsehoods on *a priori* matters. This includes falsehoods about morality ("The unjust can be happy"), which Plato ranks as the worst lies (Mahon 2017, 20).

Section 8 raises a logical caveat. Kant overestimates how much can be learned by *a posteriori* testimony given his ban on *a priori* testimony. He neglects

---

1 Those who deny that there are any *a priori* propositions vacuously agree that there are no *a priori* lies. Kant's preclusion is driven by his belief that *a priori* warrant survives all possible future experiences (since experience can only detect what is *actually* true and all *a priori* truths are necessary). The preclusion of *a priori* lies survives dilution of Kant's potent apriority as long as the residue still prevents rational belief in the lie. According to this negative remnant of aprioricity, you can be deceived into rationally suspending judgment about whether the *a priori* proposition is true, but you cannot be deceived into rationally disbelieving it.

the fact that a posteriori propositions always entail *a priori* truths. In the last section I address the economic objection that *actually* refuting an *a priori* lie is often costlier than refuting an *a posteriori* lie. This flushes out Kant's commitment to an anti-utilitarian ethics of belief.

1 Litigating your way to hell

The Scottish Calvinist James Boswell feared becoming a lawyer. He did not want to go to hell with the other liars. His father, an eminent judge, had beat into him a need for truth and integrity:

> I do not recollect having had any other valuable principle impressed upon me by my father except a strict regard for truth, which he impressed upon my mind by a hearty beating at an early age when I lied, and then talking of the dishonour of lying.   (Boswell 2001, 363)

Paternal punishment combined with maternal religiosity to yield an adolescent with a morbid fear of lying. This honesty is widely credited with making James Boswell an excellent biographer. When reviewers asked whether it was proper to reveal Samuel Johnson's follies, whims and private conversations, Boswell answered, "Authenticity is my chief boast".

Judge Boswell was chagrinned by his son's reluctance to follow him into the legal profession. The judge should have realized that respect for truth suggests parallel respect for validity. Any proof of an obligation to assert only what is true seems to yield a correlative obligation to infer only what is validly entailed.

Advocates have no greater entitlement to bend the truth than their clients. For a client who chooses to represent himself has not thereby diminished his range of permissible arguments. True, there is a superstition that indirect immoralities are less wrong than direct ones. But judges scoff at attempts to diminish the seriousness of perjury by hiring an intermediate to do the lying.

Frightened of the demons that would drag him to hell, young Boswell slept with a lit candle. At sixteen, he had a nervous breakdown. His father was oak but he was of "finer but softer wood."

Boswell drifted through half-hearted legal studies at a sequence of institutions. He finished his dilatory training only after meeting Samuel Johnson in Davis's London bookshop. Johnson, who had considered a legal career, defended the permissibility of deceptive advocacy. Boswell includes their discussion of legal debate in *The Life of Doctor Samuel Johnson* and in his adventure narrative *A Tour of the Hebrides*.

Boswell had previously sought refuge in the resemblance between lawyers and actors. Actors are paid to utter lines but are not engaged in serious speech.

Johnson rejects Boswell's theatrical analogy. The lawyer is not pretending to assert premises and to draw conclusions. Barristers are responsible for what they say in court. There are sanctions against contempt of court and "misstatements". When criticized, barristers offer justifications and excuses. Since they retract some comments, they must have earlier made assertions.

When addressing judge and jury, Johnson emphasizes, everyone has fair warning that an attorney is speaking *as an advocate*. The barrister's wig and gown signal to the other parties in the compartmentalized project that that they, not he, have the role of assessing the argument.

Johnson is a rugged individualist. All the arguer need do is to present the raw argument. Just as a fisherman does not need to gut the fish or cook the fish or pick a bone from your throat, the arguer does not need to facilitate your digestion of a proof.

> Johnson having argued for some time with a pertinacious gentleman; his opponent, who had talked in a very puzzling manner, happened to say, "I don't understand you, Sir;" upon which Johnson observed, "Sir, I have found you an argument; but I am not obliged to find you an understanding." (Boswell 1835, 317)

In a spirited three-way conversation with Joshua Reynolds, Boswell is stranded by a step in Johnson's reasoning. Ignoring Boswell's cry for help, Johnson turns to Reynolds: "Boswell is now like Jack in *The Tale of a Tub*, who, when he is puzzled by an argument, hangs himself. He thinks I shall cut him down, but I'll let him hang." (Boswell 1835, 278)

Johnson agrees that the testimony of paid witnesses should be ignored. So, Boswell inferred, the arguments of paid lawyers should also be ignored. Johnson retorted:

> Nay, Sir, argument is argument. You cannot help paying regard to their arguments if they are good. If it were testimony, you might disregard it, if you knew that it were purchased. There is a beautiful image in Bacon upon this subject. Testimony is like an arrow shot from a long-bow; the force depends on the strength of the hand that draws it. Argument is like an arrow from a cross-bow, which has equal force though shot by a child. (Boswell 1835, 281–2)

Learning from testimony requires empirical evidence of the competence and honesty of the speakers. Learning from argument only requires recognition that the premises support the conclusion.

Johnson's defense of wittingly invalid argumentation is limited to formal legal proceedings. He agrees with Boswell that a barrister should not argue insincerely with his family and friends.

Nor should there be insincere arguments between scholars. Johnson does not picture the editor as a judge and the readers as a jury. Johnson disapproved of patriotic historians who sophistically defended myths. In 1760 the Scottish poet James Macpherson published the English-language text *Fragments of ancient poetry, collected in the Highlands of Scotland, and translated from the Gaelic or Erse language.* Johnson charged that Macpherson was "a mountebank, a liar, and a fraud, and that the poems were forgeries". He disparaged the quality of the poems and characterized Gaelic as "the rude speech of a barbarous people who had few thoughts to express, and were content, as they conceived grossly, to be grossly understood". (1835, 95)

Johnson made the more checkable claim there were no manuscripts in Gaelic older than a century. The holdings officer of the Advocates' Library at Edinburgh responded with several Gaelic manuscripts, each five centuries old, plus one manuscript of even greater antiquity. "The contest over the authenticity of Macpherson's pseudo-Gaelic productions became a seismograph of the fragile unity within restive diversity of imperial Great Britain in the age of Johnson." (Curley 2009, 1)

Johnson was impatient with these learned rebuttals. He believed the Scottish scholars knew the manuscripts were forgeries.

> I asked a very learned Minister in Sky [sic], who had used all arts to make me believe the genuineness of the book, whether at last he believed it himself. But he would not answer. He wished me to be deceived, for the honour of his country; but would not directly and formally deceive me. Yet has this man's testimony been publickly produced, as of one that held Fingal to be the work of Ossian. (Boswell 1835, 98)

Instead of invoking the cross-bow analogy for historians, Johnson disparages those who "love Scotland more than truth". Johnson does not explain why sincerity is important for scholarly debate but not for legal argument. The double standard needs justification because the stakes for legal argument are *higher* than the stakes for the history of literature. If sincerity in argument mattered for either, it should matter more for legal argument.

Arguably, the Scottish sophists were more consistent than Johnson. What was admissible in court was fairplay in history. Their scholars contributed to Macpherson's triumphal burial among the literary giants in Westminster Abbey. Even now the Scots take pride in Macpherson perpetuating "the most successful literary falsehood in modern history" (Curley 2009, 1).

Yet Johnson permitted *a posteriori* lies on some points of literary history. He defends an author's right to lie in defense of his anonymity (1837, 307). Johnson reasons that since others can prevaricate to defend the author's anonymity, so can the author himself. After attaining fame and financial security, Johnson consistently repents his fraudulent reports of Parliamentary debates (written while an impoverished Grub Street literary hack). But Johnson did not repent deliberately misspelling the publisher of his poem "London" (so that Londoners would falsely infer the new poet was already popular enough to be pirated). The success of this mass deception delighted him. And Johnson never objected to ghostwriting. The hypochondriacal Johnson does not extend a license to lie to physicians. He bitterly complains of physicians who have told him paternalistic lies.

Johnson lacks the discipline and theoretical infrastructure to systematically develop his defense of lawyers. Fortunately, he had a Prussian contemporary who was marshalling the needed distinctions, doctrines, and architectonic vision.

2 Kant's preclusion of *a priori* lies

In *The Critique of Pure Reason*, Kant defines *a priori* knowledge as "knowledge that is absolutely independent of all experience" (B3). No observation or experiment can confirm or refute an *a priori* proposition (though perception of a counterexamples can show a generalization is not *a priori*). According to Kant, all *a priori* truths are *necessary* truths.[2] Experience can only establish what is actual. He agrees that experience is needed to acquire the concepts that formulate propositions. For instance, experience of alteration is needed to understand "Every alteration necessitates a cause". This experience enables scientists to presuppose that *a priori* truth. But the experience cannot prove `Every alteration necessitates a cause'. Consequently, any putative empirical support for the negation of an *a priori* proposition contains the seeds of its own debunking.

Kant gives no examples of *a priori* lies. My explanation is that his account of testimony precludes them. In a genuine lie, the listener is at the mercy of the speaker. In contrast, we are all forearmed against negations of *a priori* truths.

A congenitally blind man lacks the concepts to grasp the synthetic a priori truth that nothing is red all over and green all over. His handicap prevents him from identifying the absurdity of anything being red all over and green all over. Fortunately, the same handicap prevents him from being deceived into *believing* it. At worst, he is fooled into uncomprehendingly uttering, "Something can be red all over and green over".

---

2 In opposition to Kant, many contemporary defenders of *a priori* truths are open to the possibility of contingent *a priori* truths. The sophist Gorgias argues that nothing exists, and that if something exists, it could not be known to exist. And if something could be known to exist, then this fact could not be communicated. A listener, without leaving his armchair, has all he needs to know that each contingent conclusion is false.

Lying requires a *rational* expectation that the lie is true. Kant signed his letter to Elisabeth Motherby, "I am, with greatest respect, my honoured lady's most obedient servant" (February 11, 1793). If she read the statement literally, would Kant be rendered a liar? Kant was sometimes sarcastic. If a reader ignores signals that Kant intended the opposite of what he said, is Kant thereby made a liar? Kant told stone-faced jokes. If a dullard took the dead pan humor seriously, is Kant then a liar? No, no, no. Honesty is not hostage to hearers' irrationality.[3]

Alas, the crooked timber of humanity expects the speaker to accommodate to their irrationalities. They resent questions that require *a priori* labor. Intellectual laziness is abetted by paternalistic clergy, physicians, and politicians. Instead of admonishing the masses to think for themselves, these guardians seek to shepherd humanity with simple commands.

The guardians build sympathy for the innumerate customer in the bookstore by noting the difficulty of the sum in binary notation:

$$10000.1111110101110000101 + 10000.1111110101110000101 = 110010$$

Kant acknowledges there are performance errors in which a competent but stressed speaker denies *a priori* truths. He thinks the remedy is for the speaker to take *more* responsibility; study the statement with greater attention, organization, and diligence! Kant concedes,

> One can, of course, believe mathematical rational truths on the basis of witness reports, in part because error is in this case not easily possible, in part because it can be easily discovered, but one certainly cannot know them in this way. Philosophical rational truths are, however, not even able to be believed, they must only be known; for philosophy allows in itself no mere convincing. And particularly what applies to the objects of practical rational knowledge in morals . . . even so little could a mere belief occur in the case of these.
>
> (Kant [1900] IX, 68–70, Joseph Shieber's (2010, 334) translation)

Since belief requires understanding, no one can be deceived into believing the negation of a philosophical rational truth. As evident to essay graders, students are deceived into uncomprehendingly quoting the false sentence and believing that whatever was uttered is true. But this meta-linguistic belief does not disquote into first order error. The parroting students resemble someone ignorant of hexadecimal numerals who trustingly says, "'10.fd70a3d70a3d + 10.fd70a3d70a3d = 32' is true". Incompetence prevents the utterance from accurately reporting a belief in the quoted falsehood. The speaker is not even wrong!

---

3 The rejection of *a priori* lies provides a novel rationale for the rejection of lying about fiction. When constructing the tale, the author cannot lie because his narrative assertions are true by stipulation. Any insincere denial of one of his analytic truths is an *a priori* falsehood.

The impossibility of philosophical lies is preserved by those who dismiss philosophical utterances as nonsense. Ludwig Wittgenstein denies there are any philosophical propositions. So, no one can be deceived into believing a false philosophical proposition.

The negations of these falsehoods are *a priori* in the sense that they are accessible simply by virtue of linguistic competence. (Ditto for the utterances of nonsense that only seem grammatical.) But the students are incompetent. A diligent student will check whether he understands. Purchasers of *Math for Dummies* deserve credit for recognizing their innumeracy and taking reasonable measures to rectify their intellectual embarrassment.

3 Prudential objections to *a priori* lying

A priori trickery is often imprudent. Instead of blaming themselves, customers blame the cashier. For the sake of appearances, the prudent cashier accurately calculates. Shortchanging customers is short-sighted.

Although animals are outside the moral community, Kant condemns cruelty to cats, cows, and even the loud cock that forced him to move to new lodgings in 1777. For the resemblance of creatures to people makes sadism toward roosters a stepping stone to mistreatment of rational beings. *A priori* lying resembles *a posteriori* lying.

Accordingly, one genuinely moral motive for sincere public calculation is that deliberate miscalculation leads to genuine lies such as misreporting prices to inexperienced customers (Kant 2012, 12). The false report of a price is the negation of an *a posteriori* truth. The impression that the cashier *lied* with an *a priori* assertion may arise from mistaking what she said as a report of her *believing* the mathematical proposition.

James Mahon (2008, 220) defines lying disjunctively as asserting p with *either* an intent to deceive the hearer into believing p *or* an intent to deceive the hearer into believing the speaker believes that p. Since all propositions about the beliefs of contingent beings are *a posteriori*, Mahon's second disjunct would unmask all *a priori* lies as *a posteriori* lies.

4 Advocacy and *a priori* lies

The sophists were traveling advocates for hire. A local can persuade neighbors by appealing to his standing in the community. The sophist is an out-of-towner who persuades solely by his speech. To demonstrate he could make the weaker case appear the stronger, the sophist reversed the philosophers' policy of arguing for the most probable conclusion. The sophist argues for the least probable. If in town for a couple of days, sophists would first argue persuasively for one side and then, on the next day, argue just as well for the opposite. Fathers in the audience may have publicly disapproved of this two-faced performance. But they privately hired the sophist as a tutor for their sons.

Socrates draws an invidious comparison between the competitive debate of lawyers and the cooperative dialogue of philosophers (*Theaetetus* 173). Lovers of wisdom do not exploit *a priori* errors.[4] Philosophers help each other argue well. In the *Meno*, Socrates is a caretaker of a slave boy's reasoning. When the boy makes a fallacious inference, Socrates draws out the absurd consequences so that the beginner can recognize the mistake and try again. The psychic pollution of unjustified beliefs is cleansed by cross-examination (Sophist 230a-e).

Plato models philosophy on geometry. The geometers begin with shared axioms and reason together toward communal conclusions. Perhaps this intellectual communism reigned among Pythagoras' monks. But outside the monastery, geometers compete to be the first to discover a theorem. In Archimedes' preface to "On Spirals", he reports sending colleagues two pseudo-theorems ". . . so that those who claim to discover everything, but produce no proofs of the same, may be confuted as having pretended to discover the impossible" (Heath, 1912, 151). Archimedes disdainfully notes that one of the unnoticed plagiarism traps was an obvious inconsistency.

The dupe of an *a priori* lie receives rough justice for not respecting his own evidence. The *a priori* liar renders a public service. Victims are embarrassed into thinking for themselves. When an *a priori* lie and an *a posteriori* lie are equally effective means of deception, the sporting prefer the *a priori* lie. That gives listeners a fair opportunity to escape error.

My sociological thesis is that the sophist's tolerant attitude toward *a priori* deception was revitalized by Enlightenment individualism. Instead of viewing debate as akin to a friendly game of chess in which players can take back moves to nurture the game's potential, debate was regarded as a tournament. Each player aims to

---

4 This is why we are scandalized by John Beverslusis (2000). He contends that Socrates deliberately uses underhanded debating tactics. In the pseudo-Platonic dialogue "Axiochus", Socrates consoles a dying dullard by assembling a buffet of arguments and seeing which is swallowed. Tim O'Keefe (2006) groups this intellectual comfort food with Sextus Empiricus's method of equipollence. Sextus cures patients of philosophy by balancing each pro with a con. Repeated stalemates lead the patient to quit the neurotic game of reason. Cured patients tranquilly follow custom.

win and is free to play the man rather than the board. If the player lures his adversary into a blunder, the loser has no one to blame but himself. The collective aim of a tournament may be excellent chess. But the individual aim of each player is victory. Samuel Johnson articulated this tournament rationale for adversarial legal systems (which contrasted with the more straightforward epistemology of inquisitorial systems).

Despite being the secular philosopher most opposed to *a posteriori* lies, Kant minimizes his basis to oppose *a priori* lies (and maximizes his basis to classify "a priori lie" as a misnomer akin to "pretend lie"). As a byproduct, Kant vindicates the barrister's belief that he is not lying when marshalling arguments he regards as invalid. Nor is the attorney lying when insincerely asserting contradictions. At the same time, the lawyer is lying when disingenuously asserting *a posteriori* propositions. Lawyers can be disbarred for such lies (in an impressive contrast with priests who are not defrocked for their lies). Moreover, Kant is free of the unsystematic second thoughts that litter Samuel Johnson's defense of lawyers.

To summarize, *a priori* lies are precluded by the confluence of three Kantian themes. First, lying requires *rational* expectations of truth-telling actively raised by the speaker. Second, those expectations are limited to what the speaker said (not what was conveyed by conversational implicatures, not what may be surmised by accents, speech impediments, etc.). Third, all of these expectations purport to be based on an individual experience (not a proposition independently available to all human beings such as analytic implications or the pre-conditions of experience). Consequently, all lies purport to be *a posteriori* truths. All *a priori* lies are counterfeit lies (and so are not forbidden *as lies* though they are sometimes forbidden on other grounds).

5 Extending Kant's defense of telling misleading truths to *a priori* lying

According to Kant, only a good will has moral worth for its own sake. This leads some to infer Kant opposes lying because lying entails an intention to deceive. They are surprised that Kant permits much intentional deception. You are permitted to pack your bags to deceive a suspected thief into believing you will be traveling. You are permitted to tell misleading truths to evade intrusive inquiry. In Kant's preface to *The Conflict of the Faculties*, he forthrightly reports misleading King Frederich Wilhelm II. In the king's counter-enlightenment campaign, Kant was ordered to refrain from writing on religion. Kant pledged to desist "as your

Majesty's faithful subject". As Kant hoped, King Friedrich Wilhelm II soon died. Kant then returned to the topic of religion, pleased to have craftily avoided a lifetime commitment. Alasdair MacIntyre (1995, 337) summarizes:

> Kant therefore places himself among those who hold that my duty is to assert only what is true and that the mistaken inferences that others may draw from what I say or what I do are, in some cases at least, not my responsibility, but theirs. Those others, if they discover that, in such cases, what I said or did was well designed to mislead, as it was in Kant's own case, will certainly in the future treat me, and possibly others, as less trustworthy. But it is not this possible consequence of injury to trust that matters; what matters is the avoidance of the assertion of falsity.

The innocent victim of a lie infers only what has been licensed by the assertion. The victim can trace the falsehood back to the ancestral content of the lie. In the case of truthful misleading, there is no such pedigree. The falsehood is a bastard sired by the hearer's background beliefs commingled promiscuously with the speaker's truthful assertion. Jonathan Adler (1997, 144) endorses Kant's contrast:

> The underlying idea is, presumably, that each individual is a rational, autonomous being and so fully responsible for the inferences he draws, just as he is for his acts. It is deception, but not lies, that requires mistaken inferences and so the hearer's responsibility. (A plausible qualification, however, is that, if the intended victim is doing no wrong, the speaker does have an imperfect duty of assistance to correct the mistaken inferences he might draw.)

Adler's parenthetical qualification is *ad hoc* and undersells the generality of the autonomy thesis. If the speaker must mentor the innocent hearer's inferences, an imperfect custodial duty to *sometimes* assist is too weak and inefficient. If the speaker is an epistemic caretaker, he has a perfect duty not to mislead in first place! The speaker owes the hearer nothing but the truth.

Autonomy is not a burden to be borne only by the guilty. A far-sighted mother lets her innocent son commit fallacies that will wise him up. She does not wait for him to do wrong. She may realize that he would prefer not to be autonomous. Indulging that preference would disrespect his status as an agent.

In a more qualified endorsement of the moral preferability of misleading over lying, Timmerman and Viebahn (2021, 1494) note that false conversational implicatures do not foreclose all paths to the truth. A persistent inquirer can learn without contradicting speakers who told them misleading truths. When the hearer has a right to access the truth, a lie is worse than telling a misleading truth because the lie violates the hearer's right of unimpeded inquiry. By this logic, *a priori* lies are less restrictive to the truth-seeker than misleading truths. At the bottom of every *a priori* lie is the stamp FALSE. Thorough inspection always yields the correct result: exposure of the self-defeating lie.

There is a legal recognition of the plaintiff's duty to exercise due diligence. There is no forgery if the imitator of a famous artist signs his own name to the painting. The disclaimer was on the painting itself! We have contempt rather than sympathy for Faro players who make sucker bets at unfavorable odds.

The ex-Pietist/anti-Pietist Kant denies that a speaker can be blamed for inferences that exceed what he authorized as true. Kant thereby rejects the Quaker's conception of divine truthfulness in which every speaker ministers the hearer's beliefs. In addition to condemning misleading truth telling, the Quaker merchant speaks up about his customer's fallacies. Better to lose the sale than profit from innumeracy or illogicality.

Kant denounces Quaker paternalism. The boundaries between people circumscribe their responsibilities. If the murderer at your door slays the fugitive you refuse to protect with a lie, then the killing is solely the murderer's fault. Kant is not merely protecting the truthful misleader by erecting a moral fence. He is blaming the victim of the deception. Anyone who reasons negligently impairs his ability to discern whether he acts from a good will. This rationale works even better as a justification for telling *a priori* falsehoods rather than telling misleading truths. For the victim of an *a priori* falsehood is even more blameworthy; he believes *against* his evidence (not just beyond his evidence).

According to Kant, a deliberately false assertion is a lie only when there is a rational expectation of truth. There is no such expectation for actors, satirists, jokesters, and magicians. Consequently, these performers *cannot* lie, even if they are deliberately uttering falsehoods in the hope of deceiving the audience. Nor, says Kant, is there an expectation of truth-telling for coerced assertion. If a ruffian grabs you by the throat and demands the location of your treasure, he cannot rationally expect the truth because coercion is a better explanation of your utterance. If he irrationally relies on your deliberate falsehood, the ruffian cannot correctly object that you lied to him. Or so says Kant and some twentieth century experts on lying (Chisholm and Feehan 1977, 154).

Since the negation of an *a priori* proposition is never rationally believed, there are no *a priori* lies. They are like Kant's category of morally necessary lies – empty. The non-existence of necessary lies is derived with much labor by Kant. In contrast, the non-existence of *a priori* lies is so plain to Kant that he never mentions their possibility.

The expectation of truth is rational for freely given *a posteriori* testimony. The speaker purports to have knowledge based on experience unavailable to the hearer. He is offering to transfer this knowledge to the hearer. This transfer model does not apply to *a priori* propositions. For these propositions are available independent of empirical evidence. Consequently, there is no call to offer the *a priori* propositions as testimony. You may utter the proposition to *trigger* belief in it. Much of what passes as *a priori* testimony is just artful cognitive stimulation.

To secure attention to the stimuli, some mathematicians exploit the trappings of testimony. After hearing a television anchorman claim that parallel lines never meet *even at infinity*, the mathematics graduate student Steven Krantz conveyed the rebuttal on the prestigious letterhead of Princeton University (2005, 70). That got the attention of the newscaster. But the only way to get knowledge is to understand the implications of adding a point at infinity.

At times, the Sage of Königsberg seems to posture as a moral testifier. This false footing might be excused as a stimulus to consider *a priori* moral propositions (which cannot be learned by testimony). Much propaganda is non-cognitive persuasion under the guise of rational persuasion. The sham can cause *a priori* knowledge. Kant's rejection of *a priori* testimony clears away room for *a priori* propaganda. We can take a second look at the tactics of the ancient cynics who shocked on-lookers into insights.

As a convenience, you can offer an *a priori* proof so that the hearer can follow the reasoning instead of actively working out the proof for himself. But you cannot provide testimonial knowledge of an *a priori* proposition.

The function of testimony is to share beliefs that are based on experience. *A posteriori* lying is an unnatural use of the tongue. There is no corresponding perversion for *a priori* lying. *A priori* assertion is not designed to transfer empirical knowledge of the statement's content. When a mathematician titles his publication with a theorem, readers learn that the mathematician asserted the theorem (which can be useful to establish priority). But they must read through the proof to learn the theorem. Consequently, the function of mental communion, emphasized by Christians (Tollefsen 2014, 51–55), does not extend to *a priori* assertions. *A priori* testimony is a tumor that grows from the body of *a posteriori* testimony. This cancerous outgrowth does not deserve protection and should be excised.

Supposing the negation of an *a priori* falsehood resembles lying. A prudent logician might therefore avoid *reductio ad absurdum* in the presence of naïve listeners. The same goes for *asserting* the negation of an *a priori* falsehood. Neither supposing nor asserting the negations is *intrinsically* wrong.

Perhaps, disharmony between what one utters and what one believes is ugly. Jessica Pepp (2019) suggests this *aesthetic* aversion explains why people refrain from lying when they could deceive as effectively with a misleading truth. Pepp's hypothesis further predicts that people prefer to tell an *a posteriori* lie rather than an *a priori* lie. An *a posteriori* lie can be painted in profile, concealing the clash between representation and reality. The incongruity of an *a priori* lie is accessible regardless of presentation. Competent contemplation, a prerequisite for aesthetic detachment, exposes the mismatch between what is professed and what is the case.

Even contrarians who praise lies as beautiful focus on the visionary articulation of a consistent alternative: "Anybody can tell lies: there is no merit in a mere lie, it must possess art, it must exhibit a splendid & plausible & convincing probability; that is to say, it must be powerfully calculated to deceive" (Twain 2015, 403). Dilettantes contradict themselves; they have too little imagination to fabricate another possible world.

Kant, in contrast, must prefer an *a priori* lie over *a posteriori* lie. For he does not count *a priori* lies as genuine lies. *A priori* lying is another technique for the ethical deceiver, fairer than telling a misleading truth. When the murderer at Kant's door asks "Is he is hiding in your house?", Kant can permissibly answer: "He departed 12 PM noon". Since "PM" means 'post meridian', Kant's answer is a subtle contradiction. The murderer will not notice.

6 Whipsawing testimony

Let us now turn to Kant's explanation of why the victim of an *a priori* lie should not have sought *a priori* testimony and why others are not obliged to answer such inquires truthfully. From a student's notes of Kant's Warschauer logic lectures:

> If a cognition is so constituted that [its truth] can be discerned with the understanding alone, then the authority of others is no legitimate ground of holding-to-be-true.
>
> (Kant 1998, 584 as translated by Gelfert 2006, 641)

Understanding an *a priori* statement puts you in a position to know it. Suppose you initially take a logician to have *asserted* "Whatever will be, will be". Later you learn that she was merely singing it as a lyric. Warrant for the *a priori* belief survives this loss of testimonial backing. The mere utterance of an *a priori* truth *spurs* your understanding and thereby enabled you to know the proposition.

Denial of an *a priori* truth also serves as a trigger to knowledge of that truth. This explains how *a priori* lying sometimes backfires. If you deny that "unnoticeably" has the vowels aeiou in reverse order, reflection on the denial triggers knowledge of what it denies.

In contrast, knowledge *a posteriori* utterances often require testimony because your experience is limited. According to Kant, you are entitled to accept another's word on matters to which you lack empirical access (Gelfert 2006, 633–636). This presumption stems from Kant's egalitarianism about experience.

Your experience is no more authoritative than another's. Given that you are going to be guided by experience, you must defer to another's experience in the absence of your own experience.

When you do have experience, you may be tempted to ignore the experience of others. After all, if you look for yourself, you bypass concerns about the sincerity and competence of testimony. But Kant insists you must still heed the testimony of others. Kant *praises* this second-hand experience as sometimes superior to first-hand experience (1900, XXIV.1, 560). An astronomer eager to confirm his prediction wisely reposes greater confidence in the observation of an impartial colleague.

Your right to rely upon the experience of others is also a duty. For your experience has no more weight than the experience of other people (Gelfert 2006, 637). Testimonial knowledge is "neither in degree nor in kind in any way to be distinguished" (Kant 1900, XVI, 501) from knowledge based on one's own experience. The default response to anyone's testimony is belief. (Of course, the default can be overridden by supplementary evidence of incompetence or mendacity.)

Respect for other rational agents is borne out by deference to their experience. Whereas John Locke trusts people only in the way that he trusts clocks, Kant trusts people as autonomous agents. If Kant refuses to believe them, they are insulted. If Kant discovers a lie, Kant resents the betrayal. Testimony rests on a norm of respect for fellow rational agents.

The function of the language organ is to communicate the truth as one believes it, "truthfulness" rather than objective truth. We are designed to exchange opinions as a means of improving them. This is the basis for Kant's defense of free speech. The ruler can demand obedience. But the ruler is not entitled to the disrupt the communion of free agents.

In addition to this threat from above, there is a threat from below. According to Kant, the liar is a degenerate who communicates what he secretly believes to be false. Telling only what one believes to be true is essential to language as a social institution. Lying is therefore an offense against humanity. A beneficent lie that is welcomed by all is still wrong. A counterfeiter is not exonerated by the fact that the only effect was to please a miser who will never spend his secret hoard of pseudo-money.

The Scholastic practice of relying on authority makes one particularly vulnerable to linguistic deviants who speak beyond what they believe or even *contrary* to what they believe. One extreme solution is to follow the Royal Society's motto "Nullius in verba" (translated as 'Take nobody's word for it'). A scientist adhering to this slogan would never rely on second-hand experience (not even for the slogan!). The Society's experiments are demonstrations that allow onlookers to have first-hand experience.

The same goes for proofs. John Locke summarizes,

> For I think we may as rationally hope to see with other men's eyes, as to know by other men's understandings. So much as we ourselves consider and comprehend of truth and reason, so much we possess of real and true knowledge. The floating of other men's opinions in our brains, makes us not one jot the more knowing, though they happen to be true. What in them was science, is in us but opiniatrety; whilst we give up our assent only to reverend names, and do not, as they did, employ our own reason to understand those truths which gave them reputation. (Locke 1690, I, IV, 23)

More tersely, "Warrant is *cognizer-sensitive:* whether a particular cognizer is warranted in believing that p depends (at least in part) on features of that particular cognizer, such as that particular cognizer's experiences, beliefs, and intellectual capacities" (Casullo 2012, 185). If you and I encounter a proof that pi is a transcendental number, and only you understand it, then the proof warrants *your* belief that pi is transcendental without warranting my belief. Telling me that the proof is sound does not make the proof warrant *my* belief. That is why my continued study of the proof remains germane to attaining warrant from it. Tyler Burge (1998) optimistically compares testimony to memory because memory preserves warrant for the theorem after the proof is forgotten. The hitch is that your memory performs this service only if based on *your* previous grasp of the proof.

Nevertheless, Kant thinks the Royal Society's "Take no one's word for it" over-reacts to testimonial incompetence and insincerity. In addition to subverting the empiricism it seeks to serve, the motto is immoral. Not only are we permitted to accept *a posteriori* testimony, we are *obliged* to accept testimony on *a posteriori* matters!

Kant's divergence from the Royal Society is restricted to *a posteriori* testimony. He agrees with John Locke that no *a priori* proposition can be learned by testimony (Gelfert 2006, 628, 637, 649). Belief in philosophical propositions requires an understanding that is a pre-condition for belief. That understanding only comes from reasoning. Since philosophical testimony cannot furnish belief, it is impossible to lie to a neophyte with philosophical testimony.

Soliciting testimony from a logician will not yield knowledge of a subtle tautology. Given this impossibility, one ought not testify to *a priori* propositions. If logical testimony is offered, you should decline this futile invitation to take the logician's word for it.

Logicians and mathematicians agree with Kant's prohibition on *a priori* testimony. They insist on proof for any proposition that is not obvious. If you try to learn *a priori* propositions by trust, you gratuitously expose yourself to testimonial over-reach. Trusting students who hope to become mathematicians are hazed with pranks that humiliate them into consistency checks.

On his deathbed, P. E. B. Jourdain hoped he had proved the axiom of choice. He summoned J. E. Littlewood (1953, 129). Littlewood realized Jourdain would die a happy man if told the proof was valid. But Littlewood did not tell this *a priori* lie. Instead, he told the *a posteriori* lie that he needed more time to ascertain whether the proof was valid. On Kant's principles, Littlewood responded immorally to a morally unchallenging situation. Jourdain was asking the impossible from Littlewood. Littlewood's testimony on the validity of a proof cannot be the basis of knowledge of the proof's validity. If Littlewood had pronounced the proof valid and thereby deceived Jourdain, *Jourdain* would have been responsible for the deception.

To learn, you must think for yourself when doing mathematics. In the case of morality and metaphysics, thinking for yourself is a pre-condition for belief. This is daunting but also empowering. Credentials for an *a priori* statement, such as 2 + 2 = 4, trump any challenge by an authority.

7 Paternalism about the *a priori*

By popular reckoning, total responsibility for not believing the negations of *a priori* propositions is an overly demanding epistemic requirement. Adults, even Full Professors of Mathematics, routinely seek testimony on *a priori* matters. The German algebraist Ernst Eduard Kummer (1810–1893) was slow with simple arithmetic and so relied upon students. "Seven times nine," he began, "Seven times nine is er – ah – ah – seven times nine is . . . ." A mischievous pupil piped up, "Sixty-one". Kummer wrote 61 on the board. Another student objected; the answer should be sixty-nine. Kummer was not deceived, "Come, come, gentlemen, it can't be both. It must be one or the other."

Questions with *a priori* answers are asked with the same expectations of truth-telling as questions with *a posteriori* answers. Listeners bet their lives on the accuracy of calculations they themselves have not performed. They presume that any moral support for *a posteriori* testimony extends equally to *a priori* testimony.

Kant does not have moral grounds for enforcing reliable *a priori* testimony. After all, the seeker of *a priori* testimony asks for the impossible. If the testimony leads him astray, it is the asker's fault, not the teller's. If there is any normative support for the reliability of *a priori* assertions, it is not a *moral* norm.

This amorality creates working space for lawyers. Consider the many definitional questions that arise for breaking laws such as "Thou shalt not commit adultery". If a bachelor copulates with a married woman, is he thereby an adulterer? An attorney for the bachelor affirms that only spouses are eligible for adultery. When the defense lawyer becomes a prosecutor, she expands the franchise to anyone who copulates with a spouse. (Counsel from the editor of this volume: "If they are in Maryland, where the penalty for adultery is $10, the parties might be well advised to settle out of court.") Lawyers are permitted to lie about *a priori* propositions. Their duty to be zealous advocates may even *oblige* some *a priori* lies.

Volunteers for answering *a priori* questions are common beyond legal settings. A considerate cashier will compute aloud so that you can be a passive *a priori* thinker rather than an active *a priori* thinker. Kant's motto, "Think for yourself!", permits passive thinking in response to a proof. Kant would have reservations about the cashier's second step: Instead of publicly stating the arithmetic, the cashier computes privately. Now she invites you to believe the conclusion of her calculation is competent and sincere. This entices the customer into surrendering his autonomy.

When a customer learns that he has been deliberately short-changed, he will blame his false belief on the sophistical cashier. Many on-lookers treat the *a priori* lie on a par on with an *a posteriori* lie. This may be because these bystanders realize they are as lazy as the customer. They have no standing to rebuke the unvigilant customer for negligence. In contrast, high-minded shoppers note that reason fully arms customers against *a priori* lies. That is what makes the cashier a daring liar. Diligent purchasers deny the customer was justified in relying upon the cashier's calculation.

Autonomous shoppers are apt to be equally severe on cashiers who miscalculate in favor of the shopper. The beneficiary of the error may deny any obligation to correct the cashier's error. By Kant's lights, a daring customer who repeats the erroneous calculation back to the cashier is not lying. Nor is the daring customer lying if he insincerely says, "I verify the accuracy of your calculation". For given that `I verify that p' entails p, the falsehood of the "verified" equation makes the negation of the verification claim an *a priori* truth. What the daring customer must not say is "I believe your calculation is correct". That would be *a posteriori* testimony and so a genuine lie *about the customer's mental state.*

8 *A priori* limits on *a posteriori* testimony

Historians of arithmetic report that Nicomachus knew 8128 is a perfect number. This *a posteriori* report entails the *a priori* proposition 8128 is a perfect number. So, a ban on *a priori* testimony also bans *a posteriori* testimony that entails *a priori* propositions.

Given contemporary classical logic (first order predicate logic with identity), every proposition entails many *a priori* propositions. For all classical tautologies are entailed by every proposition. Specifically, any material conditional with a logically necessary consequent, such as "Everything is identical to itself", is itself logically necessary. Thus, Kant's ban on *a priori* testimony forces a ban on *a posteriori* testimony.

Kant is deaf to the collapse because he listens to Aristotle's theory of the syllogism – which Kant explicitly claims, in the *Critique of Pure Reason* (B viii) to be complete and only subject to refinements in presentation. Aristotle has relevance requirements that prevent trivial entailment of tautologies.

There remains a second source of collapse based on factive verbs (*know, discover, remember, forget, prove, regret*). Propositions constructed with these verbs imply the truth of their complements. Consequently, any history of an *a priori* field will be comprised of *a posteriori* propositions that entail *a priori* propositions: *The Pythagoreans regretted that the hypotenuse of an isosceles right triangle is incommensurable with its sides, The Chinese proved the Pythagorean theorem before Pythagoras, When gas was rationed by license plate numbers, New Jersey police officers forgot that 0 is an even number.* Kant must deny that one can learn mathematics by reading the history of mathematics. Further, he must deny one can even form true beliefs about philosophical propositions by reading the history of philosophy. One can learn the history of an *a priori* field only after one has learned the *a priori* propositions through *a priori* means.

Symmetric theories of testimony either accept both *a priori* and *a posteriori* testimony or reject both. For instance, until recently, Kant was interpreted as rejecting both (Schmitt 1987, 47). Thomas Reid accepts both *a priori* testimony and *a posteriori* testimony. Plato sometimes sounds like an asymmetric theorist who accepts *a priori* testimony but rejects *a posteriori* testimony (because nothing can be learned by perception). The most unstable asymmetric theory is to accept *a posteriori* testimony while rejecting *a priori* testimony. For all *a posteriori* propositions entail *a priori* propositions (and never *vice versa*).

9 Economics of *a priori* research

Kant describes mathematical mistakes as if they are always due to stress, time pressure, memory overload, or distraction. But by contemporary standards, Kant himself makes mathematical mistakes that are not shallow performance errors. In correspondence with Wilhelm Rehberg, Kant denies that the square root of a negative quantity is possible (Kant 1900, vol.11, 208).

In the *Critique of Pure Reason*, Kant admits that philosophical *a priori* errors are difficult to avoid. For instance, transcendental illusion is the source of especially tempting *a priori* errors. Kant admires metaphysicians who try to transcend the limits of experience. They remind him of caged birds whose drive for free flight makes them flutter against the bars.

Where there is *a priori* uncertainty, there is opportunity for a *priori* lying. There is also opportunity for *a posteriori* lying about propositions with *a priori* entailments. Before a stormy North Sea voyage from Denmark to England, a frightened G. H. Hardy sent a postcard to his colleague Harold Bohr: "I proved the Riemann hypothesis. G. H. Hardy" (Polya 1969, 752). Upon safe arrival, Hardy explained that the postcard was an insurance policy: God would not let the boat sink because it would have made the atheist Hardy as famous as Fermat. (The Riemann hypothesis is still unproved.)

The *a priori* liar accepts a risk of detection by armchair methods. But this is no more of a deterrent than the possibility of detection of an *a posteriori* lie through empirical means. *A posteriori* propositions are often easier to investigate than *a priori* propositions. In these circumstances, the safer lie will be the *a priori* lie.

Contrary to Kant, there can be empirical evidence for *a priori* propositions. For instance, an abacus is designed to facilitate *a priori* calculation. If the abacus is rigged to give false results, there will be *posteriori evidence* against an *a priori* truth. The lie is *a priori* but its basis is *a posteriori*. A lying mathematician resembles the corrupt abacus. His testimony is *a posteriori* evidence for the negation of an *a priori* proposition.

Admittedly, the *a priori* nature of the lie guarantees the victim an opportunity to detect the lie. But why is this necessary opportunity more significant than the contingent opportunities to detect *a posteriori* lies?

Kant's anti-utilitarian answer is that it is intrinsically wrong to believe contrary to the evidence. In "The Ethics of Belief", W. K. Clifford relies on the stronger premise that it is always immoral to believe beyond what is warranted. Kant thinks this too restrictive. He is among the orthodox defenders of faith who permit belief to sometimes fill in evidential gaps – especially when there is a need for action. The orthodox defenders of faith only forbid believing *against* the evidence. This milder restriction puts them in the same stern position as Kant. They are all committed to morally blaming the victim of *a priori* deception.

References

Adler, Jonathan. 1997. Lying, deceiving, or falsely implicating. *Journal of Philosophy* 94: 435–452.

Beversluis, John. 2000. *Cross-Examining Socrates: A Defense of the Interlocutors in Plato's Early Dialogues*. Cambridge: Cambridge University Press.

Boswell, James. 1835. *Life of Johnson*, Including *Boswell's Journal of a Tour to the Hebrides and Johnson's Diary of a Journey into North Wales*, edited by George Birkbeck Hill. Oxford: Clarendon Press, 1887.

Boswell, James. 2001. *Boswell's Edinburgh Journals 1767–1786*, edited by Hugh M. Milne. Edinburgh: Mercat Press.

Burge, Tyler. 1998. Computer proof, apriori knowledge, and other minds: The Sixth Philosophical Perspectives Lecture. *Noûs* 32 (S12): 1–37.

Casullo, Albert. 2012. *Essays on a Priori Knowledge and Justification*. New York: Oxford University Press.

Chisholm, Roderick and Thomas Feehan. 1977. The intent to deceive. *Journal of Philosophy* 74: 143–159.

Clifford, W.K. 1877. The ethics of belief. In *The Ethics of Belief and Other Essays*, 70–96. Amherst, MA: Prometheus, 1999.

Curley, Thomas. 2009. *Samuel Johnson, the Ossian Fraud, and the Celtic Revival in Great Britain and Ireland*. Cambridge: Cambridge University Press.

Friedman, Michael. 1990. *Kant and the Exact Sciences*. Cambridge, MA: Harvard University Press.

Gelfert, Axel. 2006. Kant on Testimony. *British Journal for the History of Philosophy* 14 (4): 627–652.

Heath, C.L. (ed. and trans.). 1912. *The Works of Archimedes*. New York: Dover.

Kant, Immanuel. 1900. *Gesammelte Schriften* (Berlin: Koniglich-Preussichen Akademie der Wissenschaften zu Berlin [now De Gruyter]).

Kant, Immanuel. 1910. *Kant's gesammelte Schriften* (Academy Edition), edited by G. Reimer (ed.). Berlin: Verlag von Georg Reimer.

Kant, Immanuel. 1965. *Critique of Pure Reason*, trans. N. K. Smith. London: Macmillan.

Kant, Immanuel. 1996. An Answer to the Question: What is Enlightenment?, trans. Mary Gregor. In Allen Wood and Mary Gregor (eds.), *Immanuel Kant, Practical Philosophy*. Cambridge: Cambridge University Press.

Kant, Immanuel. 1998. *Unveroffentlichte Nachschriften II. Logik Hechsel. Warschauer Logik*, edited by Tillmann Pinder. Hamburg: Felix Meiner.

Kant, Immanuel. 2012. *Groundwork of the Metaphysics of Morals*. Cambridge: Cambridge University Press.

Krantz, Steven. 2005. *Mathematical Apocrypha Redux*. Providence: American Mathematical Society.

Littlewood, J. E. 1953. *A Mathematician's Miscellany*. London: Methuen.

Locke, John. 1690. *An Essay Concerning Human Understanding*. London: The Basset.

MacIntyre, Alasdair. 1995. Truthfulness, lies, and moral philosophers: What can we learn from Mill and Kant? *The Tanner Lectures on Human Values*, 16: 307–361. Salt Lake City: University of Utah Press.

Mahon, James. 2008. Two Definitions of Lying. *International Journal of Applied Philosophy*, 22: 211–230.

Mahon, James. 2019. Classical philosophical approaches to lying and deception. In Jörg Meibauer (ed.), *The Oxford Handbook of Lying*, 13–31. Oxford: Oxford University Press,

O'Keefe, Tim. 2006. Socrates' therapeutic use of inconsistency in the Axiochus. *Phronesis* 51 (4): 388–407.

Pepp, Jessica. 2019. The aesthetic significance of the lying-misleading distinction. *British Journal of Aesthetics* 59 (3): 289–304.

Polya, George. 1969. Some mathematicians I have known. *The American Mathematical Monthly 76/7*: 746–753.

Schmitt, F. F. 1987. Justification, sociality, and autonomy. *Synthese* 73 (1987): 43–85.

Shieber, Joseph. 2010. Between autonomy and authority: Kant on the epistemic status of testimony. *Philosophy and Phenomenological Research* 80 (2): 327–348.

Smith, Philippa. 1995. "A self-indulgent misuse of leisure and writing"? How not to write philosophy: Did Cicero get it right? In J.G.F. Powell (ed.), *Cicero the Philosopher*, 301–323. Oxford: Clarendon Press.

Timmermann, Felix and Emanuel Viebahn. 2021. To lie or to mislead? *Philosophical Studies* 178: 1481–1501.

Tollefsen, Christopher. 2014. *Lying and Christian Ethics*. New York: Cambridge University Press.

Twain, Mark. 2015. *Autobiography of Mark Twain*. Berkeley: University of California Press.

II Lying, deception, and speaker commitment: Empirical evidence

Alex Wiegmann, Neele Engelmann

Is lying morally different from misleading? An empirical investigation

Abstract: Consider the following case:

> Dennis is going to Paul's party tonight. He has a long day of work ahead of him before that, but he is very excited and can't wait to get there. Dennis's annoying friend Rebecca comes up to him and starts talking about the party. Dennis is fairly sure that Rebecca won't go unless she thinks he's going, too.
>
> Rebecca: Are you going to Paul's party?
>
> (1) Dennis: No, I'm not going to Paul's party.
>
> (2) Dennis: I have to work.
>
> Rebecca comes to believe that Dennis is not going to Paul's party.

In (1), Dennis tricks Rebecca into a false belief by explicitly expressing a falsehood. By contrast, in (2) Dennis achieves his aim in a less direct way, namely by means of a conversational implicature. Cases of the first kind are usually described as cases of lying, while cases of the second kind are characterized as merely misleading. Philosophers have discussed such pairs of cases with regard to the question of whether lying is morally different from misleading. In this paper, we report the results of approaching this question empirically, by presenting 761 participants with ten matched cases of lying versus misleading in separate as well as joint evaluation designs. By and large, we found that cases of lying and misleading were judged to be morally on a par, to have roughly the same consequences for future trust, and to elicit roughly the same inferences about the speaker's moral character. When asked what kind of deception participants would choose if they had to deceive another person, the clear majority preferred misleading over lying. We discuss the relevance of our findings for the philosophical debate about lying and misleading, and outline avenues for further empirical research.

Note: Joint first authors.

Alex Wiegmann, Ruhr-Universität Bochum, e-mail: Alexander.Wiegmann@ruhr-uni-bochum.de
Neele Engelmann, Georg-August-Universität Göttingen, e-mail: neele.engelmann@uni-goettingen.de

https://doi.org/10.1515/9783110733730-005

1 Is lying morally different from misleading? An empirical investigation

Lying is a familiar and morally important phenomenon. No matter if it is in election battles, in personal relationships, or in the form of fake news – lying affects us all, and almost every day. Lying is also a classic as well as a currently prominent topic in philosophy. Two specific lines of research have been of particular interest to philosophers. The first line concerns the question of what it means to lie, i.e. how lying should be defined. From Augustine (395/1887) to Aquinas (1273), up to Williams (2002), Carson (2010), Saul (2012) and Stokke (2018), countless proposals have been put forward. According to the most prominent view in the philosophical literature, lying entails that, in order to lie, speakers have to assert something they believe to be false. This requirement may be spelled out as follows (cf. Stokke 2018; Viebahn 2021):

A lies to B if and only if there is a proposition p such that
1. A asserts to B that p, and
2. A believes that p is false.[1]

The second line of research concerns the morality of lying. Here, a central question is whether there is a significant moral difference between lying and merely misleading (see also Chapter 3). A prominent anecdote in this discussion concerns St. Athanasius (MacIntyre, 1994, p. 336):

> Persecutors, dispatched by the emperor Julian, were pursuing him up the Nile. They came on him traveling downstream, failed to recognize him, and enquired of him: "Is

---

[1] This general account has been endorsed by many authors, e.g. Chisholm & Feehan (1977), Adler (1997), Carson (2006, 2010), Sorensen (2007), Fallis (2009), Saul (2012), Stokke (2018). However, the proposals often differ from each other in relying on different accounts of assertion. Stokke (2018), for instance, bases assertions on the notion of common ground, which results in the following definition:

> A lies to B if and only if there is a proposition p such that
> (L1) A says that p to B, and
> (L2) A proposes to make it common ground that p, and
> (L3) A believes that p is false.

(L2) might be considered to be a replacement condition for a prominent but contested requirement, namely the intention-to-deceive condition (cf. Mahon 2016). Although virtually all lies are not only believed to be false but actually false, only very few authors claim that falsity is required (e.g., Carson, 2010).

Athanasius close at hand?" He replied: "He is not far from here." The persecutors hurried on and Athanasius thus successfully evaded them without telling a lie.

Here, as in similar examples to come, the speaker avoided lying because he did not assert anything that he believed to be false. By saying "He is not far from here" Athanasius only conversationally implicated – and did not say – that the person that the persecutors were searching was not standing right in front of them.[2]

When philosophers discuss the question of whether lying and misleading differ morally, they make heavy use of hypothetical cases designed to elicit intuitions one way or the other. These intuitions are then taken as evidence for or against a certain view. To present only one example at this point, Saul (2012) uses the following case to argue against the claim that lying is always worse than merely misleading:

> Charla is HIV positive, but she does not yet have AIDS, and she knows both of these facts. Dave is about to have sex with Charla for the first time, and, cautiously but imprecisely, he asks: Do you have AIDS?
> Charla replies: No, I don't have AIDS.
> Charla and Dave have unprotected sex, and Dave becomes infected with HIV.

Here, again, Charla did not say anything she believes to be false, but only conversationally implicated that it would be safe to have unprotected sex with her. However, what she did does not seem to be any morally better than if she had lied to Dave (and possibly worse since she shielded herself against a lawsuit by avoiding an overt lie and allowing herself recourse to the literal truth defense[3]).

Given the importance of case-based intuitions in this debate, and the fact that lying is deeply embedded in our social life, it is quite surprising that, to the best of our knowledge, there have not been any empirical studies that have systematically tested lay people's intuitions on the question of whether lying and misleading differ in terms of morality. The aim of this paper is to start filling this gap.

---

[2] Recently, there has been some discussion about whether a speaker can lie by means of certain deceptive conversational implicatures (see Meibauer, 2005, 2014; Viebahn, 2021, for non-empirical defences of the view that one can lie by deceptive implicatures; Reins & Wiegmann, 2021, and Wiegmann, Willemsen, & Meibauer, 2021, for empirical evidence for this view, and Weissman & Terkourafi, 2019, for empirical evidence to the contrary; see Wiegmann & Meibauer, 2019, for an overview of the empirical studies). Since this question is not important for our current enterprise, we will just assume the standard view according to which one cannot lie with conversational implicatures.

[3] We thank Laurence Horn for the observation that misleading might even be considered to be morally worse in this case.

Before we turn to our experiments, let us consider some views in this debate and their proponents. The classic and predominant view is that lying is generally worse than merely misleading. Several arguments have been put forward for this position. For instance, it has been argued that lying is worse because the liar bears the whole responsibility for the addressee's ending up with a false belief. By contrast, when merely misled, the addressee shares some responsibility, as it is the addressee who draws the inference that results in the false belief (for discussions of this view, see Adler, 1997, Green, 2001; MacIntyre, 1994; Mahon, 2003, Saul, 2012). Another argument (Strudler, 2010) is that lying destroys more trust than deceiving through conversational implicatures, due to the following asymmetry: While someone deceived by means of a deceptive implicature might still rely on what the other person *says*, a person who is lied to has nothing left to rely on.

Interestingly, Rees (2014) uses related arguments to arrive at the exact opposite conclusion, i.e. that lying is generally morally better than merely misleading (a minority position). Since misleading requires the addressee to more actively participate in their own deception, and relies on them trusting the speaker's assertions *and* implicatures, Rees holds that misleading is actually a more serious violation of trust than lying (see also Horn, 2017).

Saul forcefully argues that lying and misleading are morally on a par, by resisting the arguments put forward for the claim that lying is morally worse. For instance, she shows that the fact that a misled addressee shares some responsibility for arriving at a false belief does not entail that misleading is better than lying. To support this claim, she invites us to compare two cases of theft. In one version, we have a person (A) who is always very careful and only walks in safe parts of town in full daylight. In the other version, the person (B) is rather reckless and walks around in dangerous areas in the middle of the night. Now let us suppose that both of these people get beaten up and have their wallets stolen. Despite the fact that B's behaviour increased the risk of getting hurt and robbed, and that B, in this sense, shares some responsibility for the bad things that happened, the beating and robbing of B is just as bad as the beating and robbing of A. Regarding the argument that lying destroys more trust than misleading, Saul argues that revealing a lie does not always have fatal effects on trust. Admitting a lie can even increase future trust: she offers the example of a person who is too embarrassed to admit that her business is going badly and lies about it, but then later admits to the lie after her interlocutor states that her own business has dried up as well.

Timmermann and Viebahn (2021) defend the position that there is no general answer to the question of how lying morally compares to merely misleading. They hold that when lying, a speaker commits themself to a proposition that they believe to be false. In contrast, a misleading speaker avoids this commitment and the responsibilities it entails. This difference in commitment usually comes with a difference in how reliably the false belief is produced in the addressee. In contexts where it is morally right to deceive the hearer (e.g. Kant's murderer asking where his innocent victims are), lying is morally better than merely misleading because it will more reliably cause the intended false belief in the addressee. In situations where deceiving is morally wrong, however, the same difference in reliability speaks in favour of merely misleading, as it potentially leaves the door more open for the addressee to find out the truth.[4] Lastly, there are situations in which merely misleading is as reliable as lying, and the two means of deception therefore do not differ in terms of morality.

2 Experiment 1

The aim of this experiment was to test whether laypeople draw a moral distinction between lying and misleading in otherwise identical situations. We selected ten cases of merely misleading from the literature, created parallel versions in which the agent lied instead, and presented them to participants, asking for their moral evaluation as well as their intuitions about lying and misleading.

2.1 Design and participants

We used a 2 (type of deception: lying vs. misleading, between-subjects) x 2 (query: moral judgement vs. judgements about lying and misleading, between-subjects) x 10 (scenario: *cars* vs. *dying woman* vs. *ex-boyfriend* vs. *improper relationship* vs. *mate and captain* vs. *party* vs. *peanuts* vs. *salesman* vs. *smoking* vs. *xbox*, within-subjects) mixed design. We planned for a sample size of 400 valid responses, which yields a power of >90% to detect a small main effect (Cohen's $f = 0.16$) of the between-subject manipulation "type of deception" on moral judgements in a mixed ANOVA (determined using G*Power, Faul et al., 2007). We invited 450 participants on Prolific (*prolific.co*) to take part in the experiment; the inclusion criteria were being a native English speaker and having a minimum age of 18 years. Participants were paid £0.88 for an estimated 7 minutes of their time.

---

4 This observation usually holds but unconvincing lies leave the door open for the addressee to find the truth, and convincing instances of misleading/deception may not do so. We thank Laurence Horn for pointing this out.

2.2 Material and procedure

Participants were randomly allocated to one of the four between-subjects conditions. Within each condition, they were presented with ten scenarios taken from the literature on lying and misleading, in random order. Occasionally, we adapted or updated the scenarios to make them more suitable for experimentation. In the following, we present the full wording for all scenarios in both versions (*L*: lying, *M*: misleading; participants saw only either the *L*-version or the *M*-version of all scenarios):

Dying Woman (based on Saul, 2012, p. 70)
A dying woman asks the doctor whether her son is well. The doctor saw him yesterday, when he was well, but knows that he was later killed in an accident. The doctor says:

[L] Your son is fine.
[M] I saw your son yesterday and he was fine.

The woman comes to believe that her son is fine.

Improper relationship (based on Saul 2012, p. vii)
A politician has been accused of having an affair with his assistant. In an interview, a journalist asks the politician: "Did you have an improper relationship with your assistant?" The politician did have an affair with his assistant, but at the time of the interview the affair is over. The politician wants the journalist to come to believe that he never had an improper relationship with his assistant and replies:

[L] There was no improper relationship between me and my assistant.
[M] There is no improper relationship between me and my assistant.

The journalist comes to believe that the politician never had an improper relationship with his assistant.

Cars (based on Green 2006, p. 146)
Will is on vacation in a hotel. One day, he has some errands to run and doesn't want to walk. However, he didn't bring his car. For this reason, he asks a fellow vacationer, Carl, if he has a car that Will could borrow. Unbeknownst to Will, Carl owns four cars, all of which are currently in the parking garage of the hotel: a Mercedes, a Porsche, a Ferrari, and a Bugatti. Carl will need only the Mercedes later in the evening. However, Carl doesn't want to lend any car to Will. Carl says:

[L] I have only one car in the parking garage, a Mercedes, but I will need it later tonight.

[M] I have a Mercedes in the parking garage, but I will need it later tonight.

Will comes to believe that Carl only has one car in the parking garage, which he will need later that night.

Smoking (based on Levinson 2000, 38)
Kevin is shocked when he finds a couple of smoked cigarettes in the trash, which were all smoked by his wife, Sally. Sally does not want Kevin to know that she secretly smokes from time to time. Sally's friend Evelyn visited her yesterday. Evelyn is a smoker, but she did not smoke at Sally's house yesterday.

When Kevin confronts her, Sally says:

[L] Evelyn was here yesterday, she smoked the cigarettes.

[M] Evelyn was here yesterday, and she is a smoker.

Kevin comes to believe that Evelyn smoked the cigarettes.

Mate and Captain (based on Posner, 1980; cf. Meibauer, 2005)
On a cruise ship, a First Mate does not get along well with his Captain. One day, the mate spilled sauce on his uniform at dinner, and the captain, making fun of the mate, writes in the log: "Today, March 23rd, the Mate spilled sauce on his uniform."

A few days later, when the Mate himself has the watch, he discovers the Captain's entry in the log. He wants to embarrass the Captain in return by making their boss (the head of the shipping company they work for) believe that the Captain regularly spilled sauce on his uniform at dinners during the journey. In fact, the captain never spilled sauce on his uniform, which the Mate knows. The Mate also makes an entry in the log, which reads:

[L] The captain usually spills sauce on his uniform at dinner.

[M] Today, March 26th, the Captain did not spill sauce on his uniform at dinner.

The next day, the ship returns to its port, and the head of the shipping company reads the log. After reading the Mate's entry, he comes to believe that the Captain usually spilled sauce on his uniform at dinners during the journey.

Ex-boyfriend (Coleman and Kay, 1981)
John and Mary have recently started going together. Valentino is Mary's ex-boyfriend. One evening John asks Mary, "Have you seen Valentino this

week?" Valentino has been sick with mononucleosis for the past two weeks, but it is also the case that Mary had a date with Valentino the night before. Mary answers:

[L] I have not met with Valentino in the past two weeks.
[M] Valentino's been sick with mononucleosis for the past two weeks.

John comes to believe that Mary has not met with Valentino in the past two weeks.

Xbox (based on Levinson 2000, 122–127)
Jane recently found out that her son, Alex, often does not do his homework. For this reason, Jane and Alex agree that Alex will do his homework when he comes home from school before he does anything else, such as playing Xbox. When Jane comes home from work one evening, she asks Alex what he did that day. In fact, Alex came home, played Xbox, and only later did his homework, but he does not want his mother to know. So he says:

[L] I came home, did my homework, and afterwards I played some Xbox.
[M] I came home, did my homework, and I played some Xbox.

Jane comes to believe that Alex first did his homework and then played Xbox.

Paul's Party (based on Davis, 2005; cf. Stokke, 2018; Viebahn, 2020)
Dennis is going to Paul's party tonight. He has a long day of work ahead of him before that, but he is very excited and can't wait to get there. Dennis's annoying colleague Rebecca comes up to him and starts talking about the party. Dennis is fairly sure that Rebecca won't go unless she thinks he's going, too. Rebecca asks Dennis: "Are you going to Paul's party?" Dennis replies:

[L] I'm not going to Paul's party.
[M] I have to work.

Rebecca comes to believe that Dennis is not going to Paul's party.

Salesman (based on the anecdote about Athanasius; see MacIntyre, 1994, 336; cf. Saul, 2012, 2)
Anna Baker is tired of door-to-door salespeople ringing her doorbell and bothering her. Her address seems to be in some company's database, and the company frequently sends people over with targeted sales offers for her. She even removed her name from the front door to stop them from finding her. One day, however, the doorbell is ringing again. As she opens the door, a salesman greets her and asks: "Does Mrs Anna Baker live here?" She replies:

[L] "No, but she lives somewhere in this neighbourhood", and closes the door again.
[M] "She lives not far from here", and closes the door again.

The salesman comes to believe that Anna Baker lives somewhere in the neighbourhood, but not in that apartment.

The Peanut Attack (Timmermann & Viebahn, 2020, based on Saul, 2012)
George is cooking dinner for Frieda. He knows that Frieda has a moderate peanut allergy and that peanuts will give her a stomach ache. He wants Frieda to have a stomach ache and has asked his flatmate Hans to sneak some peanuts into the meal while George and Frieda are on the balcony. Frieda asks whether George has put any peanuts in the meal.

George replies:

[L] There are no peanuts in the meal.
[M] I didn't put any peanuts in.

Frieda comes to believe that there are no peanuts in the meal.

In the "moral judgement" conditions, participants were then asked: "How would you morally evaluate [the protagonist's] reply in the conversation above?"[5] on a seven-point scale ranging from "not bad at all" to "very bad". In the "lying and misleading" conditions, they were asked to rate their agreement with each of the following claims:

"[Protagonist] lied to [addressee]"
"[Protagonist] misled [addressee]"

on a seven-point scale ranging from "strongly disagree" to "strongly agree".

2.3 Results and discussion

383 participants completed the survey, seven of which were excluded for failing an attention check in form of a simple transitive reasoning task. ("If Peter is taller than Alex, and Alex is taller than Max, who is the shortest among them?"). Thus, data from 376 participants remained for all analyses (mean age = 35.13 years, SD = 20.1 years, 63% women, 35% men, 2% non-binary or no answer).

---

5 In *mate and captain*, the moral question was: "How would you *morally* evaluate the mate's log entry?".

Data and code for all analyses and figures are available at https://osf.io/ 67gr8/. All analyses were conducted using R (R core Team, 2019) and RStudio (RStudio Team, 2016), in combination with the following packages (in alphabetical order): *effsize* (Torchiano, 2020), *ez* (Lawrence, 2016), *MBESS* (Kelley, 2019), *reshape2* (Wickham, 2007), *showtext* (Qiu, 2020), and the *tidyverse* (Wickham et al., 2019).

Moral judgements. See Figure 1 for an overview of the results. Across scenarios, participants evaluated prototypical lies as being slightly morally worse (M = 4.43, SD = 2.06) than false implicatures (M =4.18, SD = 1.98, $F_{(1,194)}$ = 5.37, p = .02, η^2_p = .03 [.002;.07][6]). However, this effect was moderated by scenario (interaction: $F_{(9,1746)}$ = 2.51, p = .007, η^2_p = .01 [.002;.02]). Following up on this interaction by conducting separate t-tests per scenario revealed that participants discerned the largest moral difference between the prototypical lie and the false implicature in the *mate and captain* scenario ($t_{179.89}$ = 3.96, p = .001, d = 0.57 [0.28; 0.86]). Differences in other scenarios did not remain significant after Bonferroni-correcting for multiple testing. There was also a main effect of scenario, indicating that deception was generally seen as worse in some scenarios than in others, independently of linguistic type ($F_{(9,1746)}$ = 164.02, p < .001, η^2_p = .46 [.43; .48]).

How would you morally evaluate A's reply in the conversation above?

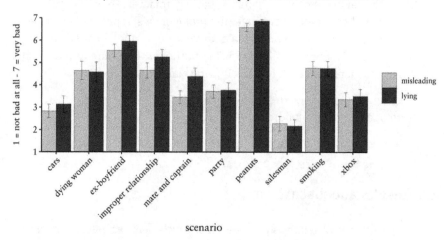

Figure 1: Means of moral evaluations of prototypical lies and false implicatures per scenario in Experiment 1. Error bars are 95% confidence intervals.

Judgements about lying and misleading. See Figure 2 (lying) and Figure 3 (misleading) for an overview of results. Participants clearly perceived all

---

6 We report 90% confidence intervals for all η^2_p; see Steiger (2004).

prototypical lies as such ($M = 6.35$, $SD = 1.33$), and differentiated them from false implicatures, which, in line with previous findings (e.g. Reins & Wiegmann, 2021, Weissman & Terkourafi 2019; Viebahn, Wiegmann, Engelmann, & Willemsen, 2021; Wiegmann et al., 2021), received clearly lower ratings ($M = 4.41$, $SD = 2.19$, $F_{(1,178)} = 145.07$, $p = <.001$, $\eta^2_p = .45$ [.36;.52]). In half of the scenarios, lie ratings for false implicatures (*cars, dying woman, ex-boyfriend, improper relationship, smoking*) did not differ from the scale midpoint of four after correcting for multiple comparisons, suggesting that participants were not sure whether the utterances in these scenarios should count as lies. In four of the remaining scenarios, false implicatures tended to be considered as lies, with ratings significantly higher than the neutral mid-point (*party*: $t_{88} = 3.24$, $p = .017$, $d = 0.34$ [0.13; 0.55], *peanuts*: $t_{(88)} = 3.51$, $p = .007$, $d = 0.37$ [0.16; 0.58], *salesman*: $t_{(88)} = 4.16$, $p < .001$, $d = 0.44$ [0.22; 0.66], *xbox*: $t_{(88)} = 3.17$, $p = 0.021$, $d = 0.34$ [0.13; 0.55]). In the *mate and captain* scenario, the false implicature was on average not seen as a lie, with a rating significantly below the mid-point ($t_{(88)} = 3.01$, $p = 0.035$, $d = 0.32$ [0.11; 0.53]).

Participants clearly judged all utterances to be instances of misleading, largely independently of linguistic type ($F_{(1,178)} = 3.30$, $p = .07$), but with some slight general differences between scenarios ($F_{(1,178)} = 6.60$, $p < .001$, $\eta^2_p = .04$ [.02; .05]). Again, the *mate and captain* scenario stood out, in which the false implicature was seen as somewhat less misleading than the prototypical lie (interaction: $F_{(9,1602)} = 5.89$, $p<.001$, $\eta^2_p = .02$ [.003; .02]).

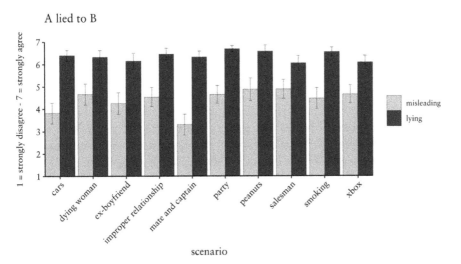

Figure 2: Means of lie ratings for prototypical lies and false implicatures per scenario in Experiment 1. Error bars are 95% confidence intervals.

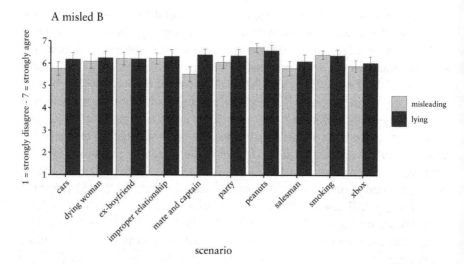

Figure 3: Means of misleading ratings for prototypical lies and false implicatures per scenario in Experiment 1. Error bars are 95% confidence intervals.

To sum up, we found that lying was on average evaluated as being slightly morally worse than misleading. However, this effect was small and mainly driven by one scenario (*mate and captain*), while no difference was found in the nine other scenarios. Accordingly, we conclude from the findings of our first experiment that lying and misleading mostly do not differ in their moral status in the eyes of laypeople.

3 Experiment 2

In the first experiment, the crucial manipulation of lying versus misleading was implemented between-subjects, i.e. each participant either saw only cases of lying or exclusively cases of misleading. It is well known (e.g., Hsee et al., 1999) that effects found in such "separate evaluation" designs often differ from findings obtained in a "joint evaluation" design, i.e. when each participant is presented with both levels of the experimental manipulation. To illustrate the effect of such design differences, let us consider a study on moral luck by Kneer and Machery (2019). They used a scenario in which the protagonist is preparing a bath for her 2-year-old son. When her phone rings, she tells her son to stand near the tub for a few minutes while she answers the phone. In the "bad luck" variant of the scenario, she finds her son dead in the tub when she returns. In the "good luck" variant, her son is still standing near the tub. The agent in the bad luck condition always received harsher moral evaluations than the agent in the "good luck" condition, but the difference was smaller when participants could compare both versions (joint evaluation). In other studies, joint evaluation even reversed the direction of effects observed in separate evaluation conditions (Hsee et al., 1999). In general, a joint evaluation design is assumed to favour reflective deliberation, as it allows one to recognize the differences between two otherwise identical cases and to explicitly assess the relevance of these differences to the judgement that is required (for example, their moral relevance). As such, joint evaluation is also the format in which philosophers typically evaluate pairs of cases and probe their intuitions about the relevance of differences between them. Arguably, the separate evaluation format of Experiment 1 is more similar to people's everyday experience. Typically, we encounter instances of lying and misleading one at a time and in otherwise very different situations, rather than side-by-side and closely matched. We have seen that in separate evaluation, people regarded lying and misleading as roughly equally good or bad. But different results are conceivable when participants are alerted to the subtle differences between the two modes of deception in a joint evaluation design.

Moreover, we wanted to address some potential causes of different moral evaluations for lying versus misleading that have been proposed in the discussion (see Introduction). So in addition to asking participants whether lying or misleading was morally worse (or equally good or bad) in each scenario, we also asked whether a lying or a misleading agent will be trusted more in the future if their deception is revealed, and which of them, if any, has a better character (cf. Saul, 2012). Finally, we were interested in which mode of deception (lying versus misleading) participants would prefer to use themselves, if they had to choose. The goal of this question was to find out whether laypeople's moral commitments are consistent with their own assumed behaviour in this regard.

3.1 Design and participants

We used a 4 (query: moral judgement vs. character judgement vs. trust judgement vs. personal preference, between-subjects) x 10 (scenario: cars vs. *dying woman* vs. *ex-boyfriend* vs. *improper relationship* vs. *mate and captain* vs. *party* vs. *peanuts* vs. *salesman* vs. *smoking* vs. *xbox*, within-subjects) mixed design. We planned for a sample size of 400 valid responses. With 100 participants per between-subject condition, we are able to detect a medium-sized effect of $w = 0.36$ in a χ^2-test against chance with 90% power (Faul et al., 2007) in conditions where three response options are offered (moral judgement, character judgement, trust judgement), and an effect of $w = 0.32$ with 90% power in the personal preference condition, where only two response options were offered. 403 participants completed the survey on Prolific (*prolific.co*) and inclusion criteria were identical to Experiment 1 (plus respondents' not having participated in Experiment 1). Participants were paid £0.90 for an estimated 9 minutes of their time.

3.2 Material and procedure

We used the same cover stories and specific scenarios for lying and misleading as in Experiment 1. However, we adapted them by contrasting the initial case description with a closely matched alternative case about another agent. This second agent found themself in the same situation as the first, but if the first agent lied, the second merely misled (and vice versa). Here's an example using the *dying woman* scenario:

> **A dying woman** asks the doctor whether her son is well. The doctor saw him yesterday, when he was well, but knows that he was later killed in an accident. The doctor says:
>
> > [L] Your son is fine.
>
> The woman comes to believe that her son is fine.
> Now please imagine that somewhere, in another hospital, the exact same situation occurs. The only difference is that here, the doctor replies:
>
> > [M] I saw your son yesterday and he was fine.
>
> The woman comes to believe that her son is fine.

The order of the types of deception (lying first vs. misleading first) was counterbalanced between participants. In the moral judgement condition, we asked: "Who behaved morally better?", with three response options: the agent who lied ("The doctor who said: Your son is fine"), the agent who misled ("The doctor who said: I saw your son yesterday and he was fine"), or neither ("Both behaved equally morally good or bad"). In the character judgement conditions, the question was: "Based on your impression of the two conversations above, who do you think has the better character?", again with the option of picking either the agent who lied, the agent who misled, or neither ("Both seem equally good or bad"). In the trust conditions we asked: "If the deception was uncovered, which doctor do you think would be trusted more by other patients in the future?"

(adapted to each scenario, see https://osf.io/67gr8/ for the full material). The both/neither option here read: "Both would be trusted or mistrusted equally." Finally, participants in the personal preference conditions were asked: "If you were in the position of the doctors in the situation described above, which reply would you have chosen? Please assume that these two replies were the only ones available to you." Here, participants were forced to choose one of the two replies. In all conditions and for each scenario, we offered participants the option to enter an explanation of their judgement in a text field if they wished to do so.

3.3 Results and discussion

18 participants were excluded for failing a simple attention check, resulting in a final sample size of 385 participants (mean age = 31.14, *SD* = 11.55 years, 64% women, 34% men, 1% non-binary, 1% another identity or no answer). See Figure 4 for the global distribution of responses in all conditions, and see Figures 5–8 for the distribution of responses per scenario within each condition.

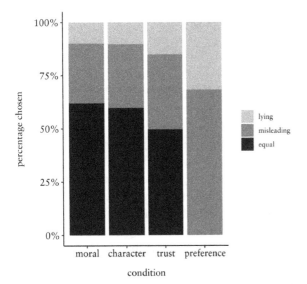

Figure 4: Global distribution of responses per condition in Experiment 2 (N = 385).

Moral judgements. Across scenarios, 62% of participants indicated that lying and misleading were equally good or bad, 28% found misleading to be morally better than lying, and 10% considered lying to be morally better than misleading (test against chance:[7] χ^2 = 43.18, df = 2, p <.001, w = 0.67). For individual scenarios (see Figure 5), the percentage of respondents who regarded both modes of deception as morally equal ranged from 47% (*cars*) to 79% (*peanuts*), but this was the largest group in all scenarios.

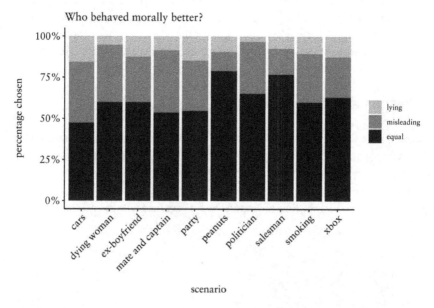

Figure 5: Distribution of responses per scenario for the moral judgement condition of Experiment 2 (N = 95).

Character judgements. Overall, 60% of participants held that speakers who lied had an equally good or bad character as speakers who misled, 30% held that misleading revealed a better character than lying, and 10% were of the opinion that a speaker who lied had the better character (test against chance: χ^2 = 34.45, df = 2, p <.001, w = 0.61). Again, the percentage of participants who regarded both speakers as equal varied between scenarios (see Figure 6), ranging from 44% (*mate and captain*) to 81% (*peanuts*), but this was the largest group in all scenarios.

Trust judgements. Again, the majority opinion among participants was that agents who lied would be trusted or mistrusted in future interactions to the same extent as speakers who misled (50%). 35% of respondents held that

---

7 Adjusted for the number of scenarios by dividing response frequencies by 10 before testing (here and for all subsequent χ2 tests).

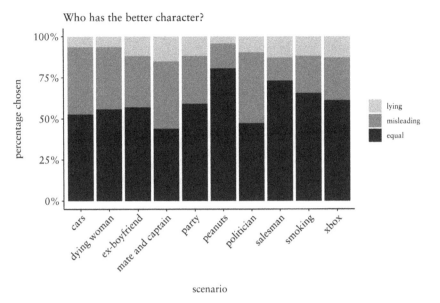

Figure 6: Distribution of responses per scenario for the character judgement condition of Experiment 2 (N = 93).

speakers who misled would be trusted more, and 15% indicated that speakers who lied would be trusted more (test against chance: $\chi^2 = 17.87$, df = 2, $p <.001$, $w = 0.42$). In some scenarios, however, participants also predominantly indicated that the misleading agent would be trusted more (see Figure 7). These were *mate and captain* (equal: 36%, misleading: 38%), and *politician* (equal: 43%, misleading: 47%). In the *party* case, participants were split (equal: 42%, misleading: 42%).

Personal preference. When faced with the question of whether they themselves would have lied or misled in the position of the agents in our scenarios, participants predominantly indicated that they would have misled rather than lied (68%, test against chance: $\chi^2 = 13.08$, df = 2, $p <.001$, $w = 0.37$). The only scenario in which the majority indicated that they would have opted for the lie was the *salesman* scenario (lying: 54%, misleading: 46%, see Figure 8).

To sum up, the findings of Experiment 2, too, suggest that most participants consider lying and misleading to be mostly on a par in terms of morality. The same holds when it comes to questions of trust and character. When we only compare lying and misleading in these latter respects, misleading fares better than lying, which is also reflected in the finding that if participants had to choose between one of these two kinds of deception, the clear majority would go

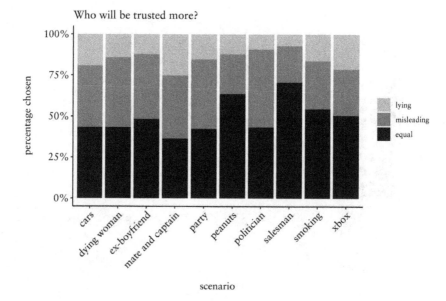

Figure 7: Distribution of responses per scenario for the trust judgement condition of Experiment 2 (N = 99).

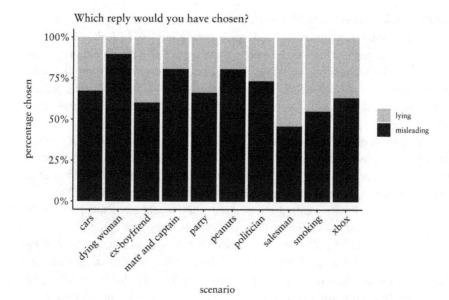

Figure 8: Distribution of responses per scenario for the personal preference condition of Experiment 2 (N = 98).

for the misleading option. Thus, even in a joint evaluation format where lying and misleading can be directly compared and reflected on, people overall do not seem to perceive any substantial moral difference between these two modes of deception. However, most people indicated that they would personally go for misleading rather than lying if forced to choose. As misleading was the second most popular option in almost all scenarios (after "equal"), this preference is consistent with their moral evaluations.

4 General discussion

In this paper, we have approached the morality of lying versus misleading empirically. To this end, we created ten pairs of scenarios, each pair consisting of a case of lying and a closely matched case of misleading. In the first experiment, we employed a separate evaluation design, i.e. participants were either presented with ten cases of lying or with ten cases of misleading, and then asked to evaluate the morality of the protagonist's utterance on a 7-point scale. While the difference between the average ratings was statistically significant, the size of this effect was small, with cases of misleading being evaluated only slightly more positively than cases of lying. Moreover, when looking at the individual cases, the difference between lying and misleading was only significant for one scenario (*mate and captain*). In the second experiment, we employed a joint evaluation design, i.e. each participant saw both the lying and the misleading variant of each of the ten cases. By and large, these results were consistent with the findings of Experiment 1. The most frequent answer was that lying and misleading did not differ morally, followed by a preference for misleading, with only a few people indicating that lying was morally better. Roughly the same general pattern was found concerning the question of which agent will be trusted more in in the future and which agent has the better character (even though the preference for the "equal"-option was not as clear-cut as it was for morality). Moreover, when we asked participants which option – lying or misleading – they would choose if they had to deceive another person, we found that a clear majority opted for misleading.

Before we discuss the implications of these findings for the various positions in the literature, let us say a few words on why we think that empirical findings matter to this debate. As we stated in our introduction, case-based intuitions play an important role in the philosophical debate about lying and misleading. To provide just one example, Saul (2012) considers the cases of *AIDS* and *Peanuts* as sufficient intuitive evidence for the claim that misleading is not always better than lying. When philosophers cite case-based intuitions as evidence for or against certain claims, the (often tacit) assumption is furthermore that not only the author or even only philosophers share their evaluation, but that most ordinary people would agree. In contrast to many other philosophical topics, lying is also deeply anchored in our social life and most people are quite familiar with it. Hence, it seems at least prima facie plausible to think that people's intuitions on this matter should be taken seriously.

The aim of the present work was, however, not to strictly test competing theories about the moral status of lying and misleading against each other. The data we present here rather offer some initial insights into people's moral attitudes about lying versus misleading, and are aimed at laying the groundwork for further investigation. Nevertheless, some tentative conclusions can be drawn. First of all, the view that lying is morally better than misleading (Rees, 2014) is clearly not supported by our results. Across all experiments and scenarios, the view that lying was better than misleading was a minority position. Other prominent views are that misleading is morally better than lying (e.g. Strudler, 2010; Adler, 1997), that lying and misleading are morally on a par (Saul, 2012), and that what is better depends on whether deceiving the addressee is morally justified in a given situation (Timmermann & Viebahn, 2021). Here, the picture is not clear cut and one might argue that there is evidence for all three views. On the one hand, proponents of the view that lying and misleading do not differ morally can point to the findings that the average ratings for lying and misleading did not differ in almost all the scenarios in Experiment 1, and that most participants in Experiment 2 indicated that lying and misleading are morally on a par. On the other hand, the view that lying is worse than misleading is supported by the overall difference found in Experiment 1 (although it was small), and by the fact that misleading was consistently evaluated as better than lying when only these two options are compared (Experiment 2). Lastly, some observations support the mixed view defended by Timmermann and Viebahn (2020). For instance, they claim that there should be no moral difference between lying and misleading in the *peanuts* scenario because both utterances are equally reliable for tricking the addressee into a false belief. And indeed, the number of participants evaluating the two utterances as morally equal was highest in this scenario (Experiment 2). In contrast, if lying is the right thing to do, then one should opt for the more reliable option for deceiving the addressee, according to this view. The only scenario in which participants had a slight preference for lying over misleading (when asked what they would do, Experiment 2) was in the *salesman* scenario, in which both modes of deception were considered to be morally acceptable (Experiment 1).

Our experiments did not include scenarios like Kant's murderer at the door, where deception might even be regarded as obligatory by participants.

It is conceivable that in such a case, lying would actually be seen as morally better than misleading. Future studies should systematically manipulate the moral status of deception in scenarios, independently of the specific linguistic devices that are used to deceive. Furthermore, the perceived reliability of specific utterances in causing a false belief should be assessed. While the present experiments only tell against the view that lying is in general morally better than misleading (Rees, 2014), such future experiments could differentiate between other views.

References

Adler, Jonathan E. 1997. Lying, deceiving, or falsely implicating. *The Journal of Philosophy*, *94*(9), 435–452.

Aquinas, Thomas. 1273. Summa theologica. (Fathers of the English Dominican Province, Trans.) Amazon Digital Services.

Augustine. 395/1887. On lying. In P. Schaff, & K. Knight (Eds.), Nicene and post-Nicene fathers, first series (Vol. 3). Christian Literature Publishing. <http://www.newadvent.org/fathers/1312.htm.

Carson, Thomas L. 2006. The definition of lying. *Nous, 40*(2), 284306.

Carson, Thomas L. 2010. *Lying and Deception: Theory and Practice*. Oxford University Press.

Chisholm, Roderick M. and Thomas D. Feehan. 1977. The intent to deceive. *The Journal of Philosophy*, *74*(3), 143–159.

Coleman, Linda, & Paul Kay. 1981. Prototype semantics: The English word lie. *Language*, *57*(1), 26–44.

Davis, Wayne. 2005. Implicatures. In E. Zalta (Ed.), *The Stanford Encyclopedia of Philosophy*. https://plato.stanford.edu/archives/sum2005/entries/implicature/.

Fallis, Don. 2009. What is lying? *The Journal of Philosophy*, *106*(1), 29–56.

Faul, Franz, Edgar Erdfelder, Albert-Georg Lang & Axel Buchner. 2007. G* Power 3: A flexible statistical power analysis program for the social, behavioral, and biomedical sciences. *Behavior Research Methods*, *39*(2), 175–191.

Green, Stuart P. 2001. Lying, misleading, and falsely denying: How moral concepts inform the law of perjury, fraud, and false statements. *Hastings Law Journal*, *53*, 157–212.

Green, Stuart P. 2006. *Lying, cheating, and stealing: a moral theory of white-collar crime*. Oxford University Press.

Horn, Laurence R. 2017. What lies beyond: Untangling the web. In R. Giora & M. Haugh (Eds.), *Doing Pragmatics Interculturally. Cognitive, Philosophical and Sociopragmatic perspectives* (151–174). Berlin, Boston: De Gruyter.

Hsee, Christopher K., George F. Loewenstein, Sally Blount, & Max H. Bazerman. 1999. Preference reversals between joint and separate evaluations of options: A review and theoretical analysis. *Psychological Bulletin*, *125*(5), 576–590.

Kelley, Ken. 2019. *MBESS: the MBESS R package*. Retrieved from https://CRAN.R-project.org/package=MBESS

Lawrence, Michael A. 2016. *ez: easy analysis and visualization of factorial experiments*. Retrieved from https://CRAN.R-project.org/package=ez

Levinson, Stephen C. 2000. *Presumptive meanings: The theory of generalized conversational implicature*. MIT Press.

Macintyre, Alasdair. 1994. Truthfulness, lies, and moral philosophers: What can we learn from Mill and Kant? *The Tanner Lectures* (delivered at Princeton) Retrieved from http://www.tannerlectures.utah.edu/lectures/macintyre_1994.pdf

Mahon, James E. 2003. Kant on lies, candour and reticence. *Kantian Review*, 7(1), 102–133.

Mahon, James E. 2016. The definition of lying and deception. In Ed Zalta (Ed.), *The Stanford Encyclopedia of Philosophy* (Winter 2016 ed.). Retrieved from https://plato.stanford.edu/entries/lying-definition/

Meibauer, Jörg. 2005. Lying and falsely implicating. *Journal of Pragmatics*, 37(9), 1373–1399.

Meibauer, Jörg. 2014. *Lying at the semantics-pragmatics interface*. De Gruyter Mouton.

Posner, Roland. 1980. Semantics and pragmatics of sentence connectives in natural language. In J. R. Searle, F. Kiefer, & M. Bierwisch (Eds.), *Speech Act Theory and Pragmatics* (169–203). Springer Netherlands.

Qiu, Yixuan. 2020. *showtext: using fonts more easily in R graphs*. https://CRAN.R-project.org/package=showtext

R Core Team. 2019. *R: A language and environment for statistical computing*. Vienna, Austria. Retrieved from https://www.R-project.org/

Rees, Clea. F. 2014. Better lie! *Analysis*, 74(1), 59–64.

Reins, Louisa M. & Alex Wiegmann. 2021. Is lying bound to commitment? Empirically investigating deceptive presuppositions, implicatures, and actions. *Cognitive Science*, 45(2), e12936.

RStudio Team. 2016. *RStudio: Integrated development environment for R*. Boston, MA. Retrieved from http://www.rstudio.com/

Saul, Jennifer M. 2012. *Lying, misleading, and what is said: An exploration in philosophy of language and in ethics*. Oxford University Press.

Sorensen, Roy. 2007. Bald-faced lies! Lying without the intent to deceive. *Pacific Philosophical Quarterly*, 88, 251–264.

Stokke, Andreas. 2018. *Lying and insincerity*. Oxford University Press.

Steiger, James H. 2004. Beyond the F test: effect size confidence intervals and tests of close fit in the analysis of variance and contrast analysis. *Psychological Methods*, 9 (2),164.

Stokke, Andreas. 2018. *Lying and insincerity*. Oxford University Press.

Strudler, Alan. 2010. The distinctive wrong in lying. *Ethical Theory and Moral Practice*, 13(2), 171–179.

Timmermann, Felix & Emanuel Viebahn. 2021. To lie or to mislead? *Philosophical Studies*, 178(5), 1481–1501.

Torchiano, Marco. 2020. *effsize: efficient effect size computation*. Retrieved from https://CRAN.R-project.org/package=effsize

Viebahn, Emanuel. 2020. Lying with presuppositions. *Noûs*, 54(3), 731–751.

Viebahn, Emanuel. 2021. The lying-misleading distinction: a commitment-based approach. *The Journal of Philosophy*, 118(6), 289–319.

Viebahn, Emanuel, Wiegmann, Alex, Engelmann, Neele, & Willemsen, Pascale. 2021. Can a question be a lie? An empirical investigation. *Ergo*, 8(7).

Wickham, Hadley. 2007. Reshaping data with the reshape package. *Journal of Statistical Software*, 21(12), 1–20.

Wickham, Hadley et al. 2019. Welcome to the tidyverse. *Journal of Open Source Software*, 4(43), 1686.

Weissman, Benjamin & Marina Terkourafi. 2019. Are false implicatures lies? An empirical investigation. *Mind & Language*, *34*(2), 221–246.

Wiegmann, Alex & Jörg Meibauer. 2019. The folk concept of lying. *Philosophy Compass*, *14*(8), e12620.

Wiegmann, Alex, Pascale Willemsen & Jörg Meibauer (forthcoming). Lying, deceptive implicatures, and commitment. *Ergo*.

Williams, Bernard. 2002. *Truth & truthfulness: an essay in genealogy* Princeton University Press.

Williams, Bernard. 2010. *Truth and truthfulness*. Princeton University Press.

Ronny Boogaart, Henrike Jansen, Maarten van Leeuwen
"I was only quoting": Shifting viewpoint and speaker commitment

Abstract: When people are accused of having said something objectionable, for instance because it is considered false or inappropriate, various strategies are available for denying or diminishing the speaker's commitment to the contested utterance (Boogaart, Jansen & van Leeuwen 2021). In this chapter we take a closer look at one of these strategies, i.e. the so-called "viewpoint defence", in which an arguer denies that the contested words were their own by attributing them to someone else. A typical instance is the claim that one was "just quoting". Our goal is first, to provide an overview of the different forms the viewpoint defence may take and second, to provide criteria for determining if and when such a defence is a reasonable strategy or may be assessed as untruthful. We show that the very act of quoting triggers implicatures that are not easy to deny. Specifically, we argue in favour of a generalized implicature to the effect that the quoter is accountable for the contents of the quote – unless a convincing alternative purpose for the quote is provided or may be inferred from the context.

1 Introduction

At the start of a debate in the Dutch parliament in June 2011 on the integrity of the central bank of Curaçao and St. Maarten,[1] Eric Lucassen of the Party for Freedom (PVV) opened his contribution with the following statements:

> (1) The country of Curaçao is being ruled by a corrupt crook. The central bank of Curaçao and Sint-Maarten is led by a fraud, and a blackmailing fraud at that.[2]

---

[1] These islands in the Caribbean constitute independent countries within the kingdom of The Netherlands.

[2] https://zoek.officielebekendmakingen.nl/h-tk-20102011-99-12.odt. Unless indicated otherwise, translations from Dutch are ours (RB/HJ/MvL).

Ronny Boogaart, Leiden University, e-mail: r.j.u.boogaart@hum.leidenuniv.nl
Henrike Jansen, Leiden University, e-mail: h.jansen@hum.leidenuniv.nl
Maarten van Leeuwen, Leiden University, e-mail: m.van.leeuwen@hum.leidenuniv.nl

https://doi.org/10.1515/9783110733730-006

Lucassen's speech was interrupted by the then Minister of the Interior and Kingdom Relations, Piet Hein Donner, who threatened to leave the room since he did not want to attend a debate where this kind of language was used with respect to ministers of other countries. However, in the ensuing argument, Lucassen defended himself by saying that the words in (1) were not *his* words, but that he was actually quoting statements made by Curaçao officials. He had intended to make this clear in the subsequent sentences of his speech. Other members of parliament backed up Lucassen by saying that "quoting is always allowed" and "never before has a Minister left the parliament because of a quote". Even though this defence appeased Donner, who decided to stay on for the debate, it may be questioned if quoting is as non-committal an act as the reactions of the parliamentarians suggest.

In this paper we take a close look at the particular defence strategy of denying commitment to something you said by pointing out that the contested words were not *your* words but the words of others. Defending oneself by claiming that one was "just quoting" the words of another speaker is a typical form of this defence. However, in this paper we regard this defence as an instantiation of the broader category of the "viewpoint defence" – as we call it – which includes all instances in which speaker commitment is denied by shifting responsibility for an utterance to another source. Our goal is not only to provide an overview of different forms the viewpoint defence may take, but also to determine if and when such a defence is a reasonable strategy or should be considered untruthful.

The viewpoint defence is a specific strategy that arguers can use when being accused of having said something objectionable. In section 2.1, we present our model of defence lines against accusations concerning controversial utterances, which is based on the general classification of types of defence from classical rhetorical status theory (Boogaart, Jansen & van Leeuwen 2021). Classifying the viewpoint defence in terms of our model brings to light some complexities of this defence strategy and the evaluation thereof, having to do mainly with the inherently polyphonic nature of viewpoint shifts. After all, the accusation as well as the defence may concern either the original utterance being quoted or the act of quoting itself.[3] For instance, in (1), we may accept that Lucassen was conveying the opinion of officials from Curaçao, but this begs the question of whether he agrees with them and, if not, why he is quoting them to begin with. From the discussion in section 2.1, two critical questions

---

3 Since our definition of the viewpoint defence includes appeals to viewpoint shifts that are not explicitly marked as such (see the cases studies in section 3.2), we are using the terms *quoting*, *quoter* and *quoted utterance* in a broad sense (cf. fn. 9).

emerge in section 2.2 on evaluating whether a specific instance of the viewpoint defence counts as a reasonable one.

Our analysis of the viewpoint defence will be illustrated and elaborated in section 3 by discussing two groups of case studies from our corpus of viewpoint defences from Dutch politics and media. In the first group, it is undisputed that the contested words are not the speaker's own since this is explicitly indicated, as in the case of direct and indirect speech or retweeting a message (section 3.1). In the second group (section 3.2), the speaker appeals to an implicit shift of viewpoint in the utterance under dispute, which raises questions about whether and how such a shift can be identified. On the basis of the case studies in section 3, we further specify, in section 4, our critical questions. In section 5 we mention a way to extend the analysis in future research.

2 Analyzing and assessing the viewpoint defence

When people are accused of having said something objectionable, either be-cause it is considered false or inappropriate, various strategies are available for denying or diminishing the speaker's commitment to the contested utterance. In Boogaart, Jansen and van Leeuwen (2021), we categorized these options by combining the four defence lines from classical status theory with the pragma-linguistic categories of explicitly and implicitly communicated meaning. We re-peat the main points of our analysis here and then determine the position of the viewpoint defence with respect to these distinctions (2.1). We conclude this discussion with two main critical questions that are instrumental in determin-ing when the viewpoint defence is reasonable (2.2).

2.1 Defence strategies and implicatures

Classical rhetorical status theory deals with the basic issues that may be addressed in a legal case, making a general distinction between four "statuses" (Leeman & Braet 1987: 76–90; Kienpointner 1997: 229; Braet 2007: 221–227). These are listed in the leftmost column of Figure 1. From the specific perspective of the defendant, they generate the four defence lines in the middle column. In the right-hand col-umn, we illustrate how we applied the four defence lines to cases where the accu-sation concerns a controversial utterance (Boogaart, Jansen & van Leeuwen 2021).

Status	Defence strategy used by the defendant	Applied to the denial of speaker commitment
(1) *Coniectura*: was the action performed?	**Denying** having performed the act	*I did not say that*
(2) *Definitio*: how should the act be defined?	**Redefining** the act in such a way that it is not illegal anymore	*I meant that differently*
(3) *Qualitas*: was the act justified?	Appeal to mitigating **circumstances**	*I was drunk* *I said it in the heat of the moment* *It was a slip of the tongue* etc.
(4) *Translatio*: has the (legal) procedure been executed correctly?	Appeal to a **wrong judge**	*Who are you to judge me?*

Figure 1: Main lines of defence according to status theory as applied to controversial utterances (summary of Boogaart, Jansen & van Leeuwen 2021: 230–231).

In classical rhetoric, the four defence lines of status theory were used to describe how one can defend oneself in response to an accusation of having committed an illicit act. In Boogaart, Jansen & van Leeuwen (2021) we showed that if the accusation concerns an utterance rather than an act the defendant may exploit the difference between the literal content of what was said and the implicatures thereof (in the sense of Grice 1975). They may, for instance, only accept accountability for the literal content of their utterance and not for the disputed implicature that the accuser "read into" their words ("Those are your words, not mine!").[4] If the accusation concerns the literal content of the utterance, the accused may choose the reverse strategy of claiming that they did not mean their words literally but, for instance, were exaggerating or being ironic or sarcastic. Thus, in our data consisting of denials of speaker commitment, we made a further distinction between these two scenarios, as has been represented in Figure 2.

The viewpoint defence is a specific way to defend oneself against an accusation of saying something false or inappropriate; in terms of Figure 2 it may be defined as the third step in (2).

---

4 In some respects, this itself is a kind of viewpoint defence since it shifts the responsibility for the contested implicature to the accuser, who is basically being accused of using straw man argumentation (Macagno and Walton 2017). We will not delve into this any further in the present paper, which deals with step 3 rather than step 2 in Figure 2.

	Scenario 1: *Denying literal meaning*	Scenario 2: *Denying an implicature*[5]
Step 1	Speaker says p and may implicate q	Speaker says p and may implicate q
Step 2	Hearer holds speaker accountable for p	Hearer holds speaker accountable for q
Step 3	Speaker denies commitment to p	Speaker denies commitment to q

Figure 2: Two scenarios for commitment denial based on the distinction between literal meaning and implicatures (Boogaart, Jansen & van Leeuwen 2021: 212).

(2) *The viewpoint defence*
    Step 1: speaker says p and may implicate q
    Step 2: hearer holds speaker accountable for p and/or q
    Step 3: speaker (S1) denies commitment to p and/or q by suggesting that another
        speaker (S2) is accountable for p and/or q[6]

The case of Lucassen's statements cited in (1) is an example of this, as is Nancy Pelosi's defence, cited in (3), after she was attacked for calling Donald Trump "morbidly obese":[7]

---

5 More precisely, we argued that Gricean *conventional implicatures* and *generalized conversational implicatures* behave like literal content when it comes to denial of commitment, so our category of implicatures is constituted by what Grice calls *particularized conversational implicatures* that are entirely context-dependent.

6 An interesting variant of the viewpoint defence that is not, strictly speaking, covered by this definition is where speakers refer to themselves at a younger age ("I said that when I did not know better") or in a different role ("I said that as a private person"). An example of the latter is the way in which Paul Abels, the Dutch National Coordinator for Counterterrorism and Security, defended himself after a Dutch television program revealed that Abels had posted negative tweets about Dutch member of parliament Pieter Omtzigt. Abels was fiercely criticized for these tweets, since he was responsible for the security of politicians and for assessing the extent to which they are being threatened. Abels defended himself by stating (among other things): "I assume that *as a politically engaged private person* I have the right to intervene on Twitter or otherwise in discussions with and about politicians and political issues." [our italics; RB/HJ/MvL] (https://www.parool.nl/nederland/topambtenaar-onder-vuur-na-ontoelaatbare-tweets-over-pieter-omtzigt~bf27f45b/?referrer=https%3A%2F%2Fwww.google.com%2F).

7 https://edition.cnn.com/2020/05/19/politics/nancy-pelosi-donald-trump-fat-hydroxychloroquine/index.html.

(3) I was only quoting what doctors had said about him so I was being factual in a very sympathetic way.

In her original statement, given in (4), Pelosi indeed suggested that there was some source for the phrase *morbidly obese*, while remaining vague about its precise origin, by adding *they say*:

(4) I would rather he not be taking something that has not been approved by the scientists, especially in his age group and in his, shall we say, weight group. *He's morbidly obese, they say*. [our italics and underlining; RB/HJ/MvL]

It is, however, not always as explicit as in (4) that contested words were meant as a quote.[8] For many cases, the term "quotative" is too restrictive because they concern less explicit forms of perspective taking. For instance, speakers may argue that they were not presenting their own thoughts on some matter, but rather temporarily taking someone else's perspective, as in forms of free indirect speech and thoughts that are less clearly marked as quoting. We do include such appeals to implicit "represented speech and thought" in our concept of the viewpoint defence because, as is well known from the literature on indirect and free indirect speech (e.g. Fludernik 1993; Sanders 2010; Verhagen 2012), there is no clear cut-off point between explicit and implicit viewpoint shifts.[9] Such in-between cases constitute important examples in view of assessing the reasonableness of this defence strategy, as will be witnessed by the case studies in 3.2.[10]

So how can the viewpoint defence ("These are not *my* words" / "I was just quoting") be characterized in terms of the four defence lines in Figure 1? The

---

8 The form of the quote used by Pelosi is already quite a bit removed from typical cases of direct and indirect speech. Leech and Short (1981: 333) consider instances where the reporting clause (*they say*) follows the represented utterance to be somewhere in between indirect speech and free indirect speech; Verhagen (2012) argues that such cases are much more like free indirect speech. For *they say*, in particular, it might also be argued that it functions as a lexical expression of "hearsay" evidentiality in English (e.g. Langacker 2017: 38), but in any case it does explicitly indicate a source of information other than Pelosi herself.
9 These cases of viewpoint shifts that are not explicitly marked as such indicate that we are using the terms *quoting, quoter* and *quoted utterance* in a broad sense (cf. fn. 3).
10 A famous example of a more or less implicit viewpoint shift can be found in the speech held by Philipp Jenninger in the German parliament in 1988, to remember the 1938 *Kristallnacht* ('Cristal Night'). To try and explain the thoughts of the Nazis in the period leading up to Kristallnacht, Jenninger made use of free indirect speech. Within 24 hours after giving the speech, Jenninger had to resign from his position as chair because he had been too sympathetic towards the Nazi ideas. Still, there were many cues in the speech making clear both that Jenninger was "quoting" and that he was distancing himself from the thoughts being represented (Ensink 1992).

brief answer to this question is that this defence strategy is an example of the second defence line, i.e. redefinition. The defendant cannot and does not deny having uttered the contested words (i.e. defence line 1), but appeals to the way in which these words should be interpreted; a parallel in the domain of acts would be arguing that a killing is a case of manslaughter rather than murder. More precisely, the viewpoint defence can often be regarded as denying an implicature, as in step 3 of scenario 2 (Figure 2). Claiming that one is "just quoting" boils down to denying that the quote as such has certain implicatures, e.g. denying that the speaker agrees with the content of the quote and that they are using it to trigger new implicatures.

However, this analysis may need some elucidation since the viewpoint defence strategy is complex: there are always two utterances involved. On the one hand, there is the original (quoted) utterance; on the other hand, there is the "quote as quote" uttered by the accused speaker at a later time. Now, both utterances may have their own implicatures attached to them. Specifically, when a speaker claims to be referring to the views of someone else, they must have their reasons for doing so at this particular point in the discourse. If one were, literally, "just quoting", this would be at odds with Grice's (1975) maxim of Relation. The act of quoting by S1 will have implicatures in the new context other than, or in addition to, the original utterance by S2 did. The complex structure of quoting may thus be represented as in (5):[11]

(5) S1 quotes [S2 says p and implicates q1] and implicates q2

As an illustration, let us look more precisely at the Pelosi defence that was cited in (3). After Pelosi had been called "a sick woman" by Trump for labeling him "morbidly obese", she claimed that she was "only quoting what doctors had said about him". In terms of (5):

S1: Nancy Pelosi
S2: "doctors"
p: Donald Trump is *morbidly obese*

What could be the different implicatures referred to as *q1* and *q2* in (5)? The term "morbidly obese" as used by S2 ("doctors") is a term of science that is well defined

---

11 The representation in (5) is not meant to suggest that, in addition to what S2 "said", S1 is also quoting the implicatures of the quoted utterance (q1). In fact, the Pelosi case makes clear that q2 may be quite different from q1. We do regard it as a generalized and thus defeasible implicature of the very act of quoting that S1 agrees with the contents of the quote and the implicatures thereof (see below). This, however, is itself an implicature at the utterance level (q2), to be distinguished from cases of "embedded implicature" discussed in the literature (e.g. Huang 2017: 167–171).

in terms of a person's BMI and that may be used to describe the condition of a patient. As such, it may for instance suggest that the patient should improve his condition and lose weight (q1). However, in the non-medical context in which Pelosi used the quote, p has different implicatures. In (4), Pelosi refers to Trump's condition to suggest that he should not be taking the antimalaria drug hydroxychloroquine to prevent getting Covid. So the latter is at least one of the implicatures of the quote as quote (q2), but there clearly are others: ultimately, in using the term, Pelosi may even be trying to suggest that Trump is not fit to be President. This becomes clearer if we look at Pelosi's full statement in (6).

> (6) "I gave him a dose of his own medicine. He's called women one thing or another over time, and I thought he thinks that passes off as humor in certain cultures," Pelosi told reporters at her weekly press conference. "I was only quoting what doctors had said about him, so I was being factual in a very sympathetic way."[12]

From (6), it appears that by calling him "morbidly obese" Pelosi was trying to get back at Trump for having offended women for their weight in the past, which is an instantiation of the fourth defence line ("who are you to accuse me!"), and she was clearly trying to be funny, which may be regarded as an instance of "redefinition" (second defence line), but a different one from appealing to a shift of viewpoint. In actual fact, Pelosi's inconsistent defence in (6) shows that her claim that she was "just quoting" and, therefore, "being factual" is simply a false one.

The Pelosi case illustrates quite well the viewpoint defence that may thus be analyzed as redefining the interpretation of the contested utterance (defence line 2) by denying an implicature (scenario 2), more specifically one or more of the implicatures attached to the act of quoting itself (q2 in (5)). In addition to particularized implicatures that are entirely context-dependent, there seems to be a generalized implicature q2 to the effect that quoters subscribe to the contents of a quote unless the context makes clear that they do not, or they explicitly distance themselves from it. For instance, even after it was clarified that Lucassen in (1) was quoting, there was no doubt that these quotes were reflecting and supporting Lucassen's own view on the administration of Curaçao and Sint-Maarten. In principle, people may of course quote all kinds of utterances they do not agree with, e.g. to ridicule someone else's opinion, but then either the context or the speaker will make clear that this is the case. This is precisely what a *generalized* conversational implicature is: in the absence of clues to the contrary, the hearer is entitled to assume that it holds. If it is not can-

---

12 https://apnews.com/article/virus-outbreak-donald-trump-nancy-pelosi-politics-6f658ee91 cc411eb2ed54b535d35314f.

celled immediately, such a generalized implicature is very hard to deny in response to an accusation at a later stage (Boogaart, Jansen & van Leeuwen 2021: 213).[13]

We conclude this section by taking a brief look at the other defence strategies in Figure 1 from the perspective of the viewpoint defence. The first defence line, i.e. denial, we restrict to cases in which the defendant claims not to have uttered (nor quoted) the contested words at all. This is what happens, for instance, if, in response to an accusation of having used abusive language in a tweet, the defendant claims that their twitter account was hacked (Boogaart, Jansen & van Leeuwen 2021: 223). If this really happened, the defendant does have a point that someone else is entirely responsible for the abusive statements since they never wrote them, or even quoted them, so this rather exceptional type of viewpoint defence would constitute an instance of the first defence line.[14] As for the third and fourth defence lines, we certainly do encounter them in our case studies to be discussed in section 3, but they do not concern the viewpoint defence as such. Of course, defendants may appeal to the circumstances to explain *why* they used an 'objectionable' quote (defence line 3), or they might launch a counterattack against the accuser (defence line 4), as Pelosi did in (6), but such supporting arguments do not themselves constitute instances of the viewpoint defence.

2.2 Evaluating the viewpoint defence

When using the viewpoint defence, a speaker is denying commitment to something they said by shifting responsibility to another speaker who is supposed to have uttered these words and/or to have held these views at some earlier time. In (5), we represented the "quotative" context that is being appealed to in the viewpoint defence as follows:

(5) S1 quotes [S2 says p and implicates q1] and implicates q2

To assess if a specific use of the viewpoint defence constitutes a reasonable defence, there are, then, two main critical questions that both need to be answered positively:

---

13 Boogaart, Jansen and van Leeuwen (2021) provide a discussion of the distinction between *cancellation* and *denial* (cf. Mazzarella, Reinecke, Noveck & Mercier 2018).

14 In our previous paper, we stated that defending oneself by claiming that you were hacked is an instance of the third defence line (Boogaart, Jansen & van Leeuwen 2021: 223). This is incorrect: an appeal to the third status boils down to an acknowledgement of having committed the act and pointing out extenuating circumstances. Saying that your account was hacked implies that you were not the acting person in the first place (i.e. denying being the wrongdoer).

CQ1: Are the contested words (p) of S1 really a quote?

CQ2: Is it reasonable to assume that S1 is not committed to the quote, including its implicatures (q1 and q2)?

As for CQ1, the case of Lucassen's statements cited in (1) already illustrated that the viewpoint defence is sometimes used in cases in which the alleged shift of viewpoint in the contested utterance is not fully explicit (see also the case studies in section 3.2), so in such instances in particular the question will be raised if there were sufficient linguistic or contextual clues for the hearer to identify the shift of viewpoint. This is not to say that CQ1 is irrelevant if the viewpoint shift in the disputed utterance is more clearly marked, because more specific questions may be raised about the identity of S2 and the reliability of the quote, i.e. did S2 actually say or think that p?

CQ2 pertains to issues that were discussed in some detail in section 2.1. Whenever a speaker (S1) quotes another speaker (S2), or, more generally, shifts perspective to some other source, they must have a reason for conveying the represented utterance or thought. Otherwise, the speaker is violating the maxim of Relation and the resulting discourse will be incoherent. In the default case, we may assume that S1 is committed to the quote (p) and its implicatures (q1) and is using it to trigger other or additional implicatures (q2). However, on the viewpoint defence, S1 is denying some aspect of this, depending on what the accusation pertains to (either p or q1 may already be controversial by itself, or the accusation may pertain to implicatures triggered by the quote as quote, i.e. q2). Such a defence is certainly not unreasonable a priori since there may be sufficient indications in the communicative context that S1 is not committed to the quote and, instead, had another purpose (q2) with the act of quoting. In that case CQ2 gets an affirmative answer.

In section 3, we will apply the theoretical notions from this section to actual instances of the viewpoint defence, taken from Dutch politics and media. This will also enable us, in section 4, to specify the critical questions CQ1 and CQ2.

3 Case studies

In this section we illustrate and elaborate our analysis by discussing two groups of case studies from our corpus of viewpoint defences that have been used in Dutch politics and media. In the first group, the viewpoint shift in the contested utterance is indicated explicitly (section 3.1). In the second group (section 3.2), the speaker appeals to a shift of viewpoint in the utterance under dispute that is more implicit.

3.1 Explicit viewpoint shifts

In this section, we present instances in which CQ1 does not arise since it is unambiguous that the contested utterance contained a quote, nor is it being disputed that the quote is reliable. The viewpoint defence in these cases consists of denying commitment to the contents of the quote, i.e. denying the generalized implicature (q2) that the quoter agrees with what they are quoting – either in blunt form, or by providing an alternative implicature.

Denying an implicature

Our first example case illustrates that a viewpoint defence in blunt form ("These are not *my* words" / "I was only quoting") is normally not reasonable. The Dutch politician Geert Wilders, leader of the right-wing Party for Freedom (PVV), defended himself in this way in 2010 in a parliamentary debate after being interrupted by his Green Left colleague Femke Halsema:

(6) **Mr Wilders (PVV):**
However, even established political parties are waking up. This is something new. Christian Democrats in Germany are starting to understand it more and more. (. . .) The party leader of the CSU, Horst Seehofer, actually goes even further. He wants a complete halt to the immigration of Turks and Arabs to Germany. (. . .) He says: multiculti is dead. Even the German Chancellor, Mrs Merkel, says that the multicultural society has proved to be an absolute failure. Not a slight failure, but an absolute failure. *If she says that, it is saying quite a lot.* (. . .)

Ms Halsema (GL):
You are saying: Islam does not belong in our country. At least, that is what I assume.

Mr Wilders (PVV):
No, I did not say that.

Ms Halsema (GL):
No, but you quote German politicians and I assume that this is what you mean.

Mr Wilders (PVV):
No, I only quoted them.

Ms Halsema (GL):
Are you now standing there, quoting all those German politicians because they are so brave and dare to say all that, and then concluding that you do not dare to say it yourself?

Mr Wilders (PVV):
I used a quotation, nothing more.[15]
[all italics are ours, RB/HJ/MvL]

It is clear that Wilders is referring to other viewpoints here (cf. CQ1 in section 2.2): he explicitly refers to the words of Horst Seehofer and Angela Merkel by quoting them in the form of direct and indirect speech ("He says . . ."; "Mrs Merkel says that . . ."). Knowing Wilders' political background it is obvious that he agrees with the content of these quotes: warning of problems related to what Wilders calls "the Islamification of the Netherlands" has been the central point of Wilders' political agenda over the last decade. In other words, in terms of (5), the discussion between Halsema and Wilders concerns the question of what Wilders' is implicating by quoting these German politicians in the "here and now" of the current debate setting (i.e. q2): Wilders is vague about this ("If she says that, it is saying quite a lot"). Halsema tries to explicate the implicature ("You are saying: Islam does not belong to our country"), which is denied by Wilders ("no, I did not say that"), who bluntly appeals to a viewpoint defence twice ("I only quoted them", "I used a quotation, nothing more").[16] This is clearly uncooperative and unreasonable: denying a plausible implicature of the quotations in the current context without providing any alternative interpretation is at odds with Grice's maxim of Relation, and unconvincing.[17] Why use these quotes if you are not agreeing? What other reasons could you have for using them?

---

15 https://zoek.officielebekendmakingen.nl/h-tk-20102011-13-7.html; translation taken from van Haaften and van Leeuwen (2020).

16 When saying "I did not say that", Wilders is also exploiting the polysemy of the Dutch verb *zeggen* ('to say'), which resembles its English counterpart in this respect. Whereas Halsema is clearly using the verb to indicate what Wilders *meant* to say (indicating speaker intended meaning), Wilders can deny, literally, having spoken these words; the distinction essentially boils down to Austin's distinction between locutionary and illocutionary meaning. Wilders' defence is also an example of defence line 2 (cf. Figure 1) and in fact equally comes down to denying an implicature by appealing to a strictly literal reading.

17 From a different perspective, example (6) is also discussed by van Haaften and van Leeuwen (2020: 123–124). They observe that Wilders' debating style in general is characterized by an unwillingness to provide answers to questions and to be involved in a 'real' debate. Van Haaften and van Leeuwen (2020) argue that this is part of a more overarching strategy of using "bad manners" in parliamentary debate in order to create and maintain an anti-elitist image.

Providing an alternative implicature

In section 2.2 we stated that the viewpoint defence is certainly not unreasonable per se. A case in point is the Dutch journalist Bert Brussen who had to defend himself for reposting a tweet in which Geert Wilders was threatened with death because of his anti-Islamic standpoints. Brussen posted a screenshot of this tweet on his weblog, and accompanied the tweet (given in (7) below) with the title "This is how you threaten Wilders with death":

> (7) Handsome reward for the one who cuts Wilders' throat. Preferably from right to left, but from left to right is okay too![18]

The Dutch public Prosecutor began an investigation for prosecuting Brussen, because, according to the Prosecutor, reposting this threat was punishable. In other words, although Brussen was clearly "quoting" (by reposting a tweet it was obvious that he referred to the viewpoint of someone else, namely: the tweeter of (7)), Brussen was still held accountable for the content of the tweet. Brussen, however, defended himself by saying that he had quoted the tweet because he found it newsworthy, and that he had reposted it with an ironic intention.[19] In other words, in terms of (5), Brussen denied commitment to the content (p) of the original tweet by suggesting that he had used this tweet to trigger alternative implicatures (q2) with this post on his weblog.

Brussen's defence seems reasonable. He did not distance himself explicitly from the controversial content of the tweet, but a media expert who investigated the case for the Prosecutor concluded that Brussen published his message on a platform aimed at a general audience where Brussen frequently wrote about newsworthy political and societal topics. According to the media expert, it was sufficiently clear for the intended audience that Brussen did not agree with the content of the tweet, especially since he had written critically elsewhere on his blog more than once about negative treatments of Wilders in the media.[20] As such, Brussen's defence that the tweet was "newsworthy", and that the title of his web post was intended to be "ironic" seems reasonable indeed. In the end, this was also the conclusion of the Prosecutor, who decided to drop the case.[21]

---

18 https://www.slideshare.net/socialmediadna/dreigtweet-op-het-weblog-van-bert-brussen.
19 https://www.nu.nl/internet/2306487/journalist-moet-retweet-verantwoorden-bij-politie.html.
20 https://www.slideshare.net/socialmediadna/dreigtweet-op-het-weblog-van-bert-brussen.
21 https://www.ad.nl/show/blogger-bert-brussen-niet-vervolgd-voor-bedreiging-wilders~a8328424/.

As Brussen's case illustrates, a viewpoint defence can be reasonable if an alternative implicature q2 is explicitly provided or clear from the context. It goes without saying, however, that giving such alternative reasons for quoting does not necessarily lead to a reasonable defence: this depends completely on the plausibility of the alternative interpretation of the quoted words being suggested. For instance, the Dutch member of parliament Thierry Baudet, leader of the right-wing populist party Forum for Democracy, once retweeted a message from Erkenbrand, a Dutch study group linked to the alt-right movement. The message consisted of a photo of Baudet lying on his piano with the photoshopped head and hands of Pepe the Frog – a symbol of the alt-right movement and white supremacists.[22] Baudet retweeted this photo adding the text 'LOL'. After being criticized for flirting with the alt-right movement and showing sympathy for its ideas, Baudet defended himself by saying that he did not know the associations connected to the meme of Pepe the Frog, and that he had posted it just because he thought it was a funny picture: "I found the tweet hilarious; I see the frog as an energetic comic character expressing a positive campaigning energy."[23] In other words, Baudet defended himself by appealing to an implicature (q2) other than the implicatures attached to the original photo (q1). Although the text that he added ("LOL") supports Baudet's interpretation, his defence does not seem very credible: at many occasions, Baudet had been showing off his intellectual background (indicated, for instance, by a PhD in philosophy of law) and claimed to be a man of wide reading. This makes it hard to believe that he was not familiar with the associations connected to the photo that he retweeted and at the very least making it dubious whether Baudet could deny accountability for the implicatures connected to the photo of the original tweet.[24]

---

22 https://twitter.com/thierrybaudet/status/817494829882048513.

23 https://www.nrc.nl/nieuws/2017/03/16/alt-right-beweging-juicht-op-het-web-hard-voor-baudet-7415174-a1550730.

24 As such, Baudet's tweet seems to be an instance of a "dog whistle" or "calculated ambivalence", i.e. a case in which one can deny a possible interpretation while at the same time conveying exactly that message for those who are happy to receive it (cf. Boogaart, Jansen & van Leeuwen 2021: 222 and the references mentioned there; see also Chapter 7). Scott (2021) provides a more elaborate discussion of the communicative intentions that can lie behind retweeting information.

3.2 Implicit viewpoint shifts

In our examples of the viewpoint defence discussed thus far, there is no discussion about the question of whose words the contested utterance resonates with: whenever a speaker is retweeting a message or using direct or indirect speech, it is evident that the viewpoint of another party besides S1 plays a role. However, as we mentioned in section 2.1, there are also instances in which this is less evident. In this section, we discuss three such cases, in which a shift of viewpoint is being appealed to, but this shift is actually difficult to identify in the contested utterance.

Representing the view of the novelist

Our first example case in this section again involves the Dutch politician Thierry Baudet, who became the subject of a controversy because of a review by his hand of *Sérotonine*, a novel written by the French writer Michel Houellebecq.[25] Immediately after the publication of the review, which Baudet had announced in a tweet,[26] he was attacked for his negative views on abortion, euthanasia and the right of women to work.[27] Baudet responded with another tweet, saying that this news coverage was partisan and that fellow politicians had displayed fake indignation in their search for voters (defence line 4).[28] Playing the victim and blaming the critics remained his strategy also in an interview, the next day, with journalists of the provocative medium GeenStijl.[29] However, when the interviewer asked him if he had put some of his own feelings in the review, Baudet appealed to the viewpoint defence:

> (8) I feel kindred to Houellebecq and his analysis of society and criticisms of the late-liberal community we are in, where we wonder about the fulfillment of life (. . .). I recognize that, so I did put something in it of myself, *but it still remains an analysis of the way Houellebecq sees the world.* [our italics; RB/HJ/MvL]

---

25 The review can be found at https://americanaffairsjournal.org/2019/05/houellebecqs-unfinished-critique-of-liberal-modernity/#.XOJe0g0yP_c.twitter.

26 https://twitter.com/thierrybaudet/status/1130383296561647616.

27 https://www.parool.nl/nieuws/baudet-oogst-kritiek-met-houellebecq-essay-over-individualisme~b9e25fca/.

28 https://twitter.com/thierrybaudet/status/1130510303316336642.

29 See: https://www.youtube.com/watch?v=W9-8XewBlP8. He also made use of an outright denial: ". . . they are making an issue now of women's rights, but [my article] is not about them and I certainly do not call them into question."

Did Baudet put forward a reasonable defence by claiming that the review presented Houellebecq's views? If you read the review, it is striking that it is full of viewpoint ambiguities. Baudet makes clever use of a mix of paragraphs containing literal quotations of Houellebecq's novel, and paragraphs that describe Houellebecq's views in a more paraphrasing fashion. The literal quotations are marked in an unconventional way (i.e. not with quotation marks but with a footnote reference providing the original French wording). In contrast, the latter paragraphs often describe views and ideas without mentioning an author they can be ascribed to; in these cases it is not clear until the next paragraph that Houellebecq is intended to be their source.

Despite several of such viewpoint ambiguities in Baudet's review, it is exactly the views for which he was criticized – on abortion, euthanasia and women's right to work – that are his own without any doubt. It is true that Baudet's stance on abortion and euthanasia can be found in a section that starts with Houellebecq's view that the liberation of the modern world, with its emphasis on the "emancipation of the individual", "has left our lives empty, without purpose, and, above all, extremely lonely". But immediately after this representation of Houellebecq's point of view, Baudet proposes a further elaboration, claiming that the "maximization of individual autonomy" lies at the heart of both social-democratic and liberal ideology.[30] No mentioning of Houellebecq can be found in this further exploration; on the contrary, Baudet refers to Dutch legislation on abortion and euthanasia, not mentioned by Houellebecq in his novel, in order to illustrate his point:

> (9) Today, even new life (in the womb) may be extinguished to avoid disturbing the individual's freedom. In the Netherlands (where I live), suicide is facilitated to ensure that here, too, no constraints – such as the duty to care for your parents – are placed on the individual.

Baudet's contested stance on women with jobs is also clearly his own. It is introduced with the statement that Houellebecq's view of the world "is validated all around us" – a claim that Baudet supports with reference to "the emancipation of women and the feminist ideology" and the prediction of a gloomy future that will result from this:[31]

---

30 Baudet claims that these movements hold maximum individual autonomy as their ultimate principle to which everything else is subordinate. Moreover, he predicts that this principle will weaken, "over time, all such institutions that the individual requires to fully actualize a meaningful existence – such as a family and a connection to generations past and future, a nation, a tradition, perhaps a church", and make it "eventually disappear".

31 According to Baudet, these ideologies expect women "to reject the traditional role of supporting a husband and strive instead for an 'equal' relationship in which 'gender roles' are interchangeable".

(10) An inevitable result of all this is the demographic decline of Europe. Another outcome is constant conflict, constant competition – and in the end, fighting, divorce, and social isolation – and a new generation of boys and girls growing up in such disfigured settings.

From the above we conclude that Baudet's viewpoint defence fails on the first critical question, i.e. the question whether he was really quoting. As our analysis reveals, there are textual indications that unambiguously show that the controversial remarks represent Baudet's subscription to and elaboration of Houellebecq's views. Because of the negative answer to the first main critical question, the second main question – whether the arguer could be held committed to the quote's content – does not require examination. Baudet can be held committed to the views he had been criticized for and his denial thereof in (8) seems insincere.

Representing the view of the victims

Our second example in which it may be questioned if the viewpoint of another party than S1 is involved, concerns a case involving Selçuk Öztürk, member of the political party DENK, in a debate in Dutch parliament, in November 2019, about the killing of 70 citizens of the Iraqi city Hawija by a bomb fired by a Dutch jet fighter.[32] That incident had taken place in June 2015, but parliament had never been informed about it; the case was disclosed through the work of journalists.[33] In his interruptions of other speakers Öztürk kept talking about "murder" in relation to the casualties, and also called the responsible Minister a *lijkenverstopper*, i.e. someone who hides dead bodies.[34]

Öztürk's choice of words caused a lot of commotion, both within parliament and outside. In his response a day later, Öztürk came up with a viewpoint defence: he said that he only had been representing the views of the people in Iraq who had been hurt by the bombing.[35] In a way, this defence made sense, as it is true that Öztürk's own contribution to the debate – which came after his interruptions of others – emphasized the feelings of the Iraqi victims. In particular, he posed a rhetorical question by asking how these victims must have felt when hearing the

---

32 See https://www.reuters.com/article/uk-mideast-crisis-idUKKBN0OK27A20150604.
33 See https://spectator.clingendael.org/nl/publicatie/de-kleine-en-grote-leugens-de-geheime-oorlog-tegen.
34 See https://zoek.officielebekendmakingen.nl/h-tk-20192020-19-23.html.
35 In Dutch: 'Het enige wat ik heb gedaan, is de gevoelens van de mensen daar, die slachtoffer zijn geworden, hier in ons parlement weergeven'. See: https://zoek.officielebekendmakingen.nl/h-tk-20192020-19-23.html.

words that the Defense Minister had used to indicate the casualties: "unintentional collateral damage" [in Dutch: *onbedoelde nevenschade*]:

> (11) This Minister does not see the seriousness of this murder. This Minister says that 22 dead women and 26 dead children are 'collateral damage'. Imagine: you are a victim who lost his wife, child or parents, and then you hear the Minister say, ice-cold, that this was unintentional collateral damage. What does this do to these people?[36]

And when he was criticised for using – again – the word "murder" in this contribution, he explicitly ascribed the use of the verb "murder" to the victims by using the form of indirect speech ("People say that . . .").

> (12) People say that their loved ones, their dear ones have been murdered. This is an emotion that should be expressed as well.

The phrases in (11) and (12) do indeed indicate that by using the word "murder" Özturk (S1) shifts the viewpoint to the victims (S2). Nevertheless, his appeal to this viewpoint shift may already fail on the first main critical question, i.e. whether the word "murder" was indeed used by S2. Even though it is likely that the word accurately describes the victims' feelings, Özturk does not mention a source so we cannot be sure whether a word like this was used by the victims with respect to the incident (and we have not been able to find such a source). Moreover, the reasonableness of Öztürk's viewpoint defence surely fails when applying the second main critical question, i.e. whether he may be assumed to agree with the appropriateness of using the word "murder" in the context. We draw this conclusion on the basis of what Öztürk also said in the debate:

> (13) People have been murdered over there. *I* cannot [think of] other words for this.

These words ("*I* cannot . . .") leave very little room for thinking that Öztürk was "only quoting". Even if we accept the viewpoint shift as based on an actual source, the example above shows that his shift to the viewpoint of the victims in some utterances does not discharge him from his own commitment to the use of the word in other utterances, such as (13). Öztürk was probably aware of this himself, as he came up with the following remarks:

> (14) I did not say that our soldiers and our Minister knowingly and deliberately killed people. I did not say that. What I did say, is that people have the feeling that they haven't known for four and a half years what exactly happened. It should be allowed to express those words and emotions here as well. In this way I try to put right what has been misunderstood in my opinion.

---

36 https://zoek.officielebekendmakingen.nl/h-tk-20192020-20-8.html. Examples (12)-(15) have also been taken from this source.

In (14), Öztürk suggests that, since he was expressing "those words and emotions" (S2), he himself (S1) is not committed to the conventional meaning of the word "murder" (i.e. "knowingly and deliberately killing people"). This defence was not accepted. As a fellow member of parliament maintained, it is just not possible to retract commitment to this conventional interpretation:[37]

> (15) (. . .) the gentleman Öztürk should realise that if he uses the word 'murder' in this house, it deals with 'premeditation', 'knowingly and deliberately' and 'intentionally'. Take back those words if you are a big boy.

Representing the view of the general audience

While Öztürk – falsely or not – attributed the represented thought to an identifiable group of people, a viewpoint defence can also appeal to a more general or abstract source. In such cases the hearer could not reasonably have recognized the viewpoint shift, which means that it has to be characterized as such afterwards, i.e. after the arguer has been criticized for what they said. An example of this rather vague type of viewpoint defence arose in a clash in a popular Dutch television talk show, in May 2015, between co-host Sylvana Simons and guest Martin Šimek. While talking about his experiences helping African boat refugees coming ashore nearby Šimek's residence in South Italy, he used the term *zwartjes* – a diminutive term that literally means "little black people".

Simons, being a black woman herself, asked Šimek what was his motivation for using the word: was it meant to be funny, or cynical? Šimek replied that it was neither of these two and took recourse to a rather general viewpoint defence by responding "it is just the way we talk about them." By thus implying that this phrasing represents the conventional way of talking about people with an African background, Šimek shifted the viewpoint to the supposed general audience.[38] In the course of the discussion he also mentioned some other

---

37 In this respect, the word *murder* is emotionally charged and has a specific legal meaning, as well as behaving a bit like an expressive element in the sense that, even if it is used within a quote, the responsibility for using it is attributed to the quoter (Potts 2007); see also the case of racial slurs in the next case study (and fn. 43).

38 https://www.bnnvara.nl/dewerelddraaitdoor/videos/269066 at 8:44; see also https://www.parool.nl/nieuws/sylvana-simons-de-vrouw-die-zich-niet-laat-beteugelen~bb43c00b/. A more benevolent interpretation is that Šimek is using the controversial word when he is "quoting" what people, including himself, say to themselves. This interpretation of his words, suggested by Šimek's statement that he was "ridiculing himself", is potentially more reasonable since one does not have to fully agree with all the things a "voice in your head" is saying.

arguments why he should not be held committed to have used a derogatory word to indicate black people, namely that his wife and children are persons of colour, that he has been kind to black people (i.e. that he had been helping the refugees), and that a discussion about the word "zwartjes" distracts from the more important message he wants to convey, namely that the strict immigration policies of Europe put refugees into the arms of criminals.[39]

Although Simons seemed to accept Šimek's defence, her acceptance does not imply that it is a sound one. Let us first have a look at the reasonableness of his viewpoint defence. It immediately fails on the first main critical question that addresses the quote's reliability. Presenting the use of *zwartjes* as "quoting" the general audience is simply presenting a falsehood, as it is not "just the way we talk about them". On the contrary: *zwartjes* has a derogatory connotation and is a term avoided in public discourse for exactly this reason. It may be the case that the word is still in use in some circles, but if you do not want to get into trouble, you do not use it.[40]

As Šimek's viewpoint defence was based on a falsehood, it is only natural that he also made use of additional arguments. However, they were not reasonable either. That Šimek's wife is a person of colour does not make it reasonable for him to use a derogatory term for people with a black skin: this is irrelevant argumentation. The same refutation holds for the claim of his having been helpful to refugees. Moreover, Šimek's defence that a discussion about the alleged word distracts from the more important issue of how these people are maltreated by Europe seems a distraction itself and creates a false dichotomy as one can pay attention to both issues at the same time.

---

39 Referring to his kindness, Šimek literally used the word *lief* ("sweet"). Its more accurate translation *sweet* may sound strange in this context, and arguably derogatory as well, but it also does in Dutch.

40 It should be noted that the general opinion about potentially racist or discriminatory wording has changed rapidly the last couple of years, given the ongoing reconsideration of many words that imply biased ideas about minorities of all kinds. In the Netherlands sensitivity regarding racial minorities may be reinforced by the vehement discussion about *Zwarte Piet* (Black Pete), initiated by Dutch people with an African background who have been demanding their position in Dutch society for some time now. The Šimek/Simons case dates from 2015 and from today's perspective it is incomprehensible that talk show host Matthijs van Nieuwkerk behaved very casually about the matter ("We are here to discuss something else") and that many Dutch people in fact were critical of Simons and found it disgraceful that she had even dared to ask her question.

4 Specifying the assessment criteria for the viewpoint defence

In section 2.2 we distinguished two main critical questions for assessing the reasonableness of the viewpoint defence (CQ1 and CQ2, repeated below). In order to be reasonable, both questions need to be answered positively. Based on the case studies discussed in the previous sections, we can now specify these critical questions by formulating several sub-questions that serve as guidelines for answering CQ1 and CQ2.

In order to answer CQ 1, three critical sub-questions can be distinguished:

CQ1: Are the contested words of S1 really a quote?

(a) Can the shift of viewpoint be recognized on the basis of textual evidence?[41]
(b) Is there a source for the represented thought or utterance, i.e. is S2 identifiable?
(c) Is the quote a reliable representation of S2's point of view, i.e. did S2 actually say p and/or implicate q1?

Sub-questions (a) and (b) – about the presence of textual evidence for the quote and an identifiable source – are closely connected and play a role in each of the case studies that we have discussed. Affirmative answers can be given in cases where the viewpoint shift is presented in an explicit way, i.e. with clear textual indicators and mentioning of S2. Wilders explicitly referred to the words of Horst Seehofer and Angela Merkel in the form of direct and indirect speech and a retweet always shows the original sender (see section 3.1).

In some other cases, either of these questions or both ((a) and/or (b)) should receive a negative answer. Whereas Pelosi did use an explicit indicator of there being some S2 (question (a)), i.e. *they say*, the reference to the source was very vague (question (b)). Baudet claimed to have presented Houellebecq's viewpoint on women's rights etc., but textual indicators actually pointed to the

---

41 Some quotes can be recognized on the basis of common knowledge even in the absence of textual indicators or a specific, identifiable source. For instance, it is not very plausible that one can quote the German (!) phrase "Wir haben es nicht gewusst" ('We didn't know') without having any association with World War II and the Holocaust, as Thierry Baudet did claim once (https://nos.nl/artikel/2285628-baudet-heeft-bij-habe-es-gewusst-geen-associatie-met-wo-ii). This denial, in fact, resembles his insistence that he was not aware of the associations triggered by the Pepe the Frog meme (section 3.1).

contrary (question (a)). Öztürk did ascribe the word "murder" to a source (question (b)), i.e. the Hawija victims, but did so only after being criticized, and therefore failed to satisfy question (a).[42] Šimek, finally, never provided any textual indicator of shifting viewpoint when using the word *zwartjes* ('little black people') (question (a)).[43]

The relevance of sub-question (c), i.e. the question of whether the quote really represents S2's point of view, can be illustrated with all the cases discussed in 3.2 and the case of Pelosi from section 2. In the case of Baudet's book review, it can easily be demonstrated that some of the views that Baudet, in his defence, ascribes to Houellebecq are not in fact to be found in the novel he is reviewing. In Öztürk's case it may be questioned whether the Hawija victims actually used a word like *murder* for the incident, since we did not find evidence for this on international websites.[44] As for the racial slur used by Šimek, it is quite clear that "the general public" does not normally use this diminutive form to refer to black people. In Pelosi's case, one can doubt whether *morbidly obese* is a generally accepted category in medicine with regard to gradations of obesity: WebMD does use it, but the Centers for Disease Control and Prevention do not. Moreover, Trump's weight does not tip the scale of WebMD's interpretation of the "morbidly obese" category, which makes it unlikely that "doctors" qualified him as such.

A positive answer to CQ1 is a necessary condition for a reasonable appeal to the viewpoint defence, but not a sufficient one: CQ2 needs to be answered positively as well. In order to find out whether this is the case, again three sub-questions can be formulated:

---

42 Also in the case of using a meme like Pepe the Frog, it is clear, at least for part of the audience, that the meme is being re-used (question (a)), but its origin is hard to pin down precisely (question (b)); see Dynel (2021) on the "epistemological complexity and ambiguity" of memes, "the voices behind which cannot always be categorically known". (Dynel revisits this issue in Chapter 6.)

43 Interestingly, even if he had attributed the use of the word to an explicit source, he might still have been accused of using racist language. In this respect, racial slurs seem to behave like other expressive elements that the speaker is accountable for, even if the expressive is embedded in another viewpoint (Potts 2007). Nowadays it is, for instance, not uncommon to be accused of racism when singing along with a song containing the n-word, or discussing the n-word as such, i.e. 'mentioning' rather than 'using' it (Herbert, under submission). For another view on the use of slurs in citation, see Allan (2016) who argues that "unintentional offence should be forgiven" (p. 226).

44 Nevertheless, evidence might be found on non-English websites.

CQ2: Is it reasonable to assume that S1 is not committed to the "quote", including its implicatures (q1 and q2)?

(d) Can a meaningful purpose be assigned to S1 representing the words or views of S2, other than the generalized implicature (q2) that S1 agrees with the contents of the quote (p and q1)?

(e) Is S1 known as someone generally subscribing to views expressed by S2 in p?

(f) Can S1 be held committed to p, q1 or q2 on the basis of other contributions to the communicative context?

In sub-question (d) we capture the more general phenomenon that, on defence line 2 (see Figure 1), providing an alternative implicature always constitutes a stronger defence than simply denying one (Boogaart, Jansen & van Leeuwen 2021: 224–229). This is certainly true in the case of the viewpoint defence, as was clearly illustrated by the first two cases discussed in section 3.1: Wilders' claim that he was "just quoting" without committing himself to any implicature failed, while Brussen's defence succeeded since he provided an alternative reading (irony) and justification (newsworthiness) for retweeting the message. The appeal to a "meaningful purpose" such as newsworthiness in sub-question (d), moreover, enables us to account for the use of quotations in journalism, and interviews more generally, that are non-committal in the sense that the journalist/interviewer is not responsible for what the interviewee is saying.

The relevance of question (e) comes to the fore in the case of Lucassen that we started out this paper with. His political party (the Party for Freedom) had been very critical about the political leaders of Curacao and Sint-Maarten, using similar phrasings as the one that Lucassen claimed to be "just quoting".[45] But this sub-question may also work the other way round: in the case of Brussen retweeting the Wilders death threat, the views expressed by Brussen elsewhere on his blog were an important argument for the Prosecutor to drop the case against him since it made an ironic reading quite plausible.

Sub-question (f), finally, plays a role, for instance, in the assessment of Nancy Pelosi "quoting" doctors. As we discussed in section 2.1, from other statements by her it could be inferred that Pelosi was trying to get back at Trump for having offended women for their weight in the past, so she was not "just quoting" without any additional intention. The relevance of sub-question (f) is also indicated by the Öztürk case (*murder*) discussed in section 3.2: one of his other contributions in the relevant debate was a clear signal that even if Öztürk had actually been quoting in

---

45 See van Leeuwen (2015: 122) for a discussion of concrete examples.

some instances (i.e. CQ1 could be answered positively), he had already expressed his own commitment to exactly this word.

5 Conclusion

Representing the words or views of another speaker is not always without consequences. If you are accused of having said something objectionable and you simply defend yourself by saying that the view you expressed was not your own, i.e. by using the viewpoint defence, this hardly ever constitutes a reasonable defence. To begin with, there should be sufficient indications in the contested utterance that the speaker was, indeed, shifting viewpoint to another speaker. In addition, the words used should of course be a reliable representation of the other speaker's viewpoint. In our case studies in section 3.2 these two conditions were not always fulfilled. But even if they are, speakers may still be committed to the words and views they are representing. This follows from the very act of "quoting", broadly defined, which triggers implicatures that are not always easy to deny. In particular, we argued in favour of a generalized implicature that, in principle, quoters agree with the contents of the quote. From our case studies in section 3.1 it appears that this implicature may be overruled only by contexual or situational information providing a convincing alternative purpose for the quote.

We showed that the viewpoint defence is used not only in cases in which it is undisputed that the contested words are not the speaker's own (as in the case of direct and indirect speech or retweeting a message), but also in cases in which the speaker appeals to a more implicit shift of viewpoint in the utterance under dispute. These more implicit viewpoint shifts turned out to have the same commitment issues as the explicit cases and played an important additional role in finding criteria for the evaluation of the viewpoint defence. Nevertheless, we think that assessing the relation between commitment and viewpoint could benefit from extending the analysis to cases that depart even more from literal quotes than our implicit cases in section 3.2. For instance: to what extent are you committed to an alleged homophobic petition that you signed, for singing along – as a white man – with a song that contains the n-word, or for banners at a demonstration you are attending? We aim to explore such questions in future research, in order to provide a more complete picture of the assessment criteria for deciding when exactly you can be held accountable for words that are not your own.

References

Allan, Keith. 2016. The reporting of slurs. In Alessandro Capone, Ferenc Kiefer & Franco Lo Piparo (eds.), *Indirect reports and pragmatics: Interdisciplinary perspectives*, 211–232. Cham: Springer.

Boogaart, Ronny, Henrike Jansen & Maarten van Leeuwen. 2021. "Those are your words, not mine!" Defense strategies for denying speaker commitment. *Argumentation* 35 (2). 209–235.

Braet, Antoine. 2007. *De redelijkheid van de klassieke retorica* [Reasonableness in classical rhetoric]. Leiden: Leiden University Press.

Dynel, Marta. 2021. COVID-19 memes going viral: On the multiple multimodal voices behind face masks. *Discourse & Society* 32 (2). 175–195.

Ensink, Titus. 1992. *Jenninger: De ontvangst van een Duitse rede in Nederland* [Jenninger: The reception of a German speech in The Netherlands]. Amsterdam: Thesis Publishers.

Fludernik, Monika. 1993. *The fictions of language and the languages of fiction*. London: Routledge.

Grice, H.P. 1975. Logic and conversation. In Peter Cole & Jerry L. Morgan (eds.), *Syntax and Semantics, vol 3: Speech acts*, 41–58. New York: Academic Press.

Herbert, Cassie. Under submission. Talking about slurs.

Huang, Yan. 2017. Implicature. In Yan Huang (ed.), *The Oxford Handbook of Pragmatics*, 155–179. Oxford: Oxford University Press.

Kienpointner, Manfred. 1997. On the art of finding arguments: What ancient and modern masters of invention have to tell us about the 'Ars Inveniendi'. *Argumentation* 11 (2). 225–236.

Langacker, Ronald W. 2017. Evidentiality in Cognitive Grammar. In Juana I. Marín-Arrese, Gerda Haβler & Marta Carretero (eds.), *Evidentiality revisited*, 13–55. Amsterdam/ Philadelphia: John Benjamins.

Leech, Geoffrey N. & Short, Michael H. 1981. *Style in fiction*. London/New York: Longman.

Leeman, Anton D. & Antoine C. Braet. 1987. *Klassieke retorica* [Classical Rhetoric]. Groningen: Wolters-Noordhoff.

Macagno, Fabrizio & Douglas Walton. 2017. *Interpreting straw man argumentation. The pragmatics of quotation and reporting*. Cham, Switzerland: Springer.

Mazzarella, Diana, Robert Reinecke, Ira Noveck & Hugo Mercier. 2018. Saying, presupposing and implicating: How pragmatics modulates commitment. *Journal of Pragmatics* 13. 15–27.

Potts, Christopher. 2007. The expressive dimension. *Theoretical Linguistics* 33 (2). 165–197.

Sanders, José. 2010. Intertwined voices. Journalists' mode of representing source information in journalistic subgenres. *English Text Construction* 3 (2). 226–249.

Scott, Kate. 2021. The pragmatics of rebroadcasting content on Twitter: How is retweeting relevant? *Journal of Pragmatics* 184. 52–60.

van Haaften, Ton & Maarten van Leeuwen. 2020. Suggesting outsider status by behaving improperly: The linguistic realization of a populist rhetorical strategy in Dutch parliament. In Ingeborg van der Geest, Henrike Jansen & Bart van Klink (eds.), *Vox Populi. Populism as a Rhetorical and Democratic Challenge*, 108–129. Cheltenham/Northampton: Edward Elgar Publishing.

van Leeuwen, Maarten. 2015. *Stijl en politiek. Een taalkundig-stilistische benadering van Nederlandse parlementaire toespraken*. [Style and politics. A Linguistic-stylistic approach to Dutch parliamentary speeches.] Dissertation, Leiden University. Utrecht: LOT Publications.

Verhagen, Arie. 2012. Construal and stylistics – within a language, across contexts, across languages. *Stylistics across disciplines: Conference proceedings*. Leiden (CD-rom). https://www.arieverhagen.nl/research/publications/publications-listing/#2012.

Marta Dynel
Memefying deception and deceptive memefication: Multimodal deception on social media

Abstract: While online deception has attracted a lot of attention in psychology and communication studies, not much has been written about the formal features of this phenomenon from the vantage point of linguistics or philosophy. Advocating the notion of *multimodal deception*, this chapter aims to address this gap and takes a pragmatic perspective on social media deception that provides humorous entertainment. The forms of overlap between humour and deception are presented as manifest in social media content. I also indicate several social media practices that are anchored in deception: trolling, scambaiting and some pranks. Next, I explore the impact of social media affordances on the workings of deception in public multi-party interactions. Most importantly, I focus on the acts of reposting and the – intentional or unwitting – repurposing of previously nondeceptive posts so that they may invite false beliefs in the receivers. This sheds new light on the fine distinctions between (purposeful) deception and (unintentional) misleading across social media contexts, as well as the epistemologically grey area. The chapter is illustrated with contemporary examples of Covid-19 mask memes.

1 Introduction

This chapter addresses the topic of deception that goes beyond the use of words, and thus the verbal modality, and is performed across modalities, which means that it cannot be equated with "nonverbal deception", i.e. deception performed without the use of words. Hardly any attention (but see Dynel 2021a on film) has been paid in theoretical scholarship to multimodal deception, i.e. the deception that operates across modes and modalities and cannot be reduced to verbalisations

Acknowledgement: I would like to thank Professor Larry Horn for his careful reading of the previous version of this chapter and his very helpful comments and suggestions.
Funding: This work was supported by the National Science Centre, Poland (Project number 2018/30/E/HS2/00644).

Marta Dynel, University of Łódź, Vilnius Gediminas Technical University,
e-mail: marta.dynel@yahoo.com

https://doi.org/10.1515/9783110733730-007

or meaningful gestures in face-to-face interactions. This chapter aims to fill this gap by examining the topic of humour-oriented deception on social media.

Multimodal deception is addressed here as an important subgenre of *online deception*, also referred to as *digital deception*, i.e. deception performed on the Internet, with a specific focus on deception on social media. Thanks to the anonymity that it affords, the Internet offers fertile ground for a whole range of deceptive practices in diverse interactional contexts, serving various purposes (cf. Utz 2005), both communicative and practical. Online deception has been amply studied by sociologists and social psychologists (e.g., Hancock, 2007; Hancock et al. 2008; Whitty and Joinson 2008; Toma et al. 2019). Previous research has concentrated primarily on romantic contexts (e.g. Toma 2017; Sharabi and Caughlin 2019) and financial activities, with the deceiver's aim being typically to trick the victim out of money (Button and Cross 2017). In the latter case, the notion of "deception" is typically superseded by other ones, such as the legal term "fraud" or the originally folk and now also academic term "scam" (see Button and Cross 2017; Beals et al. 2015). There have also been studies on "hoaxes", a term which seems to have been applied previously mainly with regard to post-truths or (false) conspiracy theories (e.g. Park and Rim 2020), as well as scamming emails (Heyd 2008).

Yet another facet of deception is prevalent on entertainment platforms, which are typically publicly available, even to unregistered users. This is, presumably, the least harmful form of deception, typically devoid of serious repercussions, apart from (potentially) affecting users' belief system about the nature of user-generated content posted mainly for the sake of humorous entertainment, and possibly damaging their own confidence and self-image when they realise they have been duped. As I will try to show in this chapter, such deception may take two different forms (and combinations thereof) and lies at the heart of different practices orientated towards online entertainment.

This chapter is divided into eight sections. Following this introduction, Section 2 depicts the notion of multimodal deception. Focusing on social media, Section 3 addresses multimodal deception as a source of humour, distinguishing the facets of the relationship between humour and deception based on the participatory framework on which public social media operate. Section 4 briefly presents the online activity types in which humorous deception may manifest itself. In turn, Section 5 sheds light on memetic cases that do not formally qualify as deception but involve the communication of false beliefs. In Section 6, I discuss the epistemological problems in the study of memes, indicating when memes may be regarded as involving deception. The chapter closes with concluding remarks in Section 7.

The data used in the present discussion are a convenience sample of examples culled from the author's corpus collected automatically in 2020 from the

Covid-19 memescape,[1] with the focus on face-mask memes, inspired by the duty of wearing masks to stop the virus from spreading. A manual analysis of the data shows that humorous memes cannot be taken at face value and necessitate in-depth investigation, given the deceptive practices and epistemological conundrums lying at their base.

2 Multimodal deception on social media

Taking stock of some of the previous definitions (e.g. Bok 1978; Fuller 1976; Chisholm and Feehan 1977; Mahon 2007, 2015; Carson 2010), it may be proposed that deception, which includes lying among other strategies, should be formally conceptualised as an intentional[2] communicative act of causing, or at least attempting to cause,[3] the *target* to (continue to) have a false belief, that is to believe to be true something the deceiver believes to be false. The successful communication of deceptive messages from the deceiver's end means that the target is deceived in accordance with the covert plan. Importantly, the target may be one or more individuals in multi-party interactions, where not all receivers need to perform the role of the target (see Dynel 2016a, 2018). This is particularly relevant to deception in social media interactions. Such deception capitalises on social media affordances, i.e. "constraints and [more importantly] possibilities for making meaning" across various modes (Kress et al. 2006: 13). Besides the various means for producing and manipulating online content, the technological affordances of social media affect the participatory framework (Dynel 2016b, 2017), which almost inevitably involves multiple receivers and can rely on the merging of various *interactional spaces*. Except for private dyadic exchanges, most social media content is available to the general public or, at least, registered participants or followers, who may be the targets themselves, or observers of acts of deception to which other individuals are being, or have been, subjected within other interactional spaces.

Deception is primarily a communicative phenomenon, which may be performed both verbally, in oral or written exchanges, and non-verbally, as noted by

---

1 The use of randomised, anonymised and publicly available data in social media research is in line with the commonly accepted ethical standards (see Franzke et al., 2020).

2 This is a narrower view of deception that does not cover *unintentional* acts of causing others to have false beliefs (for discussion and references, see Mahon 2015, Dynel 2018).

3 This proviso means that the definition encompasses acts of both successful and failed deception (i.e. where the target can see through the attempted deception and will not be fooled, thereby destroying the deceiver's communicative plan but not cancelling their intention).

many authors (e.g. Linsky 1963; Siegler 1966; Chisholm and Feehan 1977; Bok 1978; Ekman 1985; Simpson 1992; Smith 2004; Meibauer 2005; Mahon 2007, 2015), and through different modes and channels of communication in various types of human interaction, which pertains to social media among other things. Even though verbal deception is centre stage in linguistic and philosophical research, non-verbal deception is relevant to linguistic studies of deception insofar as non-verbal signs and signals may be purposefully deployed in lieu of utterances (see Dynel 2011). These should not be mistaken for non-verbal cues to deception, such as non-verbal leakage showing in a person's body language, which can be scrutinised to detect the presence of verbal deception (cf. various psychological and forensic studies).

Multimodal deception has various manifestations, similar to what has been proposed with reference to deceptive verbalisations. At least some of these can be conceptualised as multimodal counterparts of the previously recognised categories of verbal deception in the form of spoken or written utterances. These types of deception are most notably (for a detailed overview of these categories in neo-Gricean terms and references, see Dynel 2018 and references therein): *deceptively implicating* or *covertly untruthful implicature* (through the flouting of at least one of the Gricean maxims), *deceptively withholding information* (through keeping covert all/part of the believed-true meaning in order to invite a false belief in the target), *bullshit* (a message whereby the communicator produces no Gricean speaker meaning, having no concern for the truth and deceiving the receiver about his/her enterprise), *covert ambiguity* (where two alternative meanings exist, but the truthful one is covert to the target, to whom the false one is salient) and *covert irrelevance* (i.e. providing covertly irrelevant information as if it is relevant to the question under discussion).

A query remains as to whether the prototypical type of deception, that is lying, conventionally defined as asserting what one believes to be false, can be performed multimodally. While some argue that, in line with the Gricean thought, one cannot assert, and hence lie, non-verbally (e.g. Horn 2017), it seems that Grice did not really exclude "saying" through non-verbal means (see Dynel 2011), and this saying may also be mendacious. As argued previously with regard to non-verbal deception (Dynel 2018), one can lie non-verbally if this message translates directly into a verbal assertoric statement based on conventional or previously agreed on signs and symbols (and this also depends on how exactly one defines asserting) or a specific message operating in a given conversational context. While none of the examples from the corpus at hand exemplifies this, there will definitely be cases where this is possible through multimodal means. For instance, a man who is at a party with his colleagues receives a text from his wife asking, "Where are you?". In response, he sends her a previously taken selfie of himself sitting at his office desk in front of a computer screen with some document open. In this interactional context, this picture carries an assertion along the lines of: "I'm working in my office," which qualifies as a lie. Overall, given the technological affordances of social media and the complexity of the specimens of deception found there, the categories of deception proposed for language data will be subject to elaboration and expansion, with many specimens not being easily amenable merely to the categories proposed for deceptive utterances.

This is because multimodal deception involves the manipulation of various meaning-bearing semiotic resources across modalities (Baldry and Thibault 2006; Kress 2010; Jewitt et al. 2016). For example, concerning the visual modality alone, the semiotic resources include camera angle, distance, or colours, to name but a few. These elements need to be examined jointly, as stipulated by Baldry and Thibault's (2006: 18) *resource integration principle*, according to which different semiotic resources "are combined and integrated to form a complex whole which cannot be reduced to or explained in terms of the mere sum of its separate parts". Thus, the merging of the multimodal components makes for the totality of the communicated meaning (Machin and Mayr 2012; Machin 2013; see also van Leeuwen 2004 2009; Wang 2014). This analysis may also involve taking account of the cues to deception, the tell-tale signs that cause the deceptive acts to fail or that purposefully reveal to the target that they have been led up the garden path for the sake of humorous surprise. This brings the discussion to the point where the subtypes of multimodal humorous deception need to be differentiated.

3 Forms of multimodal deception-based humour on social media

As previously proposed for conversational humour, deception may be conducive to humorous effects, taking two main forms dependent on the participatory framework (and who the target of deception is), as well as whether or not the target is invited to recognise the deception a priori or post facto (Dynel 2018). Two basic sources of humorous amusement may thus be distinguished: a target taking pleasure in the surprising recognition that they have just been deceived, with no "serious" repercussions following for them, affecting their belief system or causing real-life consequences; and vicarious pleasure that an individual/individuals other than the target derive(s) from the target's being genuinely deceived, often in a witty and creative way in multi-party interactions (Dynel 2017, 2018). The same distinction can be applied to multimodal deception on social media.

Firstly, whilst the prime goal underlying prototypical deception is to cause the hearer to nurture a false belief over a period of time and act accordingly, in the humour based on deception as its primary cognitive mechanism, the humour producer purposefully reveals the deceptive act to the target. The disclosure of deception takes place immediately after it has been performed for the sake of the target's amusement consequent upon the discovery that they have just been innocuously taken in. The humour resides in the target being made privy to the fact that they have been duped.

Previously recognised humour categories, such as garden-path humour (Dynel 2009) and put-on humour, which programmatically rely on deception (see Dynel 2018 and references therein), may be realised multimodally on social media to deceive and thereby amuse the collective target, i.e. countless users. In the case of multimodal stimuli, what otherwise develops incrementally in verbal production (as more discourse is being added) may be captured in a series of images or in only one image. In either case, an online item's elements are amenable to sequential processing based on the order of images or the salience of the components in one image, as the first two examples illustrate respectively.

Figure 1: Humorous comic strip based on deception.

The comic strip in Figure 1 is a pictorial realisation of humour residing in the garden-path mechanism typically found in verbal as well as non-verbal jokes (see

Dynel 2009, 2018). Deception in the form of covert ambiguity is inherent to this humorous mechanism. The receiver is led to entertain the default interpretation only to have to cancel it at the stage of the punchline or, in the comic strip, at the stage of the final image processing. Thus, in the case at hand, the salient interpretation, the clichéd after-party scenario (escorting an intoxicated woman to a flat only to undress her and make sexual use of her), needs to be retroactively rejected to make way for another interpretation relevant at the time of Covid-19: undressing the woman in order to use her brassiere as a make-shift face mask. This surprising twist in the default but deceptive scenario that the receiver must have followed is the source of humour for them.

Figure 2: Humorous image based on deception.

At first blush, the viral image in Figure 2 seems to present Macaulay Culkin in the famous shot from *Home Alone*, which is when the boy called Kevin realises that his whole family has gone on holiday, leaving him behind. However, once the image is studied longer, it transpires that the picture shows an adult man wearing a (presumably, Photoshopped) mask with the lower part of Kevin's face and hands printed on it. Users may be duped by the covertly ambiguous image thanks to the verisimilar design on the mask, its perfect fit, as well as the terror in the man's eyes. These deceptive message components eclipse the initially less salient, tell-tale components: the sideburns, as well as the difference in the colour of the mask and the man's complexion or the clothes, notably the beaked cap with the Distro-Kid sign (i.e. a service for musicians established in 2013 for merchandising music online), that the fictional Kevin could not have worn. The user's realisation that they have been taken in and the discovery of the ingenuousness of the image are the source of humorous pleasure.

Secondly, deceptive messages in multi-party interactions on social media exert humorous effects on online users at the expense of the target, who is (meant to be) deceived but typically not amused (even if the deception should be realised mainly for entertainment purposes), remaining oblivious to the deception and developing false beliefs. This type of act realised in multi-party interactions on social media brings about different communicative effects for the various participants on the reception end, operating in different interactional spaces, in accordance with the message producer's/producers' agenda (see Dynel 2016b; Ross and Logi 2021). While in other forms of interactions, genuine deception may involve humorous effects only as an unintended by-product (or be a purposefully employed strategy for amusing the viewer in film discourse), social media acts seem to be motivated by the need to amuse online users, for instance through deception-based pranks, with the effects of the deception on the target being secondary. The humour relies on the multiple social media users' appreciation of the deceiver's dexterity and/or vicarious pleasure based on the feeling of superiority over the unknowing target subject to an act of creative deception realised in real time (e.g. in live streaming) or reported on retroactively. The duped individual(s) may thus become the "butt of the joke" thanks to their incriminating reactions, as is the case with the example in Figure 3.

Figure 3: Shots from recorded pranks based on deceiving unknowing targets for users' entertainment.

The two screenshots in Figure 3 have been extracted from a video[4] typifying one of the series of vigilante "Where is your mask?" memetic pranks from various countries,

---

4 https://www.youtube.com/watch?v=YrtoL24U6fI.

each featuring a man dressed up as a police officer (here, wearing clothes very similar to an Indian police officer's uniform) and his accomplice (here, a man wearing a red jacket). The latter is shown walking in an open public space without a face mask next to other people. All of a sudden, the pretend police officer shows up, asks his accomplice about the missing mask and attacks him physically, in some films, also reprimanding him for the lack of a mask. Meanwhile, the passers-by who are witnessing the attack and are not wearing their masks themselves either run away or start covering their faces nervously with whatever they can (masks, scarves or even their hands), scared that they will also be subject to similar corporal punishment. These targets (whether or not mask wearers) are typically shown to leave the scene unaware of the fact that the alleged officer and his victim have both been in on the deceptive scheme, nurturing a false belief that they have just witnessed a genuine act of law enforcement rather than identity deception performed by the two men staging a show. However, a few targets of deception in some of the sketches seem to be amused, which may be an indication that they actually find the surprising physical attack on someone else funny or that they have ultimately been made privy to the deceptive scheme. In the latter case, the prank may be considered to represent the combination of the two humorous-deceptive practices described in the course of this section.

On social media, the two types of deception-based humour (being the target to whom the deception is revealed or being a knowing witness to deception to which someone else is subjected) may easily mesh thanks to the merging of interactional spaces. This happens, for instance, when a prank based on put-on humour realised offline or online is (further) reported to online users so that they can also reap humorous rewards, albeit different to the target of the deception as long as they are cognisant of the deception materialising in front of their eyes. This can be illustrated with one of the series of memetic videos (cf. Figure 4),[5] each of which shows a man wearing a bizarre mask as he enters various public spaces (shops, petrol stations, or fast food restaurants) only to spark the retail workers' protests and requests that he should put his mask on. The man's underlying goal is to invite their surprised reactions and amusement when the people have found out that he is indeed wearing a mask with an image resembling his face and a surgeon's mask under his chin printed on it. The targets are shown to be deceived by what seems to be a multimodal manifestation of covert ambiguity in the offline encounters. The deception succeeds thanks to the realistic reproduction of the man's facial features (nose shape and jaw shape, as well as the beard), here further strengthened by the prop, namely a drink with a straw in his hand. Additionally, the upper line

---

5 https://www.tiktok.com/@kusteez/video/6927721390435060997?is_copy_url=1&is_from_we bapp=v1 .

of the mask is partly hidden behind the frame of the glasses. The target of this successfully executed deception fails to recognise the cues: the difference in colours between the man's skin and the mask colour, the seam on the nose, as well as the fact that the man's mouth will not move as he is saying that he is indeed wearing a mask in response to a request that he should put it on. These potential giveaways seem to be eclipsed by the otherwise convincing image. Only after the prankster takes off his mask does the target realise they have been tricked.

Figure 4: Shots from a recorded prank based on deception for the target's and users' entertainment.

Pranks such as those in Figures 3 and 4 above are duly posted and reposted online (sometimes in clusters) to entertain online users privy to the deception with the introductory shots (which present the prankster adjusting the mask) or informative captions, titles or hashtags (e.g. #prank and #maskprank), which signpost the nature of the videos. However, if the pranks are (re)posted without the contextualising information, upon the first viewing, it is also the online user that may fall prey to the deception, similar to the targets shown on screen, even if the pranksters and original posters themselves could not have had such an intention.

4 Humorous deception as online practices

The examples shown so far (as well as those to be presented further in the course of this chapter) can be collectively described as *memes*. The label "meme" is often used as a blanket term for any humorous multimodal item (see Dynel and Poppi 2020, Dynel 2021a), but formally it applies only to "digital items sharing common characteristics of content, form, and/or stance", which are "created with awareness of each other" and "circulated, imitated, and/or transformed via the Internet by many users" (Shifman 2013: 41; cf. Huntington 2015: 78). On a lower plane, the examples in Figures 2–4 report what can be considered *practical jokes* or *pranks* centred on deception, whether involving offline participants or online users as targets.

Practical jokes and pranks are typically deemed categories of humour but have not been widely described in humour research (but see Marsh 2015). A prank or a practical joke may be defined as "a scripted unilateral play performance involving two opposed parties–trickster and target–with the goal of incorporating the target into play without his or her knowledge, permission or both" (Marsh 2015: 12). What needs to be added to this definition is that the target may be not only stunned (e.g. upon finding all objects in their office wrapped in toilet paper) but also deceived, as in the examples in Figures 2–4, where offline participants or online users become the pranksters' targets.

Besides such forms of humour, there are also other practices that present humorous potential based on deception, with online users being amused at the expense of select online targets deceived in different interactional spaces. These practices include, most notably, *trolling* and *scambaiting*. Very often, scambaiting and trolling start with *identity deception*, which involves the deceiver presenting themself as someone they are not so that their deceptive scheme can materialise (Dynel 2016b). Thus, having built a fake identity, a deceiver submits false-believed content so that the target of deception should develop the required false belief(s). These interactions may sometimes be employed for humorous purposes in communication with users other than the target of deception operating in a different interactional space.

Presumably influenced by users' diversified parlance, in academic discourse, the term "trolling" is often used in reference to various aggression-based or provocative messages (see Dynel 2016b for discussion and references; Graham, 2019). However, traditionally understood, in line with the word's etymology, trolling is an online activity that inherently rests on *deception* (Dynel 2016b). As Hardaker (2013: 82) observes (only to focus on impoliteness and aggression), deception is "an almost ubiquitous, defining ingredient of trolling, involving false identities, disingenuous intentions, and outright lies", while aggression is also important, but "not necessarily always produced by the (alleged) troller." Essentially, a troll intends to come across as an individual sincerely participating in a computer-mediated interaction whilst covertly pretending to honestly have the opinions, interests or goals they are making manifest in order to deceive the target, often multiple users, and to trigger their emotional reactions. This is all for the sake of the troll's own pleasure, as well as sometimes the humorous entertainment of those who can see through the deception or who have the interactions explicitly presented as the troll's previous achievements in different interactional spaces (see Dynel 2016b). Although trolls intend their deceptive communicative intention to remain covert so that selected individuals are indeed deceived, they may make it overt to other users for the sake of their entertainment (see Bishop 2014).

Scambaiting is a practice involving prolonged interactions with scammers, i.e. individuals who commit deception with a view to extorting money from unwitting subjects. Scambaiters, however, do see through the scammers' deception and employ deception, too, as they engage in exchanges with scammers (Chia 2019; Smallridge et al. 2016). Hence, scambaiters act as if they are genuinely interested in scammers' offers (Smallridge et al. 2016), seemingly oblivious to the mischief; they do so in order to frustrate the scammers and ridicule them. Recent studies (Ross and Logi 2021; Dynel and Ross 2021) report that scambaiting has been progressing away from the vigilante function with retributive and awareness-raising goals towards pure entertainment, a feature of scambaiting observed previously but not regarded as its central definitional component (see Dynel and Ross 2021 and references therein). Scambaiters can perform complicated acts of deception with a view to sharing the fruit of their creative labour on social media platforms in order to display their creativity and amuse vast audiences with their multimodal deceptive posts testifying to the scammers having been outwitted.

The humour-oriented activities performed by pranksters, scambaiters and trolls may vary in many ways, starting from their (lack of) malicious intent towards the targets (which is what distinguishes these activities from scamming, such as phishing, catfishing and other non-humorous deception typically carried out for financial gain), through the means of offline/online realisation, to the actual repercussions for the targets. However, in any case, the targets of deception in reported pranks, scambaiting or trolling are unwitting participants (Dix 2006) unaware of the fact that they are, or will be, taking part in an online performance meant to entertain social media users (cf. Ross and Logi 2021). The humour emerging from deceptive acts relies on the change of the participatory framework and the merging of the interactional frames, which is feasible thanks to new media affordances.

5 Beyond humorous deception

Informed by social media data, two important points need to be raised about humorous communicative phenomena that seem related to deception but should not be mistaken for deception per se.

Firstly, it should be stressed that similar to verbal messages, untruthful multimodal messages need not always be deceptive at all. This is because deception consists in covert untruthfulness (see Dynel 2018), while a different variety of untruthfulness may be overt to the receiver. Overt untruthfulness of the verbal kind shows in chosen figures of speech (notably irony and hyperbole, which may but do not need to serve humorous purposes), as well as some humour (called "autotelic humour"), which does not carry any pertinent truthful meanings. Both these figures and humour are not typically announced and display absurdity or nonsense on the literal reading. The overt untruthfulness of multimodal messages on social media is realised and signposted differently, being performed with salient cues to its presence so that the fabricated/untruthful content should not deceive anybody. The fun and the humour stem from the creative or skilful means of constructing a false-believed/untruthful multimodal message to whose untruthfulness the receivers are privy. This is the case with the image manipulation done in a specific online space from which the example in Figure 5 comes: a subreddit (a part of Reddit) devoted specifically to Photoshop battles, where users exhibit their technical flair and skills through constructing fake images, with no intention to deceive other users, who are in the know, into believing that the images are genuine photographs.

Figure 5: Nondeceptive (but potentially misleading) image manipulation.

The image in Figure 5 presents a boy carrying some groceries in his hand, presumably from the shop shown in the background, and wearing a rather peculiar face mask. The image of the boy seems to be an example of the meme series involving people in private, closed spaces securing saucepan lids in front of their faces by tightening their hoodies around them for fun, rather than for genuine Covid-19 protection (see Dynel 2021b). As made overtly manifest by the context, namely both the interactional space (the specific subreddit) and the title, together with the tag at the top of the post, the image should not be taken to truthfully communicate that the subject has gone shopping with a lid covering his face, with the untruthfulness of the image being made overt from the outset. Also, based on the reflection on the lid (windows and desk with a chair), it can be surmised that the photograph of the boy must have been taken in a room and then superimposed on the photograph of the street. If someone should be taken in by this image and believe that it genuinely presents a shopper, the original poster cannot be accused of having performed an act of deception.

Secondly, what deserves to be mentioned in the context of humorous deception is the case of false beliefs being developed regardless of the communicator's lack of intent to invite those. Even though some authors claim that deception may be performed unintentionally – that is, inadvertently and/or as a result of a (communicative) mistake – it may be useful to distinguish between a purposeful act of causing false beliefs in others (deception) and *misleading*, which involves a communicator's causing false beliefs, with no intention to do so on their part (Dynel 2018). This may happen when the communicator is wrong/mistaken themselves and shares a false message bona fide as if it were true, which may be frequent on social media thanks to its affordances and the ease of reposting previously encountered content.

The meme in Figure 6 shows a woman presenting her face mask in a selfie taken in what seems to be her bathroom. At first blush, one may be inclined to believe that the woman is sincerely sharing the image of herself in a mask, presumably of her own making, that she is willing to wear it sincerely for genuine protection. However, upon closer inspection, the receiver can notice the woman's naked derrière reflected in the mirror, indicating that the mask has been cut out of the woman's leggings. Even though the humorous intent behind this selfie seems to be quite evident based on the multimodal elements, as well as the broader contextual evidence (this is, clearly, a meme presenting a new version of a viral spoof by comic author and illustrator Adam Ellis[6]), the header in the meme at hand, through seeking the author's rationale, seems to suggest the original image was posted with serious intent, rather than humorous intent. Some of the comments following this post endorse this presumption, wrongly believing that the woman must have been careless enough to accidentally capture her naked buttocks in the picture. However, this seems to be a misguided belief, representing these users

---

6 https://www.huffpost.com/entry/butt-coronavirus-mask_l_5e989875c5b6a92100e410e6.

Figure 6: Nondeceptive but potentially misleading meme.

(including the poster of the version in Figure 6) being misguided, while the original meme author, that is the meme subject, cannot have had such an intention. It is rather evident, based on the bathroom's arrangement and position that the woman has chosen, as well as the angle at which she is positioning her mobile in order to take the selfie, that she is deliberately making sure the selfie should contain her mirror reflection. It is highly unlikely that she should have meant to deceive anybody into believing that the mirror reflection captured in the picture is only accidental. The photograph must have been taken as a spoof, for the fun of it, possibly in reply to the previous post created by Adam Ellis, as many users point out in their comments on the meme in Figure 6. The only form of humorous deception – if any – in the original memetic selfie may reside in the initial inference about the mask being exhibited as if made for sincere protection, a belief which should be discontinued when the mirror reflection in the background is recognised, whereby the humour arises. Needless to say, this deception-based humour will not work for those users who first notice the mirror reflection, thus not developing any momentary false belief at all.

This last example shows that original content found online may be shared further in the same or modified form, being sometimes recontextualised and re-purposed, and intentionally or unwittingly re-interpreted, changing the original message's function and meaning (see Dynel 2020, 2021b). This may – in some cases, at least – induce the receivers to develop false beliefs, being tantamount to deception or misleading respectively. Which is the case cannot always be told because the (re)poster's beliefs and intentions cannot be determined, and the conjectures that may be made show varying degrees of certainty, as illustrated by the example in Figure 7.

Figure 7: Meme with untruthful (deceptive) content.

This Covid-19 mask meme takes as its foundation a photograph of singer and songwriter Billie Eilish wearing a translucent face mask. The caption in this meme puts her in the position of the butt of the joke, a naïve person inclined to believe that the thin lacey fabric can protect her from Covid-19. However, this can be easily recognised as a misrepresentation of the facts. The lacey mask with the golden thread was one part of the celebrity's Gucci attire that she was sporting at the Grammy Awards in January 2020,[7] when no news had been spread yet about the new coronavirus as an imminent threat anywhere outside

---

7 https://www.scmp.com/lifestyle/fashion-beauty/article/3047743/going-viral-billie-eilish-all-gucci-grammy-awards-nails.

China. Thus, the designer mask (even if inspired by the early news about a new virus) must have been nothing but decorative and very much in line with the singer's attention-grabbing style. This stands in contrast to the message that the meme communicates about Eilish's naïve beliefs. If the meme author was cognisant of when the picture had been taken, they may be thought to be responsible for an act of deception based on a lie (the verbal assertion) relative to the visual representation. However, the meme author may have been in the dark themself about the origin of the image, especially about the time when the photograph was taken, not wishing to communicate a false belief whilst intending only to share a sensational meme. Which was really the case cannot be told insofar as the meme creator's thought processes cannot be accessed anyhow.

6 Epistemological problems with social media data

In the case of memetic data on social media, the central question about the (non)deceptive status of a post concerns the poster's intentions and beliefs as they do the sharing of new or previously encountered content. Neither users nor scholars can ever categorically know that a given individual has performed an act of deception unless this deception transpires or is discovered based on the evidence at hand or a confession by the deceiver, all this on the assumption that the evidence is genuine and the confession is sincere. Thus, the examination of deceptive communication in the context of real-life data (rather than examples fabricated by scholars themselves or fictional, literary instances; cf. Dynel 2018 and references therein) is burdened with a fundamental epistemological problem: scholars do not have insight into people's mental states (beliefs, goals and intentions), which can never be established beyond a shadow of the doubt and can only be conjectured based on the available evidence with lesser or greater certainty. Whilst some of such evidence may be relatively easy to retrieve, other evidence necessitates detective work on the part of the researcher, as the example below is meant to illustrate.

The viral image in Figure 8 was amply circulated in Covid-19 memes on Reddit, Facebook and other social media pages with various contextualising captions, such as the three presented here. These seem to conceive the two subjects in the photograph as if they are representing a sincere but unwise act of mask-wearing, and hence a cautionary example of how masks should not be worn, without questioning the provenance of the peculiar image and seeking to discover the intentions of the absurdly dressed elderly man or his interlocutor.

However, in response to these memes, many users posted queries in this respect, to which no answers were provided.

Figure 8: Memes with epistemologically ambivalent conten.

However, a painstaking online search led me to a relevant tweet posted on 12th July, "This picture was taken in Vienna Austria a few days ago. The guy in the 2 masks is an old actor, quite the experimental kind. This was during a covid help for artists demonstration in Vienna. Numbers here are low so masks outside really not necessary."[8] What I duly learnt and does not seem to be widely appreciated is that the viral picture was taken during a silent march and rally held by artists in Vienna in June 2020 in order to draw attention to the difficult situation for artists at the time of the coronavirus crisis.[9] As several news websites report in German, a salient protester during this event was actor and director Hubsi Kramar, who took part in the protest wearing only a backpack, sunglasses and two face masks, one on his forehead and the other as a fig leaf. The mask on his forehead (presumably indicating the need for common sense or for protecting one's sanity) bore the sign hardly discernible in the memetic image but a bit clearer in other images reporting the event: "Kurz Sebastian" (i.e. "short Sebastian"), the reversal-based wordplay alluding to the name of the current Chancellor of Austria, Sebastian Kurz. Clearly, this

---

8 I would like to thank Prof. Liz Haas for sharing this public tweet, which I was fortunate enough to chance upon. I am also very grateful to her for referring me to online articles in German that allowed me to learn the story behind the viral picture.
9 https://wien.orf.at/stories/3056021/.

shocking attire was meant as an act of provocation in order to ridicule the importance of mask-wearing through covering what has to be covered in public with what has to be worn and in order to signpost the problematic socio-political situation. With his clothes on, Kramar duly gave a speech protesting the restrictions imposed on Austrians' rights and freedom.[10] This is the socio-political context of which the various memes seem to be stripped (pun intended), untruthfully picturing the man as the epitome of stupidity in terms of mask wearing rather than as a political rebel performing a tongue-in-cheek act of civil disobedience.

Each meme is typically a derivative of a previous meme, which the poster must have encountered only to replicate it through imitation, this being the essence of meme creation (see e.g. Shifman 2013; Dynel 2016c; Maraev et al. 2021). In many cases, it is difficult to tell whether the meme posters' (re)interpretations of the images are purposeful attempts at deception about the nature of the content they are modifying and sharing, whether the posters are labouring under a misapprehension themselves, or whether they do not care at all about the truthfulness of (parts of) the content they are sharing. Following this last interpretation, users may be driven by the need to share interesting content, as if indifferent to the deceptive, or rather misleading, potential their posts can have if someone should develop false beliefs as a result. Such users may then be considered bullshitters in Frankfurt's (2005) sense insofar as they do not care at all about the truth of the content they are sharing while deceiving the receivers about the overarching purpose of the posts, which are to earn them social media plaudits, rather than to communicate any factual information. Interestingly, on certain social media platforms, such practices may be legitimised by the user community engaged in the sharing of *autotelic humour*, i.e. humour for its own sake, with no overarching truthfulness considerations coming into play (Dynel 2018), which is also why it is difficult to talk about any (purposeful) deception. It is, however, when such content is decontextualised through reposting that it may invite false beliefs in those new receivers, sometimes in line with the reposter's intent, whereby deception does come into being.

7 Concluding remarks

This chapter has taken a pragmatic perspective on humorous multimodal deception on social media, a topic hitherto unexplored in linguistics. The forms of

---

10 https://zackzack.at/2020/07/04/hubsi-kramar-trauerrede-beim-staatsbegraebnis-der-ici/.

overlap between humour and deception manifest in social media content were presented with regard to the participant framework, distinguishing between the target of deception and the receiver of humour. Thus, a distinction was made between forms of multimodal humour involving the multimodal deception of a target who is to reap humorous rewards after being made privy to the deception in sequential processing, and genuine deception of a target that becomes the butt of a joke as the act of deception realised online or offline (but typically primarily for the sake of humorous purposes on social media) is shared with online users. Thus, the potentially humorous social media practices centred on deception, namely trolling, scambaiting and pranks, were briefly introduced. Further, I explored the impact of social media affordances on the workings of deception, notably the ease of reposting and the – intentional or unwitting – repurposing of previously nondeceptive posts to invite false beliefs in the receivers. This shed new light on the fine distinctions between (purposeful) deception and (unintentional) misleading across social media contexts, as well as the epistemologically grey area.

The thrust of this discussion is that even the seemingly trivial genre of memes cannot be taken at face value, and there is much epistemic ambivalence underlying what meme (re)posters' underlying mental states have been. Additionally, users may sometimes have little concern for the truth of the intertextual items they are reusing in their memes. As reported by previous research (Dynel 2020; Dynel and Ross 2021), oriented towards causing amusement and earning online kudos, users are prone to submit fabricated content, for instance, sometimes ascribing stupidity to others so that the latter can be disparaged for holding naïve or false beliefs.

References

Baldry, Anthony & Paul J. Thibault. 2006. *Multimodal transcription and text analysis*. London: Equinox.

Beals, Michaela, DeLiema, Marguerite, Deevy, Martha. 2015. *Framework for a taxonomy of fraud*. Stanford Center on Longevity. http://162.144.124.243/~longevl0/wp-content/up loads/2016/03/Full-Taxonomy-report.pdf

Bishop, Jonathan. 2014. Trolling for the lulz? Using media theory to understand transgressive humour and other internet trolling in online communities. In Jonathan Bishop (ed.), *Transforming politics and policy in the digital age*, 155–172. Hershey: IGI Global.

Bok, Sissela. 1978. *Lying: Moral choice in public and private life*. New York: Random House.

Button, Mark & Cassandra Cross. 2017. *Cyber frauds, scams and their victims*. London: Routledge.

Carson, Thomas L. 2010. *Lying and deception: Theory and practice*. Oxford: Oxford University Press.

Chia, Stella. 2019. Seeking justice on the web: How news media and social norms drive the practice of cyber vigilantism. *Social Science Computer Review* 36(8). 655–672. https://doi.org/10.1177/0894439319842190

Chisholm, Roderick & Thomas Feehan. 1977. The intent to deceive. *The Journal of Philosophy* 74. 143–159.

Dix, Alan, Jennifer G. Sheridan, Stuart Reeves, Steve Benford & Claire O'Malley. In S.W. Gilroy & M.D. Harrison (eds.), *Interactive systems: Design, specification, and verification*, 15–25. Berlin: Springer.

Dynel, Marta. 2009. *Humorous garden-paths: A pragmatic-cognitive study*. Newcastle: Cambridge Scholars Publishing.

Dynel, Marta. 2011. Turning speaker meaning on its head: Non-verbal communication and intended meanings. *Pragmatics and Cognition* 3. 422–447.https://doi.org/10.1075/pc.19.3.03dyn

Dynel, Marta. 2016a. Killing two birds with one deceit: Deception in multi-party interactions. *International Review of Pragmatics* 8. 179–218. https://doi.org/10.1163/18773109-00802002.

Dynel, Marta. 2016b. Trolling is not stupid: Internet trolling as the art of deception serving entertainment. *Intercultural Pragmatics* 13. 353–381.

Dynel, Marta. 2016c. 'I has seen Image Macros!' Advice Animals memes as visual-verbal jokes. *International Journal of Communication* 10. 660–688.

Dynel, Marta. 2017. Participation as audience design. In Christian R. Hoffmann & Wolfram Bublitz (eds). *Pragmatics of social media. Mouton de Gruyter handbooks of pragmatics*, volume 11, 61–82. Berlin: Mouton de Gruyter.

Dynel, Marta. 2018. *Irony, deception and humour: Seeking the truth about overt and covert untruthfulness. Mouton Series in Pragmatics*. Berlin: Mouton de Gruyter.

Dynel, Marta. 2020. Vigilante disparaging humour at r/IncelTears: Humour as critique of incel ideology. *Language & Communication* 74. 1–14. https://doi.org/10.1016/j.langcom.2020.05.001

Dynel, Marta. 2021a. When both utterances and appearances are deceptive: Deception in multimodal film narrative. In Fabrizio Macagno & Alessandro Capone (eds.), *Inquiries in philosophical pragmatics. Perspectives in pragmatics, philosophy & psychology, Vol. 28*, 205–252. Cham: Springer. https://doi.org/10.1007/978-3-030-56696-8_12

Dynel, Marta. 2021b. COVID-19 memes going viral: On the multiple multimodal voices behind face masks. *Discourse & Society* 32(2). 175–195. https://doi.org/10.1177/0957926520970385

Dynel, Marta & Fabio I. M. Poppi. 2020. Caveat emptor: Boycott through digital humour on the wave of the 2019 Hong Kong protests. *Information, Communication & Society*. 24(15). 2323–2341.https://doi.org/10.1080/1369118X.2020.1757134

Dynel, Marta & Andrew Ross. 2021. You don't fool me: On scams, scambaiting, deception and epistemological ambiguity at r/scambait on Reddit. *Social Media + Society* 7(3). 1–14. https://doi.org/10.1177/20563051211035698.

Ekman, Paul. 1985. *Telling lies: Clues to deceit in the marketplace, politics, and marriage*. New York: Norton & Company.

Frankfurt, Harry. 2005. *On bullshit*. Princeton, NJ: Princeton University.

Franzke Aline, Anja Bechmann, Michael Zimmer, Charles Ess & the Association of Internet Researchers. 2020. *Internet Research: Ethical Guidelines 3.0*. https://aoir.org/reports/ethics3.pdf

Fuller, Gary. 1976. Other-deception. *The Southwestern Journal of Philosophy* 7. 21–31.

Graham, Elyse. 2019. Boundary maintenance and the origins of trolling. *New Media & Society*. DOI: 10.1177/1461444819837561

Hancock, Jeffrey. 2007. Digital deception: Why, when and how people lie online. In A. N. Joinson, K. Y. A. McKenna, T. Postmes, & U. Reips (eds.), *The Oxford handbook of internet psychology*, 289–330. Oxford: Oxford University Press.

Hancock, Jeffrey, Lauren Curry, Saurabh Goorha & Michael Woodworth. 2008. On lying and being lied to: A linguistic analysis of deception in computer-mediated communication. *Discourse Processes* 45(1). 1–23. https://doi.org/10.1080/01638530701739181

Hardaker, Claire. 2013. "Uh . . . not to be nitpicky . . . but . . . the past tense of drag is dragged, not drug": An overview of trolling strategies. *Journal of Language Aggression and Conflict* 1. 58–86.

Heyd, Theresa. 2008. *Email hoaxes*. Amsterdam: John Benjamins.

Horn, Laurence R. 2017. Telling it slant: Toward a taxonomy of deception. In Janet Giltrow & Dieter Stein (eds.), *The pragmatic turn in law*, 23–55. Berlin: Mouton de Gruyter.

Huntington, Heidi E. 2015. Pepper spray cop and the American Dream: Using synecdoche and metaphor to unlock internet memes' visual political rhetoric. *Communication Studies* 67(1). 77–93. https://doi.org/10.1080

Jewitt, Carey, Jeff Bezemer & Kay O'Halloran. 2016. *Introducing multimodality* (1st ed.). London: Routledge. https://doi.org/10.4324/9781315638027

Kress, Gunther. 2010. *Multimodality: A social semiotic approach to contemporary communication*. London: Routledge.

Kress, Gunther, Carey Jewitt, Jon Ogborn & Tsatsarelis Charalampos. 2006. *Multimodal teaching and learning: The rhetorics of the science classroom*. London: Bloomsbury.

Linsky, Leonard. 1963. Deception. *Inquiry* 6. 157–169.

Machin, David. 2013. What is multimodal critical discourse studies? *Critical Discourse Studies*, 10(4),347–355. https://doi.org/10.1080/17405904.2013.813770

Machin, David & Andrea Mayr. 2012. *How to do critical discourse analysis: A multimodal introduction*. London: SAGE.

Mahon, James E. 2007. A definition of deceiving. *International Journal of Applied Philosophy* 21. 181–194.

Mahon, James E. 2015. The definition of lying and deception. In *The Stanford Encyclopedia of Philosophy* (Fall 2015 Edition), ed. E. N. Zalta, URL = <http://plato.stanford.edu/archives/fall2015/entries/lying-definition/>.

Maraev Vladislav, Ellen Breitholtz, Christine Howes, Staffan Larsson & Robin Cooper. 2021. Something old, something new, something borrowed, something taboo: Interaction and creativity in humour. *Frontiers in Psychology* 12. 654615. doi: 10.3389/fpsyg.2021.654615

Marsh, Moira. 2015. *Practically joking*. Logan, UT: Utah State University Press.

Meibauer, Jörg. 2005. Lying and falsely implicating. *Journal of Pragmatics* 37. 1373–1399.

Park, Keonyoung & Hyejoon Rim. 2020. "Click First!": The effects of instant activism via a hoax on social media. *Social Media + Society*. doi:10.1177/2056305120904706

Ross, Andrew S. & Lorenzo Logi. 2021. 'Hello, this is Martha': Interaction dynamics of live scambaiting on Twitch. *Convergence*. doi:10.1177/13548565211015453

Sharabi, Liesel L. & John P. Caughlin. 2019. Deception in online dating: Significance and implications for the first offline date. *New Media & Society* 21(1). 229–247. https://10.1177/1461444818792425

Shifman, Limor. 2013. *Memes in digital culture*. Cambridge, MA: MIT Press.

Siegler, Frederick. 1966. Lying. *American Philosophical Quarterly* 3. 128–136.

Simpson, David. 1992. Lying, liars and language. *Philosophy and Phenomenological Research* 52. 623–639.

Smallridge, Joshua, Philip Wagner & Justin Crowl. 2016. Understanding cyber-vigilantism: A conceptual framework. *Journal of Theoretical & Philosophical Criminolog, 8*(1). 57–70.

Smith, David L. 2004. *Why we lie: The evolutionary roots of deception and the unconscious mind.* New York: St. Martin's Press.

Toma, Catalina L. 2017. Developing online deception literacy while looking for love. *Media, Culture and Society 39*(3). 423–428. https://doi.org/10.1177/0163443716681660

Toma, Catalina L., James Alex Bonus & Lyn M. Van Swol. 2019. Lying online: Examining the production, detection, and popular beliefs surrounding interpersonal deception in technologically-mediated environments. In Tony Docan-Morgan (ed.), *The Palgrave handbook of deceptive communication*, 583–601. London: Palgrave Macmillan.

Utz, Sonja. 2005. Types of deception and underlying motivation: What people think. *Social Science Computer Review* 23. 49–56. https://doi.org/10.1177/0894439304271534

van Leeuwen, Theo. 2004. Ten reasons why linguists should pay attention to visual communication. In P. LeVine & R. Scollon (eds.), *Discourse and technology: Multimodal discourse analysis*, 7–20. Washington, WA: Georgetown University Press.

van Leeuwen, Theo. 2009. Discourse as the recontextualization of social practice: A guide. In Ruth Wodak & Michael Meyer (eds.), *Methods of critical discourse analysis, 2nd ed*, 144–161. London, UK: SAGE.

Wang, Jiayu. 2014. Criticising images: Critical discourse analysis of visual semiosis in picture news. *Critical Arts* 28(2). 264–286. https://doi.org/10.1080/02560046.2014.906344

Whitty, Monica & Adam Joinson. 2008. *Truth, trust and lies on the internet.* London: Routledge.

III Puffery, bluffery, bullshit: How to not quite lie

Tim Kenyon, Jennifer Saul

Bald-faced bullshit and authoritarian political speech: Making sense of Johnson and Trump

Abstract: Donald Trump and Boris Johnson are notoriously uninterested in truth-telling. They also often appear uninterested even in constructing plausible falsehoods. What stands out above all is the brazenness and frequency with which they repeat known falsehoods. In spite of this, they are not always greeted with incredulity. Indeed, Republicans continue to express trust in Donald Trump in remarkable numbers. The only way to properly make sense of what Trump and Johnson are doing, we argue, is to give a greater role to audience relativity – and in some cases, audience participation – in notions like bullshitting and bald-faced lying. In this paper, we develop a new understanding of bullshitting, one that includes bald-faced lying, and recognizes that different communicative acts may be directed at different audiences with a single utterance. In addition, we argue for recognition of the category of bald-faced bullshitting, a particular speciality of both Trump and Johnson, and one especially useful to authoritarian leaders.

Donald Trump and Boris Johnson are notoriously uninterested in truth-telling. They also often appear uninterested even in constructing plausible falsehoods. What stands out above all is the brazenness and frequency with which they declare their falsehoods. In spite of this, they are not always met with disbelief. Indeed, Republicans continue to trust Donald Trump in remarkable numbers. Taking all this into account, it is not so straightforward to characterize what they are up to, using the standard tools of the philosophical literature. Is it bald-faced lying? Well, Trump is believed by a great many people and he surely knows this, so it seems wrong to insist that his lies are not intended to deceive. Is it bullshitting? Harry Frankfurt insisted that a bullshitter must conceal what they are up to, and these two make no great effort at concealment.

Acknowledgement: We are very grateful to Larry Horn for insightful and enjoyable comments on an earlier draft of this paper.

Tim Kenyon, Brock University, e-mail: tkenyon@brocku.ca
Jennifer Saul, University of Waterloo and University of Sheffield,
e-mail: Jennifer.saul@uwaterloo.ca

https://doi.org/10.1515/9783110733730-008

The only way to properly make sense of what Trump and Johnson are doing, we argue, is to give a greater role to audience relativity in notions like bullshitting and bald-faced lying. In this paper, we develop a new understanding of bullshitting, one that includes bald-faced lying, and allows for different communicative acts to be directed at different audiences with a single utterance. In addition, we argue for the importance of recognizing bald-faced bullshitting, a particular speciality of both Trump and Johnson, and one especially useful to authoritarian leaders. Our definition helps us to understand what Trump and Johnson are doing, but it also, we argue, serves well to accommodate the concerns that motivated Harry Frankfurt's concept of bullshitting.

We argue that Frankfurt's notion of bullshit is not sufficiently clear as it stands, which probably accounts for the many very different understandings in the literature, most of which are presented as interpretations of Frankfurt. We suggest understanding bullshit utterances as characterized by a lack of concern for one's audience as truth seeker. We show that our understanding includes all cases that meet Frankfurt's original definition, but that it offers more precision and that there are great benefits to explicitly discussing the audience – or *audiences* – as a part of the definition. Crucially, this enables us to make sense of the multiple ways that utterances by leaders like Trump and Johnson function, and to better understand the dangers that they pose. This will also help us to draw out some differences between what Trump and Johnson are doing.

1 Boris Johnson and Donald Trump's bullshit

We will start with some of the clearest examples of bullshit from Trump and Johnson, before later discussing more complex cases.

1.1 Boris Johnson on the Brexit referendum

Although he rose to Prime Ministerial power as a Brexiter,[1] and although he is the Prime Minister who took the United Kingdom out of the European Union, it is widely believed that Johnson in fact had no view about whether Brexit was a good

---

1 We have chosen to use the word 'Brexiter' rather than 'Brexiteer' though both are in popular usage. Interestingly, 'Brexiteer' is the one favoured by Johnson and other supporters of Brexit. Although the swashbuckling connotations suggest recklessness, this is thoroughly embraced by top supporters of Brexit, who consider the term a marketing triumph. We think there may

or bad idea for the United Kingdom when the referendum was held.[2] At the time of the referendum campaign, Johnson had a newspaper column.[3] He wrote one column in favour of Brexit and one column opposed, and then decided at the last minute which one to publish and which side to back. It is generally accepted that this decision was based entirely on what he thought would serve his career best. This is not the action of a person with true convictions, determined to argue for what he believes to be true. Nor, however, is it the action of a deceptive character, determined to conceal the truth and lead his readers into falsehood. Instead, it seems to be naturally characterised as the action of a bullshitter, someone who simply had no interest in truth or falsehood. All indications are that his focus was on nothing but political, or possibly personal, expediency.

1.2 Boris Johnson on his bus-making hobby

In June 2019, while campaigning to become the next Prime Minister, Boris Johnson gave a truly remarkable interview. Asked by the interviewer what he did in his free time, Johnson began to ramble:[4]

> I like to paint. Or I make things. I have a thing where I make models of buses. What I make is, I get old, I don't know, wooden crates, and I paint them. It's a box that's been used to contain two wine bottles, right, and it will have a dividing thing. And I turn it into a bus.
>
> So I put passengers – I paint the passengers enjoying themselves on a wonderful bus – low carbon, of the kind that we brought to the streets of London, reducing CO_2, reducing nitrous oxide, reducing pollution.

Johnson's delivery was that of a man making things up – he took a few halting passes at the idea of making "things," before settling on "buses" – and the hobby was strange and implausible. Among other things, wine tends to come in crates of larger than two bottles. It also seems likely that Johnson had a pre-existing desire to mention busses whenever he could in the interview. Busses had played a very important role in his political career: both the busses that he brought in as mayor of

---

be an interesting parallel here to the embrace of obvious falsehood. For more on the term see https://www.spectator.co.uk/article/victory-of-the-swashbucklers.

2 https://www.reuters.com/article/uk-britain-eu-johnson-views-factbox-idUKKCN1TC1YD, https://www.vanityfair.com/news/2016/07/did-boris-johnson-want-to-remain-all-along.

3 https://www.theguardian.com/politics/2016/oct/16/secret-boris-johnson-column-favoured-uk-remaining-in-eu.

4 https://www.theguardian.com/politics/2019/jun/26/mesmerising-boris-johnsons-bizarre-model-buses-claim-raises-eyebrows.

London and the famous Brexit bus. Nobody seem to find it at all believable that Johnson had genuinely been recounting a hobby. Here are two representative comments.[5]

- Matt Bevan: "I have thought about this a lot and realised that this is exactly how my 3yo son would answer this question."
- Simon Blackwell: "Only just caught up with the Boris Johnson model bus interview." Feels like a screw-you status thing – "I can literally say any old unbelievable shit and still become PM." Like Trump's "I could stand in the middle of Fifth Avenue and shoot somebody and I wouldn't lose voters."

In short, Johnson appeared to be inventing a ridiculously implausible story, with no concern whatsoever for whether anyone would take it seriously. It is hard to imagine that Johnson had any real interest in the truth or falsehood of the story,[6] or in whether anyone believed him. This was all the more puzzling because there was no real need for this invention. It was in response to an easy question about how he relaxes, the sort of query that traditionally helps politicians to seem friendly and relatable – rather than a tendentious topic where he might feel the need to cover up an inconvenient truth.

1.3 Donald Trump on COVID-19 miracle cures

When Trump became a sudden and ardent advocate for hydroxychloroquine as a treatment for COVID-19, there was no clear evidence one way or the other about its efficacy (although it is now clear that it is ineffective and dangerous as a remedy for COVID-19). Trump was very unlikely to have any belief about its efficacy at the time when he said that hydroxychloroquine was a miracle cure. He was not, then, lying.[7] Nor was he saying something that he genuinely believed – it is hard to see how he would have come by this belief. However, support for hydroxycholoquine had quickly become a useful culture war issue for Trump. (He did have a small financial stake in the product.[8]) Trump was, intuitively, bullshitting: pushing a

---

5 https://www.theguardian.com/politics/2019/jun/26/mesmerising-boris-johnsons-bizarre-model-buses-claim-raises-eyebrows.

6 Interestingly, this is despite the fact that he surely knew whether the story was true or not.

7 We assume in what follows that lying requires asserting what one believes is false or not true.

8 Journalists have concluded that Trump's financial stake in hyrdoxychloroquine was likely too small to have been the reason for his endorsement. https://www.washingtonpost.com/politics/2020/04/07/trumps-promotion-hydroxychloroquine-is-almost-certainly-about-politics-not-profits/.

controversial unproven drug in such a way as to provoke his opponents and please his supporters, with no concern for the actual efficacy of the drug.

Or, take Trump's advocacy of bleach or light as treatments, documented by William J. Broad and Dan Levin in the New York Times (Broad and Levin 2020):

> After the administrator, William N. Bryan, the head of science at the Department of Home-land Security, told the briefing that the agency had tested how sunlight and disinfectants – including bleach and alcohol – can kill the coronavirus on surfaces in as little as 30 seconds, an excited Mr. Trump returned to the lectern.
>
> "Supposing we hit the body with a tremendous – whether it's ultraviolet or just very powerful light," Mr. Trump said. "And I think you said that hasn't been checked, but we're going to test it?" he added, turning to Mr. Bryan, who had returned to his seat. "And then I said, supposing you brought the light inside the body, either through the skin or some other way . . ."
>
> "And then I see the disinfectant where it knocks it out in a minute – one minute – and is there a way we can do something like that by injection inside, or almost a cleaning?" he asked. "Because you see it gets in the lungs and it does a tremendous number on the lungs, so it would be interesting to check that".

Trump did not manage to have a financial stake in light, even a small one. But he did have a stake in depicting himself as an all-around genius – someone who could weigh in as an outsider on practically any matter, surprising the resident experts with his insights. A month earlier, in March 2020, Trump had visited the Centre for Disease Control and described his knowledge and understanding in terms of his uncle, "a great super genius" who had been a university professor. "Every one of these doctors said, 'How do you know so much about this?'" Trump said. "Maybe I have a natural ability. Maybe I should have done that instead of running for President."

Even had he just been brainstorming, and not fostering an image of himself as having serious Coronavirus treatment insights, it is quite obviously an enormous mistake for an untrained, ignorant person in a position of authority to brainstorm in public about potentially deadly things that one might try in order to fight COVID-19. (And it does look like some of them followed his suggestions.[9]) This, then, seems like a particularly dangerous instance of bullshitting.

---

9 https://www.usnews.com/news/health-news/articles/2020-06-05/cdc-some-people-did-take-bleach-to-protect-from-coronavirus.

1.4 Donald Trump on Canadian trade deficit

In this incident, discussed by Quassim Cassam (2021), Donald Trump told Canadian Prime Minister Justin Trudeau that the US had a trade deficit with Canada. Afterwards, in a speech with reporters present, Trump explained that he had no idea whether that was true or not. (https://www.washingtonpost.com/news/post-politics/wp/2018/03/14/in-fundraising-speech-trump-says-he-made-up-facts-in-meeting-with-justin-trudeau/). This is a rare case in which a speaker admits to having no concern whatsoever with the truth of what he says. So it is an exceptionally clear case of bullshitting.

2 Frankfurt's account, and ours

According to Frankfurt, truth-tellers and liars alike care about the truth; the one to achieve it, the other to avoid it. Bullshitters – a paradigm case here is the used car salesman who will say anything to make the sale – don't care about the truth. They are simply saying things to make other things happen. Liars and bullshitters are therefore distinguished by their motives in speaking:

> [The bullshitter's] statement is grounded neither in a belief that it is true nor, as a lie must be, in a belief that it is not true. It is just this lack of connection to a concern with the truth – this indifference to how things really are – that I regard as of the essence of bullshit. (Frankfurt 2005: 33–34)

This is a provocative idea, and has struck many people as deep, insightful, and diagnostic of trends in public and political discourse in recent decades. But for all the attention the idea has received there are at least two significant complications regarding it. First, most discussions stray quite far from Frankfurt's definition. And second, the distinction as Frankfurt sketches it is not very clear.

2.1 Divergent understandings in the literature

We'll take the first concern first. There has been enormous enthusiasm for the idea that bullshit is something of a defining trait of the last several decades, and these discussions always take Frankfurt as their starting point. But they tend to stray quite far from Frankfurt's idea of the bullshitter as one who lacks interest in truth, whether to achieve it or to avoid it. Some assimilate bullshit to various kinds of lying. For example, Reisch and Hardcastle (2006: xi) cite the prevalence of fraud as evidence for the rise in bullshit; but fraud normally involves carefully engineered deceptions[10] by people and corporate bodies determined to make sure that the truth is not learned. Reisch and Hardcastle also cite Descartes and Hume (2006: xv-xvi) as being on a mission to root out bullshit, focusing on their efforts to eliminate unjustified beliefs. But many people who are very interested in the truth of their beliefs sadly have unjustified ones – indeed most do, according to Descartes and Hume! Finally, Douglas (2006) describes much climate scepticism as bullshit; in fact the cases she describes look much more like straightforward lies or motivated reasoning, in which people (roughly speaking) manage to construct justifications for the beliefs that they want to maintain.[11] An interest in concealing the truth is not the attitude of a bullshitter, according to Frankfurt's definition.

Others treat bullshit as an issue of semantic content, an approach that misses (or, in some cases, rejects) the role of the speaker's attitudes or intentions on Frankfurt's account. (Cohen (2006) describes this as focusing on *bullshit* rather than *bullshitting*.) Hardcastle (2006) argues that the Vienna Circle, with their concern for eliminating meaningless statements, were at heart concerned with bullshit. The sort of meaninglessness they were concerned with, however, was for the most part not apparent to the ordinary thinker, and the meaningless statements they singled out were certainly endorsed by many seekers after truth. G. A. Cohen (2006, previously 2002), in one of the best-known papers on bullshit, defines bullshit as unclarifiable unclarity, taking continental philosophy as his prime example. Again, though, it's clear that many truth-seekers believe such claims. Cohen's underlying suggestion that bullshit is essentially a kind of semantic *bafflegab* is seen also in Pennycook, Pennycook, Gordon, Cheyne, Barr, Koehler & Fugelsang's psychological studies of "bullshit detection," using examples like "Hidden meaning transforms unparalleled abstract beauty" (2015: 549). In fact, it is not difficult to think of theoretical or argumentative framing that would make this sentence informative and useful, or anyhow no worse than a bit overwrought. But even if we grant that a speaker utters it on the basis of somewhat muddled ideas, still nothing precludes that the muddled speaker asserts it with the intention of expressing and encouraging the uptake of a truth. In that case it still would not satisfy core aspects of Frankfurt's definition, bafflegab or not. Even those discussing Trump have offered some content-focused characterizations. For example, Jacquemet (2020: 127) writes that "we can characterize [bullshit] as ego-centered discourse frequently (even if not always) marked by the use of repetitions, intensifiers, superlatives, and ellipsis".

---

10 Which may or may not be *lies*. (See Saul 2012 for an extensive discussion of the difference between lying and merely misleading.)
11 See, for example, Kunda 1990.

While it's very common for scholars to offer divergent interpretations of an idea, this is an unusually wide range of interpretations.[12] A key reason for this, we think, is a certain lack of clarity in the original text. Recall Frankfurt's truly striking characterisation of the concept of bullshit.

> [The bullshitter's] statement is grounded neither in a belief that it is true nor, as a lie must be, in a belief that it is not true. It is just this lack of connection to a concern with the truth– this indifference to how things really are– that I regard as of the essence of bullshit. (Frankfurt 2005: 33–34)

Just before and after this, Frankfurt himself gives some examples that don't really fit with it at all, including a semantic content case. One of his oft-cited examples is the Fourth of July orator (originally Black's example of humbug) who refers to "our great and blessed country whose Founding Fathers under divine guidance created a new beginning for mankind" (2005: 16). As we've seen, this kind of utterance is often described as bullshit. But it's also true that very large numbers of Americans actually believe that this is an accurate description of the country. Just based on its content, Frankfurt says that it is "surely humbug", and goes on to make clear that he thinks it is bullshit. Interestingly, Frankfurt argues that it is bullshit by noting that "the orator does not really care what his audience thinks about the Founding Fathers, or about the role of the deity in our country's history, or the like" (17). This focus on indifference to what the audience thinks, rather than indifference with respect to truth or falsehood, does *not* actually fit Frankfurt's official definition. We will see, however, that it does fit well with ours.

To find examples that fit the definition, Frankfurt has to turn to a strange story (2005: 24–34) about Wittgenstein taking a speaker's hyperbole literally and thinking that she is asserting something for which she has no evidence. Although Frankfurt himself expresses doubts about the accuracy of this way of understanding the story, he notes that if it did happen that way, Wittgenstein would have thought (falsely) that the speaker was bullshitting. And that is the closest Frankfurt comes to a real-life example of bullshit. This does not amount to a great case for Frankfurt's claim that "one of the most salient features of bullshit is that there is so much of it" (2005: 1).

So why does Frankfurt think that there is so much bullshit? The reason is that it's almost irresistible, even for Frankfurt, to use the term 'bullshit' much more widely than the official definition allows. The ordinary usage of 'bullshit' is much broader – easily encompassing not just the Fourth of July orator, but all the various meanings that appear in the literature.

---

12 And we have by no means attempted a comprehensive survey.

We also think that it's easy to suppose nobody could really live the life of a Frankfurtian bullshitter, if we take as our paradigm the used car salesman who would say anything to make a sale. It's no stretch to imagine someone who would bullshit *about that* in particular, but a person genuinely and quite generally unconcerned with the truth or falsity of what they say would – we might think – surely not make it very far in life. According to this line of thought, their falsehoods and heedlessness would be found out, and surely people would cease to trust them. Their lack of concern with truth and falsehood would mean that people would find it very strange and difficult to interact with them. If they didn't mend their ways, surely they'd end up as some sort of exiles from the human community. This makes the slide to the non-technical uses of 'bullshit' all the more tempting, to round out an otherwise lean set of cases. Real examples fitting the official definition could be expected to be hard to come by, but real examples of the *broader* phenomena are everywhere.

In fact we are sceptical of this kind of reasoning, which leans so heavily on what would *surely* happen to thoroughgoing bullshitters. It rests on a highly idealized sense of the practical consequences of bullshitting and lying – one that is to some extent dangerous. It invites complacency to believe that the social, political, or legal contexts of speech will automatically exert pressure against outright bullshit when it is detected.

This also highlights a curious tension between two of Frankfurt's convictions: that there is very obviously a lot of bullshit around, and that bullshitters by definition must attempt to conceal their art.[13] If the penalties for bullshitting, in lost credibility or other social sanctions, were as severe and certain as the idealized story suggests, the need for concealment would be very clear. But then perhaps the phenomenon shouldn't be quite so *obviously* prevalent. We will argue that there are in fact pretty ordinary contexts in which being seen to bullshit does not incur many or any negative consequences, as well as contexts in which being seen to bullshit has benefits that outweigh being seen as a bullshitter, and still others in which it is actually a key means of achieving the intended outcomes.

2.2 Bullshit, lies, and attitudes towards truth

As Cohen (2006) notes, there is a fair bit of slippage in the literature between discussions focused on the *bullshitter* and their state of mind on the one hand and

---

13 We will take up the concealment claim in more detail when we discuss bald-faced bullshitting later in this paper.

those focused on the sort of *content* that counts as bullshit on the other. Our focus here, like Frankfurt's, is on the bullshitter and their state of mind. We will start with the question of whether liars really care about avoiding truth, as Frankfurt suggests, and whether this really distinguishes them from bullshitters.

If we think about truth as a universal or abstract notion, the answer must pretty clearly be that people don't care about avoiding it when they lie. That is, virtually no people tell lies out of a kind of ideological opposition to truth in the abstract.

Things are at first not much clearer if we think in terms of avoiding particular truths instead. Arguably, when people lie, they do so not from an antipathy to some specific truth in itself. They lie for more prosaic and less hifalutin reasons – because, for example, they don't want to get caught. It's fair to respond to Frankfurt that a criminal lying on the witness stand doesn't fundamentally care about the truth; they care about not going to jail. Frankfurt writes that the bullshitter in the earlier scenario, unlike the liar, "is not trying to deceive anyone concerning American history. What he cares about is what people think of *him*" (2005: 18). But the liar too is often concerned to deceive an audience in order to affect what the audience thinks of them.

If most lies (and other deceptions) arise from the overriding wish to bring about or to avoid various specific outcomes, then the distinction between lies and bullshit that initially seemed so bracing in Frankfurt's characterization of bullshit might seem threatened. Liars too say things to make other things happen. But most of the consequences the liar wants to fend off are due to other people's knowing some truth or other. Even if a speaker ultimately lies because they want to avoid the consequences of their audience's knowing the truth, the audience's knowledge will be a shared cause of a range of those consequences. For someone lying on a witness stand about having evaded corporate taxes, the jury's knowledge of what they did is an immediate outcome that could lead to a heavy fine, or to house arrest, or to jail time. Aiming to avoid any of those unwelcome consequences means aiming to avoid the jury's coming to know some awkward truths. So we can at least say that liars often attend closely to how their speech affects the truth or falsity of the beliefs their audience forms on its basis.

2.3 Our proposal

To recap: when someone lies, for the most part what they care about isn't The Truth, nor is it strictly this truth or that truth. They care about their audience's state of mind with respect to this truth or that truth, and what will result from their audience's having that state of mind. What distinguishes lying from bullshitting isn't how one treats the truth, then, but how one treats one's audience. To see this more clearly, it is helpful to divide the phenomena a bit more precisely than Frankfurt does.

Frankfurt makes the idea of bullshit vivid by contrasting it with things we wouldn't normally expect to be placed together in a single virtuous group: truth-telling and lying.

> Someone who lies and someone who tells the truth are playing on opposite sides, so to speak, in the same game. Each responds to the facts as he understands them, although the response of the one is guided by the authority of the truth, while the response of the other defies that authority and refuses to meet its demands. The bullshitter ignores these demands altogether. He does not reject the authority of the truth, as the liar does, and oppose himself to it. He pays no attention to it at all. By virtue of this, bullshit is a greater enemy of the truth than lies are. (60–61)

This way of carving things up has a striking rhetorical effect, but it can steer us wide of the underlying insight. This is because neither truth-telling nor lying captures the point Frankfurt goes on to make about the distinctiveness of bullshitting.

Truth-telling is not itself a speech-act, after all. It's an achievement, one that asserters represent themselves as aiming at. But even a good-faith asserter may fail at this, depending on the accuracy of their beliefs, without either lying or bullshitting. Someone who earnestly asserts a sincerely believed falsehood is not truth-telling; but they are trying to tell the truth, to facilitate their audience's believing the truth, and that's really what matters for Frankfurt's purposes.

Moreover, the cases of lying that contrast with bullshit, for Frankfurt, are above all cases of intended deception. The liar in these cases is concerned to inculcate a false belief, or at least to prevent a true one, in their audience; this is the respect in which even a liar pays respect to the truth. But some lies, such as bald-faced lies, do not have this aim, while some strict truth-telling does. Deceptive assertion includes both deceptive lying and the assertion of carefully phrased truths that are intended to mislead the audience, a particularly common move among lawyers, politicians, and others for whom strict truth-telling might be an effective way of avoiding sanctions for dishonesty. Speakers employing this strategy are often extremely careful and calculating truth-tellers, yet they are not good-faith asserters.[14] They join deceptive liars in Frankfurt's schema of speakers who respect the truth to at least the extent that they are concerned to direct their audience away from it.

---

14 For an extended study of this sort of speech, see Saul (2012).

So truth-telling and lying aren't quite the right categories to bundle together to effect Frankfurt's surprising contrast. A better contrast is between *good-faith assertions* and *deceptive assertions* on one hand, and bullshit on the other. Good-faith asserters and deceptive speakers alike are concerned with whether their speech will lead their audience to believe some contextually relevant truth (or perhaps to believe it with conviction, as someone might lie or palter simply to cause their audience enough doubt that they will not act in some unwelcome way). Frankfurt's insight is that a distinct type of communicative pathology is displayed when someone makes an assertion without attempting either to facilitate or to obstruct the audience's uptake of a true belief by means of the utterance. This is bullshitting.

This contrast, as we've seen, is a matter of the speaker's actions and attitude towards their audience. Without attempting to characterize this precisely, we note that the contrast clearly concerns the presence or absence of a concern with whether one's audience forms a true belief on the basis of one's assertions. Here, then, is our proposal about Frankfurtian bullshit: it is characterized by a speaker's indifference as to whether their speech provides the basis for an audience to uptake or recover truths.[15] So there *is* a sense in which bullshitters don't care about the truth, but it comes via their treatment of an audience. The bullshitter's aims, with respect to the effects of their speech on an audience, don't principally concern the truth or falsity of this audience's beliefs. It is crucial for us that this be formulated in terms of *an* audience rather than *the* audience. As we will argue, a speaker may have different attitudes toward different audiences, and thereby bullshit with respect to one audience but not another.

2.4 Implications of our proposal

Our proposal captures one of Frankfurt's most crucial illustrative cases, the used car salesman who answers every question with what he imagines the customer wants to hear, neither knowing nor caring whether it's true or false. It also fits extremely well with Frankfurt's Fourth of July orator, who may well care about saying only true things. All Frankfurt tells us is that the orator doesn't really care

---

15 This is distinct from Stokke and Fallis's view that bullshitting is indifference to the process of inquiry (2017). First, the indifference is broader, not just focused on a process of inquiry. And second, our view is particularly about attitudes toward audiences, which allows for different attitudes toward different audiences. This second feature is crucial, we insist, to modern political uses of bullshit.

what his audience thinks – and this fits our version of his definition, rather than the official one.

Our account leaves us in need of more detail about Cohen's jargon-loving impenetrable philosopher. If they are merely trying to sound deep, without any concern for getting closer to the truth, then they are bullshitting. But if they are genuinely engaged in the project of trying to understand and arrive at philosophical truths, then they are not bullshitting.

Importantly, our account includes bald-faced lying as bullshitting. Bald-faced liars deliberately assert falsehoods, but they *manifestly* have no intention of deceiving their audience.[16] Here is a classic example from the literature, discussed by Carson (2006), Stokke (2018a) and Harris (2020) among others: A student lies bald-facedly to the Dean regarding a plagiarism accusation, knowing that the Dean has a policy of punishing only those who confess. In this case, says Harris (emphasis ours),

> [t]he student . . . is not motivated by the aim of changing anyone's mind. Rather, the student's aim is to do whatever it takes to avoid punishment. The dean has a policy of not punishing anyone who doesn't confess. Knowing this, the student utters the magic words, *with little concern for any other effects that their action might have.* (Harris 2020: 15)

Harris doesn't link the example to bullshit, but it captures much of what makes bald-faced lying in general a practice that counts also as a clear and characteristic case of bullshit.

Speakers like the plagiarizing student count as bullshitters not because they are heedless as to the truth-value of their assertion, but because they make no effort to facilitate the uptake of a truth or a falsehood in their audience. Here too we will see that the existence of multiple audiences for an utterance means that the very same utterance can be both a deceptive lie and a bald-faced lie relative to different audiences – perhaps both a deception and an instance of bullshitting.

The relationship between bullshit and lies has been much discussed (see for example Carson 2010, Stokke 2018b), and Harry Frankfurt (2002) later[17] admitted that the categories are not mutually exclusive. Don Fallis and Andreas Stokke (2017: 279) also take this view, reframing Frankfurt's definition in terms of the

---

16 Philosophical discussion of bald-faced lies stems from Kenyon (2003), Carson (2006), and Sorensen (2007), although Kenyon calls these 'cynical assertions'. For an overview of the literature on bald-faced lies see Meibauer (2019).

17 A word about this 'later': Our citations for "On Bullshit" are to Frankfurt (2005), but the text was initially published in 1986. So (2002) is later, with respect to the writing of the original text.

speaker's indifference toward inquiry: "the cooperative project of incremental accumulation of true information with the aim of discovering how things are, or what the actual world is like." Some lies will count as bullshit on this view as well, but not ones in which the speaker is "concerned with her assertion being a true or a false answer to a QUD [question under discussion]" (2017: 279), which they take to be the majority of lies. Stokke and Fallis aim to accommodate cases of bullshitting offered as counterexamples to Frankfurt, in which a speaker who ought to count as a bullshitter could still be said to care about the truth, provided they deeply wished that their utterance was correct. These cases do count as bullshit on Stokke and Fallis's account, because caring about whether an inquiry gets closer to the truth means shaping your assertions to the evidence, not simply asserting and wishing it were so.

Our account too will accommodate such counterexamples by capturing them as cases of bullshit. This is a matter of some importance, too, because such examples are not limited to exotic thought experiments. There is a Trumpian case, in fact: a remarkable moment that took place when Trump tweeted claims about the projected course of a hurricane, which contradicted statements from meteorological authorities (Oprysko 2019). When the contradiction was pointed out, Trump responded by showing a map blazoned with the logo of the National Oceanic and Atmospheric Administration, which had a projected course for the storm consistent with Trump's tweet, but which had obviously been altered with a Sharpie to make it so. Of course one cannot facilitate true or false beliefs on the basis of one's assertions by faking evidence after the fact; it's supposed to work by fitting one's assertions to the evidence in the first place. So on our account too these cases straightforwardly count as bullshit. If caring about *how an inquiry goes* means caring about how the other inquirers are served by one's utterances, our proposal and that of Stokke and Fallis might be of a piece. But we think there is something more basic and straightforward in the thought that bullshitting is linked to speakers' attitudes towards audiences, rather than their attitudes about the courses of inquiry. It also allows us to distinguish attitudes toward different audiences, which is crucial to our project of understanding the kinds of falsehoods that have recently assumed so much prominence in our politics.

We think our account can moreover make progress on Frankfurt's "exercise for the reader," following his important observation that bullshit can sometimes be benign:

> In fact, people do tend to be more tolerant of bullshit than of lies, perhaps because we are less inclined to take the former as a personal affront. We may seek to distance ourselves from bullshit, but we are more likely to turn away from it with an impatient or irritated shrug than with the sense of violation or outrage that lies often inspire. The problem of understanding why our attitude toward bullshit is generally more benign than

our attitude toward lying is an important one, which I shall leave as an exercise for the
reader.   (2005: 50)

Specifically, we note that failing to offer one's audience the grounds for a presumptively true belief is consistent with offering them something else in place
of it. Sometimes what's offered is welcomed by the audience as amusing or di-
verting. Poorly concealed or even unconcealed bullshit can be genial, agreeable
bullshit. People might enjoy it for its entertainment value or for the sheer nerve
of the bullshit artist.

This can also help to explain the sort of tolerance for falsehood that has
enabled the rise of figures like Trump and Johnson. Both of them were at one
point commonly viewed as entertainment figures, not to be taken too seriously.
Boris Johnson's fame was in part due to his time on *Have I Got News For You*, a
comedy quiz show, and Donald Trump was gossip column fodder before becoming a reality TV star. In this capacity, he discussed his loose attitude toward
the truth: "I play to people's fantasies. I call it truthful hyperbole. It's an innocent form of exaggeration – and a very effective form of promotion" (Trump
and Schwartz 1987: 57–58). The Huffington Post infamously featured Trump's
candidacy on its entertainment rather than politics pages. Treating Trump and
Johnson as entertainers meant not worrying too much about the truth value implications of what they said. This enabled their rise, and also habituated the
public both to a high level of falsehood in their speech and to their bald-faced
bullshitting, as simply part of their personal brands.

On the other hand, one might wonder whether the audience-relativity of
our definition opens up the prospect that an audience could turn an otherwise
good-faith asserter into a bullshitter simply by telegraphing a complete unwillingness to believe them. By undermining any expectation or hope of facilitating
the uptake of true beliefs, could an audience in turn undermine the speaker's
ability even to form a good-faith intention to do so? The speaker would speak
the truth as they perceive it, but would do it so hopelessly as to be utterly indifferent to the effects of its uptake.

Of course weird results can follow from weird thought-experiment situations;
but in practice no such complete failure of hope could be warranted. The full effects of uptake are never exhausted in the instant of utterance, after all. Assertions
are often greeted with transparent disbelief at the point of utterance, yet are subsequently accepted, at least to a degree, anywhere from hours to decades later. This
prospect in itself enables a speaker's expectations to support good-faith assertion,
irrespective of an audience's advertised imperviousness to acceptance.

Of these and other implications of our account, we focus on two in particular.
The first we've already noted: with a single instance of speech, a speaker may be

bullshitting with respect to one audience but not with respect to another.[18] The second is that, contrary to what Frankfurt says, whether speech counts as bullshit does not depend on whether the underlying attitudes are concealed from the audience. A bullshitter may do anything from concealing their effort to advertising it in pursuit of various outcomes, and yet be a bullshitter all the same. As we will see, these features and their interactions are important to making sense of a range of cases, and especially important to making sense of Trump's and Johnson's use of bullshitting as a political tactic.

3 Different audiences and different bullshit

3.1 Audience relativity

Politicians typically address very large audiences. Thus, it is not surprising if some of these people are quite credulous. Mathiesen and Fallis (2017) have drawn our attention to a wonderful quote from Jonathan Swift: "as the vilest writer has his readers, so the greatest liar has his believers" (Swift 2004 [1710], p. 195). As they note in their important discussion, "even if a politician says something that is extremely implausible, or that can easily be shown to be false, some people will believe" (2017: 11).

Yet implausibility, extreme or otherwise, is itself relative to an audience. Our definition of 'bullshit' makes crucial reference to the audience, which allows us to better understand some complexities of real-life cases. In particular, it allows us to understand cases in which an utterance is bullshit aimed at one audience and a deception aimed at another audience, or different kinds of lies or bullshit for different audiences.[19]

---

18 To be clear, this is not a version of the view Hermann Cappelen (2005) calls *speech act pluralism*, which is focused on *locutionary pluralism*. Cappelen's view is pluralist about the different propositional contents asserted via an utterance, where this is distinguished sharply from the semantic content. Our view is instead pluralist about the types of sociolinguistic acts (misleading, deceiving, bullshitting) that a single assertion can be used to perform, depending on the audiences exposed to it. Because these are not all illocutionary acts – some depend, for example, on perlocutionary matters – our view is also not quite an illocutionary pluralist one like those of Sbisà (2013) and Lewiński (2021).

19 This has a certain structural similarity to dogwhistling (Lopez 2015, 2019; Saul 2018) and Connolly's theory of trolling (2021). It is also structurally related to the sort of view that Marina Sbisà (2013) and Marcin Lewiński (2021) call "illocutionary pluralism", which Lewinski argues is important for making sense of dogwhistles. Importantly, however, our view does not involve different *contents* being communicated to different groups (the primary focus for Cappelen

The Canadian Trade Deficit case seems at first like simple undifferentiated bullshitting: Trump was neither attempting to tell Trudeau true things nor to convince Trudeau of false things. Although Cassam (2021: 52) writes that "Trump's primary objective was presumably to induce in Trudeau the belief that America had a trade deficit", we don't think this is plausible. Any international leader meeting with Trump would obviously be advised in advance as to whether their country had a trade deficit with the U.S., and could not be expected to need or take Trump's word for it. Surely even Trump would be aware of this, and not hope to make Trudeau believe in a (non-existent) trade deficit.

Still, there are important distinctions to be drawn between different audiences in a case like the Canadian Trade Deficit. Trump had no hope of convincing Justin Trudeau that the United States had a trade deficit with Canada. However, he might well have intended to convince some of his supporters of this. Many Trump supporters take him to be extremely knowledgeable, and would simply assume that he was right in what he said to Trudeau. The confrontation would also impress them as one in which Trump spoke the truth to a problematic ally. For this audience, Trump's utterance was meant to be believed. If that's right, then what was a bald-faced lie for one audience (Trudeau, and Trump's critical domestic opposition) was a deceptive lie with some chance of success for another (Trump's supporters). It was bullshit relative to one audience, but not relative to another. We will see this theme repeated in subsequent examples.

Considering multiple audiences is also important to understanding what Boris Johnson is up to, at least in some of his utterances. Although a significant portion of the UK public take Johnson to be a bullshitter or a bald-faced liar, others find him credible, at least on some subjects. Despite the fact that Johnson's vacillation over Brexit was well known, Brexit supporters took what he said very seriously, and believed it. Indeed, despite two years of high-profile debunking, 42% of the British public still believed the claims of the infamous Brexit Bus to be true in 2018.[20] As we write this in 2021, a clear majority of the British public now takes Johnson to be untrustworthy, but 35% still trust him.[21] Utterances directed at this group may not be simply bald-faced lies. However, it is important to note also that some of Johnson's utterances clearly are not meant to be believed. His claims of a

---

2005), but rather different expectations and intentions about how the utterance will be received by different groups.

20 https://www.theneweuropean.co.uk/brexit-news/kings-college-research-finds-leave-voters-still-believe-lie-about-34688.

21 https://www.standard.co.uk/news/politics/boris-johnson-untrustworthy-poll-keir-starmer-ipsos-mori-poll-b931903.html?fbclid=IwAR2PFZAvmLrcesbAkigu7ON4oVrrRTg5FlJKrhQfihCp8aaXtXmcNyUA3mo.

bus-making hobby are a case in point: these are much more likely to fall into the category of quirky or entertaining bullshit that we discussed earlier.

3.2 Types of lies and types of bullshit

Some remarkable cases of bald-faced lies emerged from the Trump administration and their supporters. Importantly, these were often not *just* bald-faced lies. Instead, many of these utterances are to be classified differently with respect to different audiences. Such lies and bullshit are useful tools for authoritarian leaders, functioning differently with respect to different audiences. Our analysis helps us to bring out this point.

Trump's claims about the 2020 election are frequently described as bald-faced lies. And they clearly are, with respect to much of the country, who he has no hope of convincing. However, they are deceptive lies with respect to those among his base, who fervently believe him.[22] But with respect to other Republican politicians they are authoritarian lies – bald-faced lies designed to extract humiliating displays of loyalty from subordinates, who know that they are false.

These Republican officials are Trump's enablers. They repeat his lies, even if they know them to be false. These lies are what we call "compliance lies" – lies told to show loyalty to an authoritarian leader. Authoritarian rulers have a well-documented pattern of demanding lies from their subordinates, as a kind of loyalty test or commitment to shared consequences.[23] Tyler Cowen writes:[24]

> By requiring subordinates to speak untruths, a leader can undercut their independent standing, including their standing with the public, with the media and with other members of the administration. That makes those individuals grow more dependent on the leader and less likely to mount independent rebellions against the structure of command. Promoting such chains of lies is a classic tactic when a leader distrusts his subordinates and expects to continue to distrust them in the future.

---

22 https://www.theguardian.com/us-news/2021/may/24/republicans-2020-election-poll-trump-biden.

23 Grimaltos and Rosell (2017) draw a distinction within the category of deceptive lies, between doxogenic lies and falsifying lies. Doxogenic lies are told in order to get the audience to believe some particular claim, an intention which would still be satisfied if the claim turned out to be true. Falsifying lies are told in order to get the audience to believe a falsehood, where the falsity of the belief is the primary goal. We think it may be worth drawing a similar distinction within bald-faced lies, between cases where the proposition asserted is of primary purpose and cases where the obvious falseness is of primary importance. The bald-faced lies of an authoritarian leader will often be of this latter sort: for loyalty tests, the obvious falseness is vital.

24 https://www.bloomberg.com/opinion/articles/2017-01-23/why-trump-s-staff-is-lying.

We take the 2020 election lies of many Republican officials to be compliance lies. But these are also are deceptive lies for some audiences, and bullshit or even bald-faced bullshit for others.

Russell Muirhead and Nancy Rosenblum discuss this point in their book on recent conspiracist thinking, *A Lot of People Are Saying* (2019: 68–69):

> it is not clear that Trump cares whether his falsehoods are believed; he seems to care only that they are affirmed. He wants the power to make others assent to his version of reality. When Sean Spicer, Trump's first press secretary, said that Trump's inauguration had "the largest audience ever to witness an inauguration," he may not have intended for people to believe him, only to describe the world Trump wished for, and us to enact Trump's own power.[25]

Another compliance lie was Sean Spicer's insistence (described above) that Trump had had a larger inauguration crowd than Barack Obama did, despite very clear photographic evidence to the contrary. Spicer's initial utterance of this lie may have been quite generally (and optimistically) deceptive in intent. But his assertions standing by the lie when confronted about the photographic record are probably more complicated. These would be a continued attempt at deception, as directed at the portion of the public inclined to believe any spokesperson for the Trump White House. Many of Trump's supporters did believe that his inauguration crowd was bigger than Obama's, or at least were willing to say that they did (Schaffner and Luks 2017).[26] At the same time, the continued insistence on the lie looks like open, undisguised bullshitting and bald-faced lying as directed at the questioning reporters and that portion of the audience who understood or trusted the photographic record. This sort of lie, clearly uttered to show loyalty to Trump, is a compliance lie. Spicer's utterance, then, is simultaneously a deceptive lie and a bald-faced lie (which is a species of bald-faced bullshit). It is a particular kind of bald-faced lie: a compliance lie to show fealty to a lying leader. In another example of simultaneously deceptive and bald-faced lying, Trump lawyer Sidney Powell made manifestly false claims about Dominion Voting Systems voting machines as a part of her

---

25 For Muirhead and Rosenblum, this makes the utterance not a lie. But that is because they require an intention to deceive, thus ruling out bald-faced lies.

26 We recognize that there are open questions as to the psychological or doxastic attitudes underlying the politicized answers people give to surveys and polls, their voting behaviour, and other forms of assent or acceptance. Some evidence indicates that partisans endorse politicized propositions out of solidarity, or as a declaration of identity, and not as an unqualified reflection of the credence they assign to those propositions (Bullock, Gerber, Hill & Huber 2015, Berinsky 2018, Peterson and Iyengar 2021). (On the other hand, the high rates of vaccine and masking refusal, which have extremely high-stakes consequences, suggest that many of these beliefs may be genuine.) This sort of partisan performance of belief, to the extent it occurs, is also an outcome at which a political bullshitter may aim.

argument that Trump had actually won the 2020 election. In responding to a subsequent defamation lawsuit, she defended herself by insisting that no reasonable person would have believed her claims, due to their obvious falsity.

> Indeed, Plaintiffs themselves characterize the statements at issue as "wild accusations" and "outlandish claims . . . They are repeatedly labelled "inherently improbable" and even "impossible." . . . Such characterizations of the allegedly defamatory statements further support Defendants' position that reasonable people would not accept such statements as fact.   (Porterfield 2021)

Powell's defence, in essence, was to declare that she was bullshitting, telling bald-faced lies. Yet this declaration does not undo the bullshitting project, which is continued elsewhere in her response with the claim that Powell believes those same allegations.

> The Complaint . . . alleges no facts which, if proven by clear and convincing evidence, would show that Sidney Powell knew her statements were false (assuming that they were indeed false, which Defendants dispute). Nor have Plaintiffs alleged any facts showing that Powell "in fact entertained serious doubts as to the truth of h[er] publication." In fact, she believed the allegations then and she believes them now.

The claims made by Powell and others about fraudulent vote counts in the 2020 election are believed or assented to by a disturbingly high number of Americans; no doubt she made those assertions intending to secure acceptance for them where possible. For anyone bothering to read the legal response, as the excerpts indicate, the key claims are depicted fairly openly as bullshit. But then, those who make the effort to read the legal filing are only one of the audiences exposed to Powell's claims about the election, and very far from the largest. Powell too is both a deceptive liar and a bald-faced liar. In addition, her fervent persistence in telling these lies serves as a compliance lie: in her case, this seems to have been her path into Trump's inner circle, rather than something demanded of her once she was there.

The bald-faced liar does not care to hide that they are lying, and might even parade the fact. What these examples indicate is that the same is true of *bald-faced bullshitting*. A bald-faced bullshitter is one who not only does not intend to induce in an audience a belief in what they have said; they moreover make no effort to hide that they are bullshitting, and in fact might advertise or revel in it.

This is something that Frankfurt would not himself have allowed. He writes (2005: 54),

> The bullshitter may not deceive us, or even intend to do so, either about the facts or about what he takes the facts to be. What he does necessarily attempt to deceive us about is his enterprise. His only indispensably distinctive characteristic is that in a certain way he misrepresents what he is up to.

On this point, we simply disagree with Frankfurt. We suspect that he did not consider cases of bald-faced bullshitting, much as writers on lying for quite a long time did not consider cases of bald-faced lying. And it is vital to recognise bald-faced bullshitting in order to understand how some political speech pushes the boundaries of bullshit and adapts it for political ends. The existence of multiple audiences to so much political speech already means that the respects in which bullshit is bald-faced can be quite hard to perceive, and lost to analysis.

In short, some Republican politicians and followers of Trump do believe his falsehoods. For these followers, Trump's lies are deceptive lies. Others recognise them as lies. For these followers, they are bald-faced authoritarian lies and bullshit. Many of these people repeat them or act as though they believe them to demonstrate loyalty. Their lies are compliance lies, told to show loyalty to Trump. These lies, told by Trumpist politicians, have two audiences. They are deceptive lies for credulous Trump followers who hear them. But they are bald-faced lies for their most important audience: Trump. Trump wants them to assert these bald-faced lies to show their loyalty. Neglecting the way that these utterances work for multiple audiences would lead us to miss out on much of what they do.

Examining the role of authoritarian and compliance lies for Trump helps us to draw an important contrast between him and Johnson. Johnson does not seem particularly concerned to extract this kind of demonstration of loyalty from his followers. Although some joked that armies of interns would be up all night manufacturing buses from wine boxes to support Johnson's claims, nothing of the sort happened. As various high officials have broken with Johnson, he seems relatively uninterested in and impervious to their disagreements. Johnson's unique selling point has remained his use of entertaining bald-faced lies and bullshit.[27]

---

27 It has been suggested that some of Johnson's more overtly incredible utterances may have been more carefully planned than they seemed, perhaps intended to interfere with otherwise politically unwelcome Internet search engine results. So, for example, his bus-building hobby remarks could have diluted the results of searches for information about the infamous "Brexit bus," a campaign bus for the Vote Leave organization, supported by Johnson, that displayed deceptive or misleading claims. Similarly, Johnson's strangely describing himself as a "model of restraint" could have buried search engines results about his alleged affair with a model. We take no position on the plausibility of this theory. Yet the possibility alone suggests an interesting new trope of bald-faced bullshit: aiming to be recognized as bullshitting, in some comment-worthy way, in order to game search engine results. We thank Neri Marsili for bringing this to our attention. (For more on this theory, see https://www.wired.co.uk/article/boris-johnson-model-google-news?fbclid=IwAR1wtd sipGgBcTG2yHAHCDRi-o7_ye0bZu4Jku5AIaEjnHe6mlUlSvKoFAQ, consulted 28 August 2021.)

Although his regime is deeply authoritarian in other ways,[28] he has not followed Donald Trump down this path.

3.3 Power bullshit

Despite the fact that it can sometimes seem playful or benign, bald-faced bullshit can play a special role for an authoritarian leader. As Jason Stanley writes, discussing Trump, "authoritarian propagandists are attempting to convey power by defining reality" (Stanley 2016). Stanley argues that while this may be accurately described as bullshit – since Trump is unconcerned with what is actually true or false – that is a misleading description. Stanley takes this to be misleading because it places our focus on truth and falsehood rather than on power. But we have argued that bullshit is not directly about truth and falsehood; it is about how a speaker treats an audience. And when there are multiple audiences, as there so often are with political speech, it can be about how one audience is treated in front of another audience. Stanley's view misses the respects in which Trump's authoritarian speech asserts power precisely because it is bullshit, and bald-faced bullshit in particular.

Consider again Trump's altered hurricane map: here we have a lie so obvious, a drawing so amateurish, that Trump could not possibly have intended it to be believed. In telling this bald-faced lie, he certainly made no effort to provide the basis for a true belief or a false belief for his audience. But this was not an indifference to the audience's *stake* in believing the truth; it was a display of what he could do in the face of that stake. It's the sort of move made by someone who really has nothing to fear from what their audience might come to know or believe, or who at least wants to convey such impunity. It displays the attitude: not only do I not need to tell you the truth; I don't even need to bother deceiving you. We call this a *power move*.

Authoritarian political bullshitting is very often a power move: bullshitting in a way that is obvious, with no attempt at concealment. These bullshitting power moves are particularly flagrant floutings of communicative norms, openly displaying disrespect or contempt for the immediate audience. They might display dominance directly to the audience being disrespected, or for the sake of a secondary or onlooker audience. With power moves, a speaker does not just happen to reveal

---

28 https://www.independent.co.uk/news/uk/politics/boris-johnson-stop-search-police-b1891317.html; https://www.theguardian.com/world/2021/apr/28/policing-bill-will-have-chilling-effect-on-right-to-protest-mps-told.

contempt for an audience; they can *deploy* contempt as a political tool. This may be intended to cow opponents or to excite and energize supporters. They are also often used, as we have seen, to extract humiliating compliance lies from top subordinates. All of these can flow from the ostentatious use of bullshit.

Stanley worried that describing Trump as a bullshitter meant comparing him to a salesperson, saying anything to make a sale. But a salesperson would not dare to declare themselves a bullshitter as a power move, clearly communicating to their audience that they had no concern with truth, nor with the consequences of the audience's perception of their bad faith. Trump's openness about his bullshitting is in this sense similar to his declaration that he could shoot somebody on Fifth Avenue and get away with it.[29] In our view, calling attention to Trump's use of power bullshit is a crucial part of understanding the dangers of Trump's authoritarian speech, rather than a means of trivialising it.[30]

Freed from the assumption that the bullshitter aims at concealment, our account accommodates cases in which the bullshitter does not merely fail to hide their enterprise, but to some degree, to some audiences, actually advertises it. Some bald-faced bullshit has just this purpose: it advertises contempt for the audience, by advertising contempt for the audience's interest in receiving truths.

4 Power moves and politics

4.1 How do power moves succeed politically?

It is hard to avoid being struck, quite frankly over and over, by a feeling of amazement that politicians who make a habit of bullshitting, indeed bald-faced bullshitting, could have any success at all, let alone success on the scale of a Donald Trump or a Boris Johnson. We can understand this better by looking more closely at the dynamics of bald-faced political bullshit as a power move used with multiple audiences.

As we have noted, bald-faced bullshit involves making assertions that openly demonstrate an indifference to the quality of the beliefs an audience could base on the assertions. Even though this may be tolerated or laughed off as part of a shtick,

---

29 https://www.npr.org/sections/thetwo-way/2016/01/23/464129029/donald-trump-i-could-shoot-somebody-and-i-wouldnt-lose-any-voters.

30 That said, we also agree with Quassim Cassam (2021) that it is not enough to simply categorise Trump's speech as bullshit. Where it is hate speech, it must be understood as hate speech, and so on. All of these categories are important to understanding the harms done.

it is hard to see how it could inspire trust. And yet Donald Trump has retained an enormously high level of trust from his base: Even in April 2021, 66% of Republicans said they trusted Trump *for medical advice* about the COVID19 pandemic, literally a life-or-death matter.[31] So how does he maintain the trust of his followers?

A large part of the answer, we believe, is that trust isn't just about believing that someone says true things. Trust can also be about feeling that someone is on your side, which can often mean feeling that they share your enemies. This is where use of the power move in political speech comes in. In those cases, authoritarian bald-faced bullshit that openly and disrespectfully treats one audience as not even worth deceiving may be an effective way of consolidating trust with another audience that witnesses it – a more effective means than a less overt sort of bullshit would be.

The fact that there are multiple audiences for political speech is, we suggest, fairly clear to the audiences themselves. An authoritarian speaker's bullshitting power moves in this context can be an important means of communicating contempt for their opponents, consolidating support among those who perceive that the speaker's cultural and political enemies are also their cultural and political enemies. This fits remarkably well with Timothy Snyder's discussion of the use Vladimir Putin makes of obvious lies (both authoritarian and compliance lies), which confer what Snyder calls "implausible deniability" (2018: 163). He quotes Charles Clover:

> Putin has correctly surmised that lies unite rather than divide Russia's political class. The greater and more obvious the lie, the more his subjects demonstrate their loyalty by accepting it, and the more they participate in the great sacral mystery of Kremlin power.

When a speaker lies directly to one audience, but is overheard and recognized as lying by a second audience, the second audience has not necessarily been – and need not feel – lied to. They might even feel solidarity and recognize a shared purpose instead. By the same token, an awareness of multiple political audiences enables one audience to recognize bald-faced bullshit without feeling that it is directed at them. For Trump's supporters, his bald-faced bullshitting can be seen as directed at others among the total audience – specifically, those who have it coming. This is done on behalf of his supporters; they are in on the joke.

As Clover's observations suggest, they may even perceive themselves as needed participants: it is at rallies where they are the primary audiences that

31 https://www.independent.co.uk/news/world/americas/us-politics/trump-covid-medical-advice-republicans-b1829784.html. As we note in footnote 27, there are worries about whether utterances like these always express the actual beliefs of those who make them. Nonetheless, it seems that there is much more trust placed in Trump than one would expect of a bald-faced bullshitter.

much of the communication aimed at secondary audiences via media coverage will take place. Some elements of Trump's rally audiences are genuinely deceived by his lies. But other elements of the audience may well recognize that his utterances will be received as bald-faced lies by onlookers – the mainstream media, intellectuals, liberals, leftists, other enemies. By bullshitting *at* his base in such rallies, and having that bullshit received so warmly, he is *fucking with* those secondary audiences. Anyone perceiving this among his supporters can feel that they are participating in the theatre that empowers him to fuck with people who deserve it, who have it coming. Being a bullshitter in this sense – being seen as a bullshitter, being seen as *intending* to be seen as bullshitter – is consistent with being perceived as trustworthy in a broader sense.

In short: being perceived as a bald-faced bullshitter is not a liability in those circumstances, but an advantage. Many of his supporters, we suggest, admire Trump for his ability to upset or "own the libs." As a result, when the targets of his contempt or disinterested analysts complain about his speech, and spotlight his power move bullshit, his erstwhile supporters are not moved to abandon him because he's a bullshitter. They already knew he was a bullshitter – toward those who they think deserve it. That he uses bullshit to get under his opponents' skin is not a mark against him, but a sign of his superiority over the shared enemy. *They just can't handle him; they freak out; he makes them melt down.* Bald-faced bullshit of this kind does not work in spite of its overtness, then, but because of it. It may even benefit from the indignation it generates, which to a partisan onlooker audience functions as evidence of its value and efficacy.

4.2 How power moves might fail politically

More recent speeches from Trump following his electoral defeat and removal from the presidency suggest both the accuracy of this characterization, and an important vulnerability of the underlying strategy. As political commentator Amanda Marcotte observes,[32]

> In his speech at the second "annual" Conservative Political Action Conference of the year, Trump bragged about how he lies about polls and elections when he doesn't win them. "You know, they do that straw poll, right?" Trump asked, referring to the 2024 GOP nominee straw poll CPAC conducted of attendees. "If it's bad, I say it's fake. If it's good, I say, that's the most accurate poll perhaps ever."

---

32 https://www.salon.com/2021/07/12/trump-supporters-think-theyre-players–but-theyre-still-just-pawns/.

> Trump got a huge laugh from the CPAC crowd . . . As the clip went viral, there was no outraged reaction from GOP voters, no anger that he lies to them in order to enlist their support for an authoritarian coup. Trump supporters aren't mad about Trump admitting he lied for one simple reason: They don't think they're the ones being lied to.

These are not just lies, moreover. With no attempt to deceive, the lies are bald-faced, hence bullshit. And the core audience, or a significant element of it, also isn't angry about the bullshitting, because here too they don't think they're the ones being bullshitted.

On the other hand, if reduced media coverage and analysis leaves this primary audience feeling like the secondary audience of enemies is no longer bearing witness, our account suggests that this kind of authoritarian speech should become much less appealing to them. The content of the speech and its bald-facedness (both as lies and as bullshit) can largely stay the same; but without the theatre of dominance or contempt that it is intended to enact for onlookers, this will just seem tedious or disrespectful to the primary audience.

As journalist Benjamin Fearnow reported of a post-loss Trump speech to his previously most dedicated extremist base, which met with little real-time media attention,[33]

> a majority of the top QAnon user comments simply expressed their outright boredom with Trump's post-election stump speech . . . "Judging by the Trump-supporting normies I live with, they were bored with his speech," wrote another QAnon user. "I support Trump but this is getting ridiculous."
>
> "Love President Trump. But, if I'm being honest, it's a lot of the 'same old-same old,' we've all heard a thousand times before," wrote Annmarie Calabro.

Why not suppose that this boredom and frustration really does arise solely from the repetition of a tired message? In short, because Trump gave highly repetitive speeches for years, to audiences whose enthusiasm rarely waned as long as major media coverage and outraged reactions were the norm. When the secondary audience was visible just offstage, the bullshit excited the audience and kept them engaged.

---

33 https://www.newsweek.com/qanon-supporters-express-boredom-same-old-trump-speech-this-getting-ridiculous-1604489.

5 Conclusion

In this chapter, we have developed an understanding of bullshit which allows us to acknowledge and explore differences between audience subgroups for the same utterance. We have shown the utility of this by demonstrating the way that it allows us to make sense of an otherwise baffling fact: the continuing trust placed in bald-faced bullshitters like Donald Trump and Boris Johnson, despite the apparent obviousness of what they are doing. (This trust is especially startling due to its life-or-death consequences in the COVID19 pandemic.)

While little in the political speech of Trump, Johnson, and others cut from their cloth is entirely novel, we suggest that an understanding of authoritarian political bullshit helps to diagnose elements of its recent rise. These include the increasing overlap between mass entertainment media and mass political media, enabling the incubation of bullshitters who hone their craft in seemingly benign form before their move into the political sphere; and the unprecedented magnitude and ubiquity of secondary audiences, through news media and social media reportage and analysis, for bald-faced bullshit to be leveraged in the ways we have suggested. Although some have suggested that classifying authoritarian political speech as bullshit under-rates its significance or danger, we have argued that bald-faced bullshit is characterized by a contempt for the audience and a naked assertion of power. Understood in this way, classifying a leader's utterances as bald-faced bullshit illuminates some alarming features of their communicative style, and its dependence on and exploitation of divisions and enmities within a polity.

References

Berinsky, Adam. 2018. Telling the Truth about Believing the Lies? Evidence for the Limited Prevalence of Expressive Survey Responding. *Journal of Politics* 81(1): 2011–24.

Broad, William J., and Dan Levin. 2020. Trump Muses About Light as Remedy, but Also Disinfectant, Which Is Dangerous, New York Times. April 24, 2020. https://www.nytimes.com/2020/04/24/health/sunlight-coronavirus-trump.html. Accessed November 15, 2020.

Bullock, John, Alan Gerber, Seth Hill, and Gregory Huber. 2015. Partisan Bias in Factual Beliefs about Politics. *Quarterly Journal of Political Science* 10(4): 519–78.

Cappelen, Herman. 2005. Pluralistic skepticism: Advertisement for speech act pluralism. *Philosophical Perspectives* 19.1: 15–39.

Carson, Thomas. 2006. The Definition of Lying. *Nous* 40: 284–306.

Carson, Thomas L. 2010. *Lying and Deception: Theory and Practice*. Oxford: OUP.

Cassam, Quassim. 2021. Bullshit, Post-truth, and Propaganda. In E. Edenberg and M. Hannon (eds.), *Political Epistemology*, 49–63. New York: Oxford University Press.

Cohen, G. A. 2006. "Deeper Into bullshit", in Reisch and Harcastle, *Bullshit and Philosophy* (Open Court Press), 117–136.

Connolly, P.J. 2021. Trolling as Speech Act (or, the art of trolling, with a description of all the utensils, instruments, tackling, and materials requisite thereto: With rules and directions how to use them). *Journal of Social Philosophy* (early access online).

Douglas, Heather. 2006, "Bullshit at the Interface of Science and Policy: Global Warming, Toxic Substances, and Other Pesky Problems", in Reisch and Hardcastle, *Bullshit and Philosophy* (Open Court Press), 215–228.

Frankfurt, Harry. 2002. Reply to G. A. Cohen. In Sarah Buss and Lee Overton (Eds.), *Contours of Agency: Essays on Themes from Harry Frankfurt*, 340–344. Cambridge, MA: MIT Press.

Frankfurt, Harry. 2005. *On Bullshit*. Princeton: Princeton University Press.

Grimaltos, Tobies and Sergi Rosell. More and More Lies: A New Distinction and its Consequences. *Teorema* 2017: 5–21.

Hardcastle, Gary and George Reisch. 2006. On Bullshitmania. In G. Hardcastle and G. Reisch (eds.), *Bullshit and Philosophy*, vii–xxiii. Chicago: Open Court.

Hardcastle, Gary. 2006. The Unity of Bullshit. In G. Hardcastle and G. Reisch (eds.), *Bullshit and Philosophy*, 137–150. Chicago: Open Court.

Harris, Daniel W. 2020. Intentionalism and bald-faced lies. *Inquiry*, Published online. DOI: 10.1080/0020174X.2020.1775381

Jacquemet, Marco. 2020. 45 As Bullshit Artist: Straining for Charisma. In Janet McIntosh and Norma Mendoza-Denton (eds.), *Language in the Trump Era: Scandals and Emergencies*, 124–136. Cambridge: Cambridge University Press.

Kenyon, Tim. 2003. Cynical Assertion: Convention, Pragmatics, and Saying 'Uncle'. *American Philosophical Quarterly* 40(3): 241–248.

Kunda, Ziva. 1990. The Case for Motivated Reasoning. *Psychological Bulletin* 108(3):480–98.

Lewiński, Marcin. 2021. Illocutionary Pluralism, *Synthese*. Published online, DOI https://link.springer.com/article/10.1007%2Fs11229-021-03087-7.

Lopez, Ian Haney. 2015. *Dog Whistle Politics*. Oxford: Oxford University Press.

Lopez, Ian Haney. 2019. *Merge Left*. New York: The New Press.

Lynch, Michael. Forthcoming. Power, Bald-Faced Lies, and Contempt for Truth. *Revue Internationale de Philosophie*.

Mathiesen, Kay and Don Fallis. 2017. The Greatest Liar Has His Believers: The Social Epistemology of Political Lying, in Emily Crookston, David Killoren, and Jonathan Trerise (eds.), *Ethics in Politics*, 35–53. Abingdon: Routledge. https://ssrn.com/abstract= 2937409.

Meibauer, Jorg. 2019. Bald-Faced Lies, in Jörg Meibauer (ed.), *Oxford Handbook of Lying*, 252–263. Oxford: Oxford University Press.

Muirhead, Russell and Nancy L. Rosenblum. 2019. *A Lot of People Are Saying: The New Conspiracism and the Assault on Democracy*. Princeton: Princeton University Press.

Oprysko, Caitlin. 2019. An Oval Office mystery: Who doctored the hurricane map? Politico. September 4, 2019. https://www.politico.com/story/2019/09/04/donald-trump-sharpie-hurricane-map-1481733. Accessed January 21, 2021.

Pennycook, Gordon, James Allan Cheyne, Nathaniel Barr, Derek J. Koehler, and Jonathan A. Fugelsang. 2015. On the reception and detection of pseudo-profound bullshit. *Judgment and Decision Making* 10: 549–563.

Peterson, Erik and Shanto Iyengar. 2021. Partisan Gaps in Political Information and Information-Seeking Behavior: Motivated Reasoning or Cheerleading? *American Journal of Political Science* 65(1): 133–147.

Porterfield, Carlie. 2021. Sidney Powell Argues Her Dominion Defamation Lawsuit Be Tossed Because 'No Reasonable Person' Would Believe Her. *Forbes*, March 22, 2021. https://www.forbes.com/sites/carlieporterfield/2021/03/22/sidney-powell-argues-her-domin ion-defamation-lawsuit-should-be-dropped-because-no-reasonable-person-would-be lieve-her/. Article supporting materials at: https://s3.documentcloud.org/documents/20519858/3-22-21-sidney-powell-defending-the-republic-motion-to-dismiss-dominion. pdf. Accessed March 22, 2021.

Saul, Jennifer. 2012. *Lying, Misleading, and What is Said*. Oxford: Oxford University Press.

Saul, Jennifer. 2018. Dogwhistles, Political Manipulation, and Philosophy of Language. In Daniel Fogal, Matt Cross and Daniel Harris (eds.), *New Work on Speech Acts*, 360–383. Oxford: Oxford University Press.

Sbisà, Marina. 2013. Some remarks about speech act pluralism. In Alessandro Capone, Franco Lo Piparo, and Marco Carapezza (eds.), *Perspectives on Pragmatics and Philosophy*, 227–244. Berlin: Springer.

Schaffner, Brian and Samantha Luks. 2017. This is what Trump voters said when asked to compare his inauguration crowd with Obama's. *Washington Post* 01/ 25/2017. https://www.washingtonpost.com/news/monkey-cage/wp/2017/01/25/we-asked-people-which-inauguration-crowd-was-bigger-heres-what-they-said/ Accessed June 12, 2021.

Snyder, Timothy. 2018. *The Road to Unfreedom*. New York: Tim Duggan Books.

Sorensen, Roy. 2007. Bald-Faced Lies! Lying Without the Intent to Deceive. *Pacific Philosophical Quarterly* 88 (2): 251–264.

Stanley, Jason. 2016. Beyond Lying: Trump's Authoritarian Reality, *The Stone, New York Times*, November 4, 2016. https://nyti.ms/2ea62OH, consulted December 29, 2020.

Stokke, Andreas. and Fallis, Don. 2017. Bullshitting, Lying, and Indifference Toward Truth, *Ergo* 4: 10. http://dx.doi.org/10.3998/ergo.12405314.0004.010.

Stokke, Andreas. 2018a. *Lying and Insincerity*. Oxford: Oxford University Press.

Stokke, Andreas. 2018b. Bullshit. In Jörg Meibauer (ed.), *Oxford Handbook of Lying*, 264–276. Oxford: Oxford University Press.

Swift, Jonathan. 2004. *A Modest Proposal and Other Prose*. NY: Barnes & Noble Publishing.

Trump, Donald and Tony Schwartz. 1987. *The Art of the Deal*. New York: Random House.

Laurence Horn

Practice to deceive: A natural history of the legal bluff

Abstract: After centuries of exploration, the territory between falsehood and truth remains imperfectly mapped. Among cartographers of the forensically complex and ethically dubious domain of intentional deception there is widespread recognition of the importance of the practice of bluffing but no unanimity concerning its nature. Standard treatments of the relationship between bluffing and lying by philosophers, legal scholars, and specialists in business ethics, while addressing a range of subtle yet significant distinctions in the criteria for lying, provide no definitive analysis. I address this gap here, touching along the way on the relation of bluffing to deception, false implicature, bullshit, puffery, fraud, and the literal truth defense. What distinguishes bluffing from other varieties of misrepresentation and misdirection is the perpetrator's inherent vulnerability to being called by the target of the bluff; when called, a bluff collapses. But the potential bluffee – in the courtroom or hearing room, at the negotiating table or the card table, on the field of battle or the campaign trail, in the wild – is vulnerable as well, since apparent bluffs may conceal actual knowledge or power. Consequences for victims of a successful bluff, or for those lured into injudiciously calling a non-bluff, can range from an unsatisfactory settlement or wrongful conviction (or acquittal) to bankruptcy, military defeat, or death.

Acknowledgments: Earlier versions of parts of this chapter were presented to the International Law and Language Association (September 2019) at UCLA Law School, and virtually to the Koerner Center for Retired Faculty at Yale (December 2020), the LSA Linguistics in Law workshop (January 2021), and the University of Pennsylvania ILST seminar (February 2021). I am grateful to those who attended those presentations for their comments and suggestions, as well as to my poker mentors, especially Barbara Abbott, Polly Jacobson, and the late Ellen Prince. Thanks to Roslyn Burns for pointing me to John Ohala's study of the relation of intonation contours to dominance and submission displays and its application to Elizabeth Holmes's baritone voice stylings (and her Steve Jobs-adjacent black-on-black outfit). I am particularly indebted to Jennifer Saul and Tim Kenyon for their extremely valuable comments on an earlier draft.

Laurence Horn, Yale University, e-mail: laurence.horn@yale.edu

https://doi.org/10.1515/9783110733730-009

1 What do we talk about when we talk about bluffing?

According to the *Oxford English Dictionary*, to bluff (sense 3) is "to assume a bold, big, or boastful demeanour, in order to inspire an opponent with an **exaggerated notion of one's strength**, determination to fight, etc." (Here and below, **emphasis** is added.) This is a generalization of the OED's poker-specific sense 2, "To impose upon (an opponent) as to the value of one's hand of cards, by betting heavily upon it, speaking or gesticulating or otherwise acting in such a way as to **make believe that it is stronger than it is**, so as to induce him to 'throw up' his cards and lose his stake, rather than run the risk of betting against the bluffer." At first glance, this is plausible enough: the bluffer engages in conscious misrepresentation, feigning strength, knowledge, or invulnerability in order to intimidate an adversary at the card table, in the boardroom or courtroom, on the battlefield or political arena, in the used car lot, etc. But this cannot be the whole story. While the motivation for the canonical bluff may well be to exaggerate (physical or epistemological) strength, bluffing to **conceal** strength is a robust strategy as well, as any serious poker player knows.

This is recognized in the *American Heritage Dictionary*'s entry, which makes it clear that weakness as well as strength (however defined) can be bluffed, while also correctly focusing on the nature of the bluff as a false display and not specifically a linguistic signal:

American Heritage Dictionary online, s.v. BLUFF, v. intr.

(1) To engage in a false **display** of confidence or aggression in order to deceive or intimidate someone: *The management debated if there would really be a strike or if the union was bluffing.*

(2) To make a **display** of aggression, as by charging or baring the teeth, as a means of intimidating another animal.

(3) To try to mislead opponents in a card game by heavy betting on a poor hand **or by little or no betting on a good one**.

A poker player who engages in sandbagging or "slow play" misrepresents their hand as weaker than it actually is, luring more players and chips into the pot. Perhaps the most celebrated instance of bluffing weakness is the wooden horse fashioned by Odysseus to lull the Trojan enemy into drawing that appealing gift inside Troy's walled fortress, whereupon the concealed Greek warriors emerged to open the gates of the city and facilitate their army's decisive attack. Innocuous or inviting in appearance but concealing a deadly threat within, the original equine template inspired the Trojan horse of our own day: a computer program in which malicious or harmful code concealed by apparently harmless programming or data is unleashed to gain control and do its chosen form of damage.

(4) Bluffing **strength**: appearing to pose a threat but actually harmless
Bluffing **weakness**: appearing harmless but actually posing a threat

Misleading displays of both strength and weakness are not limited to the deceptive practices of our own species. While both the OED and AHD entries invoke the bluffer's agency and conscious intention to mislead, deceive, or intimidate, this does not apply when it is an insect or amphibian that runs the bluff.[1]

2 Bluffing in the wild: Deimatic display as dishonest signaling

> 'Deimatic' comes from the Greek δειματσω, 'to frighten', and is generally used to describe behavior in which, when under attack, prey suddenly unleash unexpected defences to frighten their predators and stop the attack . . . While some animals' deimatic displays honestly indicate their unprofitability as prey, many seem to be bluffing, being, in reality, a tasty prey item.
> (Umbers, Lehtonen & Mappes 2015: R58–R59)

In the latter case, biologists refer to DISHONEST SIGNALING; standard illustrations, courtesy of praying mantises, moths, and frogs, appear in Figure 1, from the relevant wikipedia entries, including https://en.wikipedia.org/wiki/Deimatic_behaviour.

For example, some species of frogs will "inflate themselves with air and raise their hindparts, and display brightly coloured markings and eyespots to intimidate predators" (https://en.wikipedia.org/wiki/Deimatic_behaviour), while another, *Eupemphix nattereri* shown in Figure 1d, startles attackers by lifting its rear to display two false eyes.[2] (Note also the possibility of intraspecific bluffing, as illustrated by the crustaceans described by Steger & Caldwell 1983.)

---

1 Or consider E. B. White's comment in a 1973 letter to a Mr. Nadeau: "Sailors have an expression about the weather: they say, the weather is a great bluffer" – no conscious intention there either. The target of a bluff, however, must be conscious if not human, whence the tweets and Reddit posts pointing out "You can't bluff a virus".

2 Some readers may recall a similar ploy practiced by a literary tribe of mammals (de Brunhoff 1934). Babar, king of the elephants, strategized a defense against an enraged rhinoceros horde primed to overrun them by disguising his biggest soldiers in wigs, painting their tails red to serve as noses and painting "large, frightening eyes" on their rumps. The elephants presented (from the rear) as fierce monsters to the rhinos, who retreated in shock and disarray.

Figure 1: Deimatic behavior. a: Adult female *Iris oratorio* performs a bluffing threat display, rearing back, forelegs & wings spread, mouth open (https://en.wikipedia.org/wiki/Iris_orato ria); b: Brightly colored wings of the jeweled flower mantis, *Creobroter gemmatus*, suddenly open to startle predators (https://en.wikipedia.org/wiki/Creobroter); c: *Spirama helicina* moth mimics the face of a snake in a deimatic or bluffing display (https://tinyurl.com/48cfu7v9); d: *Eupemphix nattereri* (https://en.wikipedia.org/wiki/Physalaemus_nattereri).

Following the "principle of antithesis" (Darwin 1886), Colapinto (2021: 80) summarizes recent scholarship (e.g. Morton 1977, Dawkins & Krebs 1978) on dishonest signaling:

> Aggressive or angry animals stand tall and stiff, raising their hair to suggest a bigger, more formidable body – visual signals that they match the voice, through low-pitched, growling sounds that also "suggest" a bigger, more formidable body (a size bluff); submissive or loving animals win romantic partners by loosening their stance, even pushing the body against the ground like an affectionate dog, flattening the hair – and producing the high, or whining, pure vocal notes that suggest a small, submissive body (a reverse size bluff).

False signaling or bluffing evolved as a strategy for adaptation, as variously displayed in lizard postures, cricket calls, moth pheromones, butterfly wings, and human vocalization. At the same time, recipients of the deceptive signals have continuously evolved techniques for detecting false displays, leading to a "biological arms race" (Dawkins & Krebs 1979), a constant re-equilibrium of the transmission and reception of deceptive signals and their detection that Colapinto (2021: 87) describes as "co-evolutionary one-upmanship".

Sound spectrographs compiled and analyzed by Morton (1977) demonstrate the ubiquity across bird and mammal species of the tendency for low frequency growls to mark aggression and for high pitched vocalizations to signal interest in affinity and mating. As Ohala (1997: 99) remarks, "both parties put on displays, both visual and acoustic, which help to convey an impression, often a false one, of their size". The acoustic correlate in humans is the high F0 in polar questions and rising pitch in declaratives signaling deference or submission to authority, while the low F0 or falling intonation in statements and replies affirms dominance, confidence, aggression, or threat. This FREQUENCY CODE – "the association of high acoustic frequency with smallness and low acoustic frequency with largeness" (Ohala 1997: 102) – extends to the non-physiologically determined aspects of sexual dimorphism, to the tendency to rate lower-pitched men as more dominant and men using raised pitch as more deferential, and to the adoption (conscious or not) of a lowered pitch in women to express (or bluff) confidence and reliability.[3]

Bluffing weakness can be an adaptive behavior as defense mechanism to avoid predators or to disguise predatory intention. Examples of the former range from moth larvae that have evolved the appearance of twigs to dead-leaf butterflies that resemble (surprise!) dead leaves. And it's not just insects that play dead (a.k.a. exhibit "tonic immobility" or "thanatosis") as illustrated in Figure 2.

Predators may bluff weakness or otherwise misrepresent their appearance as a form of "aggressive mimicry" or of the "wolf in sheep's clothing" strategy (Wickler 1968; https://en.wikipedia.org/wiki/Aggressive_mimicry). The predator (insect, spider, fish, bird) appears as innocuous or seductive as a Trojan horse, making the true threat known when it's too late for their unsuspecting prey. Sex or food may be used as a lure. Some species have evolved the practice of kleptogamy, technically known as the "sneaky fucker strategy", in which a low-status male disguises himself as female (cf. Dawkins & Krebs 1978).

---

3 An eloquent illustration of the last strategy is the adoption by the embattled former Theranos CEO Elizabeth Holmes of a gravitas-projecting baritone, as detailed in publications (e.g. Carreyrou 2018), television news shows (e.g. the 2021 Australian 60 Minutes segment "Blood Money", https://tinyurl.com/5csvv8bj), and the 2021 podcast "The Dropout" (https://abcaudio.com/podcasts/the-dropout/).

Figure 2: A possum playing possum
(https://en.wikipedia.org/wiki/Apparent_death#/media/File:Opossum2.jpg).

In an evolutionary perspective, bluffing strategies have developed among mammals as a way of averting conflict within and across species. Non-human primates regularly engage in physical bluffing displays, while humans tend to be harder to frighten off in situations of potential conflict. Our fellow mammals have evolved physiological structures that support the bluff, in particular piloerection or horripilation (hair standing up on end to render an individual appear larger, thus frightening off predators), responses that humans have largely moved beyond. Primatologist Sherwood Washburn attributes the loss of such specific physical structures to the fact that "we have been bluffing with language for a long period of time" (discussion comment in de Reuck & Knight 1966: 39).

Humans do, however, use false displays to bluff out humans or other animals. Consider the ubiquitous "Beware of dog" signs tracing back to those found in the ruins of ancient Pompeii like the one in Figure 3.

Figure 3: *Cave canem* (public domain).

But what if there's no dog? Other bluffing alerts include fake security alarm signs, security keypads, and security cameras, all of which are readily available for purchase on Amazon (and fake watchdogs are free). One website offers "15 Ways to Fake That You Have a Home Security System and Deter Burglars" (https://tinyurl.com/yybzu2m5), and a tweeter observes "I think most 'Emergency Exit Only – Alarm Will Sound' doors are bluffing, but I'm too much of a coward to find out" (https://tinyurl.com/y3vdrklz).

A canonical strength bluff is the scarecrow, although such bluffs are often successfully called, as noted by a Canadian government site: "Because the threat that birds might associate with scarecrows is perceived rather than real, habituation is likely to occur relatively quickly unless other scare techniques are used in conjunction" (https://tinyurl.com/y2urz6lh). But crows are notoriously clever; indeed, the technique may be more successful against human Canadians:

> Some new scarecrows are popping up on the Prairies, but these ones are bluffing more than birds. Life-size metal cutouts of uniformed Mounties are being placed next to busy roads and intersections in Lloydminister, a city straddling the Alberta-Saskatchewan boundary. The city says the fake officers are part of a pilot "scarecrow initiative" aimed at discouraging speeders. (https://tinyurl.com/z3kw897p)

3 (Human) bluffing in the wild: Five specimens

As our dictionary glosses indicate, *bluff* (unlike *mislead* or *deceive*) is not a success verb, whether or not agents or targets are human; failed attempts to "engage in a false display in order to deceive or intimidate" by those who "try to mislead opponents" are still bluffs. But *bluff out* is a success verb, describing a perlocutionary act: if you bluff me out, your bluff has worked. Further, as the same dictionary entries suggest, human bluffers invariably act as intentional agents. In terms of the distinction drawn in the introduction to this volume, they are engaged not just in misleading but in **dis**leading: intentionally misrepresenting (exaggerating or concealing) their strength or knowledge to create a false inference on the part of their target.

The practice of bluffing extends over a wide range of such attempts well beyond the card table: prosecution vs. defense; buyer vs. seller; labor vs. management; antagonists on the field of military or business combat—more generally, adversaries in any social exchange who operate with imperfect trust and opposed goals. We must also situate bluffing within the landscape of related acts. What is the relation of bluffing to lying? perjury? misleading or deception? bullshit? puffery? fraud? false implicature? To address these questions, it will be helpful to look at cases in which the term is explicitly invoked or to which it implicitly applies and to attempt a characterization of the common features involved. (As elsewhere, emphasis is added.)

FLIGHT ATTENDANT

[FBI agent Hammond interrogates flight attendant Cassie about a certain passenger – her one-night stand, as Cassie and the reader know – who was later found murdered]

"And you two had a lot of interaction."

"I doubt I had more 'interaction' with him than I did with any other passengers I was serving", she said. It was a lie, but *interaction* struck her as a vague, ridiculous word that was impossible to quantify . . . She guessed it was also possible that Hammond had phrased his sentence this way because **he was bluffing: he was trying to frighten her into believing that he knew more than he did.**

(Chris Bohjalian (2018), *The Flight Attendant,* p. 148)

GUERNICA

Since the long-exiled "Guernica" finally came to Madrid on Sept. 10 [1981], Rafael Fernández Quintanilla, the witty diplomat who dealt with the Museum of Modern Art in New York and Picasso's heirs, has felt free to disclose some of the secrets of his protracted negotiations. One is **an elaborate bluff.** To demonstrate that the Spanish Government had in fact paid Picasso to paint the mural in 1937 for the Paris International Exhibition, Mr. Fernández Quintanilla had to secure documents in the archives of the late Luis Araquistain, Spain's Ambassador to France at the time. But Araquistain's son, poor and opportunistic, demanded $2 million for the archives, which Mr. Fernández Quintanilla rejected as outrageous. He managed, however, to obtain from the son photocopies of the pertinent documents, which in 1979 he presented to Roland Dumas, the Paris lawyer named by Picasso to determine when "public liberties" had been re-established in Spain, permitting delivery of the "Guernica" to the Prado.

"This changes everything," a startled Mr. Dumas told the Spanish envoy when he showed him the photocopies of the Araquistain documents. "You of course have the originals?" the lawyer asked casually. **"Not all of them,"** replied Mr. Fernández Quintanilla, **not lying but not telling the truth, either.**

(Markham 1981)

TAKING THE STAND

[In the judge's chambers, attorney Mickey Haller – defending Detective Harry Bosch against a frame-up – dramatically pulls out his phone and glances at a text from his investigator Cisco.]

> "Mr. Haller," [the judge] intoned. "Those are very significant allegations. Do you plan to offer any evidence to go with them if I allow you to present this in open court?"
> "Yes, Your Honor," Haller said. "The last witness I would present is Terrence Spencer himself. We were able to locate him over the weekend, hiding out at a home down in Laguna Beach . . . I had him served with a subpoena, and at this moment **he's out in the hallway with my investigator and ready to take the stand.**"

[At this point the conspirators throw in the towel. The case against Detective Bosch is dismissed. In the aftermath, Bosch wonders out loud what the witness Spencer will divulge about the conspiracy now that it has collapsed, but Cisco explains Spencer will plead the fifth amendment (= "take the nickel") and refuse to testify. Bosch is incredulous.]

> "Your text to Haller in the courtroom. You said he was ready to testify."
> "No, I said you can put him on the stand but he'll take the nickel. Why, what did Mick say?" [. . .] Bosch stared across the hallway at Haller. "Son of a gun," he said.
> "What?" Cisco asked.
> "I saw him read your text, and then he told the judge that Spencer was ready to go on the stand. **He didn't exactly say he would testify, only that he'd go on the stand. He tipped the whole thing with that bluff.**"
> "Smooth move."
> "Dangerous move."
> (Michael Connelly (2017), *Two Kinds of Truth*, Chapter 39)

– a dangerous move, Bosch notes, because Haller's bluff could have easily been called, and the jig would have been up.

But not all cases of bluffing are so identified. Two classic instances cited extensively in the literature on misleading and lying qualify as bluffing under the dictionary definitions: a speaker intentionally misrepresents his epistemic strength (his state of knowledge) in an attempt to deceive his target without lying, hoping – like Hammond, Quintanilla, or Haller – that his bluff will not be called.

SLIPPERY SAINT

Persecutors, dispatched by the emperor Julian, were pursuing him up the Nile. They came on him traveling downstream, failed to recognize him, and enquired of him: "Is Athanasius close at hand?" He replied: **"He is not far from here."** The persecutors hurried on and Athanasius thus successfully evaded them **without telling a lie**.

(MacIntyre 1995: 336)

SWISS BANK

[From the celebrated Supreme Court decision Bronston v. United States, 409 U.S. 352 (1973), overturning sleazy movie producer Samuel Bronston's perjury conviction; see Chapters 1, 2, 9, 13, and 14 for more on Bronston]
https://supreme.justia.com/cases/federal/us/409/352/case.html [June 10, 1966]

"Q. Do you have any bank accounts in Swiss banks, Mr. Bronston?"
"A. No, sir."
"Q. Have you ever?"
"A. The company had an account there for about six months, in Zurich."

It is undisputed that, for a period of nearly five years, between October, 1959, and June, 1964, petitioner had a personal bank account at the International Credit Bank in Geneva, Switzerland, into which he made deposits and upon which he drew checks totaling more than $180,000. It is likewise undisputed that petitioner's answers were **literally truthful**.

(a) Petitioner did not at the time of questioning have a Swiss bank account. (b) Bronston Productions, Inc., did have the account in Zurich described by petitioner.

Beyond question, petitioner's answer to the crucial question was not responsive if we assume, as we do, that the first question was directed at personal bank accounts. **There is, indeed, an implication in the answer to the second question that there was never a personal bank account**; in casual conversation, this interpretation might reasonably be drawn. But we are not dealing with casual conversation, and **the statute does not make it a criminal act for a witness to willfully state any material matter that implies any material matter that he does not believe to be true**.

– Chief Justice Warren Burger (409 U. S. 357–8)

The perjury conviction in SWISS BANK was overturned because, the Burger court ruled, Bronston had not lied (see Tiersma 2003, 2005 and Solan & Tiersma 2005 on the "literal truth defense").

This finding is consistent not only with perjury statutes (see Chapters 10–14) but with definitions of lying dating back to St. Augustine (5th c. CE). Four criteria have traditionally been invoked for what it takes for speaker S to lie to hearer H (Mahon 2015; Horn 2017a,b):

> (5) Criteria for lying
> **(C1)** S says/asserts that p
> **(C2)** S believes that p is false
> **(C3)** p is false
> **(C4)** S intends to deceive H into believing that p

To **assert** what you **correctly believe to be false** with **the intent to deceive** is surely to lie. But must all four criteria be satisfied? The OED (s.v. LIE, n., 2b) invokes **C3, C1,** and **C4**: a lie is "a false statement made with intent to deceive". But **C2** is arguably more essential than **C3** or **C4** (cf. Chisholm & Feehan 1977 and Coleman & Kay 1981 for theoretical and empirical support of this position, Turri & Turri 2015 for a defense of **C2**, and Wiegmann, Samland, and Michael Waldmann 2016 for a rebuttal). For Augustine in *The Enchiridion*, "Every liar says the opposite of what he thinks in his heart, with purpose to deceive", yielding **C1, C2,** and **C4**.[4] In stipulating that the liar in stating p must believe not-p (**C2**) without stipulating that p is actually false (**C3**), Augustine prefigures perjury laws deriving from the landmark 1911 U.K. statute:

> If any person lawfully sworn as a witness or as an interpreter in a judicial proceeding wilfully **makes a statement** material in that proceeding, **which he knows to be false** or **does not believe to be true**, he shall be guilty of perjury.
>
> (http://www.legislation.gov.uk/ukpga/Geo5/1-2/6/
> section/1/enacted [1911, §1])

Since asserting (or statement-making) is necessary for both lying and perjury, one cannot lie, but at most deceive or intentionally mislead by uttering a literally true statement with false implicatures, whether conversational or conventional.[5] Conversational implicatures are triggered by the speaker's exploitation of one or more of Grice's maxims in (6).

---

4 The status of the intention to deceive (**C4**) within the definition of lying is a matter of some dispute. Sorensen (2007), Carson (2010), and others have pointed to cases of BALD-FACED LIES in which the speaker utters a believed-false statement with no expectation that it will be believed, and there is thus no intention to deceive. Others regard such cases as non-lies. (See Meibauer 2019 and Chapter 7 for additional considerations.)

5 Adler (1997: 452) proposes "a lies-deception distinction corresponding to an assertion-implicature one", and Saul (2012) elaborates a view on which the distinction between lying and misleading as isomorphic to Grice's distinction between what is said and what is implicated. See Horn 2017a,b for defenses of this position and Weissman & Terkourafi 2019 for partial empirical support. Meibauer (2005, 2014) proposes an "extended definition" of lying on

(6) Grice's Maxims of Conversation (Grice [1967]1989: 26):
    QUALITY: Try to make your contribution one that is true.
        1. Do not say what you believe to be false.
        2. Do not say that for which you lack evidence.
    QUANTITY:
        1. Make your contribution as informative as is required . . .
        2. Do not make your contribution more informative than is required.
    RELATION: Be relevant.
    MANNER: Be perspicuous.
        1. Avoid obscurity of expression.
        2. Avoid ambiguity.
        3. Be brief. (Avoid unnecessary [sic] prolixity.)
        4. Be orderly.

On the neo-Gricean Manichean model of pragmatics I have proposed (Horn 1984, 1989, 2007), the hearer-oriented Q Principle, "Make your contribution sufficient", "Say as much as you can (given Quality and R)", is dialectically opposed to the speaker-oriented R Principle, a correlate of least effort, "Say no more than you must (given Q)". The Q Principle subsumes the first submaxim of Quantity along with the first two submaxims of Manner, while the R Principle incorporates the second submaxim of Quantity, Relation, and the Be brief and Be orderly submaxims of Manner. The cases of bluffing we have reviewed can all be glossed through the operation of Q, R, or both, given our understanding of bluffing as an instance of not lying but of intentional misleading or, as proposed in the introduction, DISLEADING.

GUERNICA
"I don't have all the originals" $+>_Q$ "I have some of the originals"
[cf. <*some, all*> scale]
SLIPPERY SAINT
"Athanasius is not far from here" $+>_Q$ "Athanasius is not here before you"
FLIGHT ATTENDANT
"You two had a lot of interaction" $+>_R$ "You two were romantically/sexually involved"
TAKING THE STAND
"Spencer is prepared to take the stand" $+>_R$ "Spencer is prepared to testify"
SWISS BANK
"The company had a Swiss bank account" $+>_{Q,R}$ "I had no personal Swiss bank account"

---

which false implicatures do count as lies. This question is further pursued in Stokke 2013, Mahon 2015, and Chapters 1–4 and 12–14.

Scalar implicature (Horn 1972), the locus classicus of Q-based inference as in GUERNICA, is responsible for many instances of bluffing in and out of the court-room.[6] Or in the hearing room, as when Monica Lewinsky was questioned in a 1999 congressional hearing by Sen. Ed Bryant (R-TN) as to whether her affidavit in the earlier Paula Jones case was partially false (http://tinyurl.com/n9xcp8t). Her literally true but misleading response was "Incomplete and misleading" – falsely scalar-implicating (via the <*misleading, false*> scale) that it wasn't false. Here again, as with Bronston, Lewinsky's bluff could have been but wasn't called by her interrogator.

4 Bluffing and the landscape of deception

Even without a consensus definition, there is substantial agreement as to the na-ture of and motivation for bluffing (by humans), but the details can be tricky to pin down. For Brenkert (2013: 587), "Bluffing is an attempt to get someone to do or to agree to something through deception." This, however, is too narrow: a bluffer can seek to affect beliefs and dispositions, not just actions. In his influential paper on bluffing in business transactions, Carr (1968: 144) depicts bluffers as using "conscious misstatements, concealment of pertinent facts or exaggeration to per-suade others to agree". But bluffs need not be statements and can indeed be en-tirely non-linguistic – even at the card table or the car lot. It suffices to "engage in a false **display**", whether or not you're a vertebrate. Unlike lying, which requires a speech act of asserting or promising, bluffing does not necessarily involve the per-formance of a speech act – or of language at all. In this respect and others, bluffing is a species not of lying but of disleading or attempted deception.

The key feature of bluffing is the vulnerability of the agent to the possibility that the target will call the bluff. In a successful bluff, the bluffee is, in the words of the OED entry, "induced to 'throw up' his cards and lose his stake, rather than run the risk of betting against the bluffer", given the very real dan-ger of guessing wrong. In one classic case, Speaker of the House Newt Gingrich (R-GA) and his fellow Republicans threatened a government shutdown in 1995 if President Clinton refused to cut funding for Medicare. Time's Man of the Year Gingrich and his Republican acolytes assumed this would make for a popular

---

6 Chisholm & Feehan (1977: 155) offer a constructed example prefiguring GUERNICA: in telling you "My leg isn't bothering me too much today" when it's not bothering me at all I may suc-ceed in misleading or deceiving you (given the scale <*not too much, not at all*>) but – absent any assertion that my leg is bothering me – I do not lie.

assault against government spending. But Clinton called their bluff, success-fully portraying Gingrich as extreme and unreasonable. The Republican revolt collapsed and Clinton was easily re-elected the next year.

But there are dangers for those who call or dismiss an apparent bluff. Parents may ignore repeated teen suicide threats as merely a bluff – until it's too late. (The "boy who cries wolf" scenario.) U.S. diplomats ignored signals from the Chinese during the Korean War, assuming their threats to intervene on the side of North Korea were bluffs. (Spoiler: they weren't; see Sartori 2005: 36–7)

In the legal realm, Aaron Sanders, PLLC ("Nashville office, Global presence") offers a useful rendition of the dangers of bluffing for both agent and target:

> If there's one thing lawyers are good at, it's bluffing. It's pretty much impossible to tell, simply from the style and tone of a letter, whether the lawyer believes in what he or she is writing. Often, lawyers aren't bluffing. They're giving you fair warning, a chance to climb down, to get out of trouble with your skin. Frustratingly, the downside of bluffing all the time is that there's no way for a lawyer to signal: "Um, I really mean it, this time."
>
> (https://www.aaronsanderslaw.com/ive-received-a-nasty-letter-from-a-lawyer/)

A bluff is an act or display that collapses if called, but the caller is vulnerable too, because the supposed bluffer (at the negotiating table, on the battlefield, in the courtroom) may not be bluffing. The risk of calling is loss – of face, bargaining position, money, freedom, or life.

In some contexts, however, the bluffer is happy to be (occasionally) caught. This is bluffing as advertising: the creation of uncertainty as a strategic ploy. The adage in poker is "Play the same hand differently; play different hands the same".

> Often the best gains come when you do not have a strong hand, but you get others to believe you are bluffing. They stay in the game, calling, perhaps even raising your bets, creating a large pot. You bluff (perhaps even get caught), not so much to fool people about the quality of your hand, but to gain a reputation as someone who does bluff. That reputation allows you to confuse and mislead opponents in other hands.
>
> (Dees & Cramton 1991: 153)

The goal is to keep the adversary off balance, a practice with putatively ancient roots: "Aristotle . . . beeyng asked what vauntage a man might get by lying, he answered: to be unbelieved when he telleth truth" (Baldwin 1571: cap. xxiii).[7] Whether the payoff for competitive bluffing is as effective for business negotiators as it is for poker players playing the long game is, however, an open

---

7 Extolling the "psychic effect" of bluffing in both poker and "every day life", Clemens France writes (1902: 386) that "it makes the man who bluffs play better and the opponent play worse."

question (cf. Friedman 1971, but also Guidice, Alder & Phelan 2009 for a sobering reconsideration).

5 Bluffing and bullshitting

For both lexicographers and philosophers, the realm of the bluff lies within the kingdom of bullshit. The OED glosses the verb *bullshit* as 'to talk nonsense (to); to **bluff one's way through** (something) by talking nonsense'. Similarly, a *bullshitter* is 'one who exaggerates or talks nonsense, **esp. to bluff or impress**'. Frankfurt (2005: 45–7) classically distinguishes both bullshit and bluffing from lying, inspired by Ezra Pound's line, "Don't bullshit *me*" (*Pisa Cantos,* 1948, cited in OED's *bullshit* entry):

> The speaker . . . demands that the claim be supported with facts. He will not accept a mere report; he insists on seeing the thing itself. In other words, he is **calling the bluff** . . . **It does seem that bullshitting involves a kind of bluff**. It is closer to bluffing, surely, than to telling a lie . . . Lying and bluffing are both modes of misrepresentation or deception. Now the concept most central to the distinctive nature of a lie is that of falsity: the liar is essentially someone who deliberately promulgates a falsehood. Bluffing, too, is typically devoted to conveying something false. Unlike plain lying, however, it is more especially a matter not of falsity but of fakery. **For the essence of bullshit is not that it is *false* but that it is *phony*.**

Bullshitting and bluffing, like lying, involve insincerity, but to different ends. Note that Frankfurt's appeal to phoniness would be applicable to the full range of "dishonest signaling" from the deimatic displays of moths and mantises to misrepresentations by negotiators and poker players. But while to "call bullshit", as in Pound's line, amounts to calling a bluff, bullshit is largely restricted to spoken or at least linguistic phoniness (it can be used for "No Parking" or "No Smoking" signs, which may evoke the angry response "That's bullshit"). What if I knowingly put down a non-existent word at Scrabble, especially if I make a habit of it – I'm certainly bluffing, but am I also bullshitting? Perhaps only if I defend my move verbally.

For Frankfurt (see also Chapter 14), bullshitters do not essentially misrepresent the facts; they essentially seek to deceive us about what they are up to. This is not so for bluffers, whose motivation is different. We bullshit to impress and for self-aggrandizement; we bluff to mislead about our knowledge, position, or future actions. And we have seen, we can bluff weakness – but it is less clear that we bullshit weakness, unless false modesty or insincere humility counts as bullshitting.

Carson argues (2010: 61–63) contra Frankfurt that "one can tell a lie while producing bullshit" and that "bullshitters can be concerned with the truth of what they say".[8] Evasive bullshit can be used to avoid the question under discussion, but if what is said is false, it's also a lie. An evasive bullshitter may indeed be concerned with the truth of what he says: a politician who dodges the question (but wants to avoid false claims in case he's found out), an unprepared student who (for partial credit) answers a different question from the one asked, or a witness who cannot afford to directly answer the question posed (but wants to avoid perjury). Thus evasive bullshitters can be seen as seeking, in the words of the OED definition, "to bluff [their] way through". Without abrogating Quality-1 ("Do not say what you believe to be false"), the evasive bullshitter flouts one or more of Grice's other maxims, through un(der)informativity, irrelevance, absence of evidence, and/or withholding of the truth.

6 Bluffing and puffing

As defined by the Federal Trade Commission, puffery is used "to denote the exaggerations reasonably to be expected of a seller as to the degree of quality of his product, the truth or falsity of which cannot be precisely determined" (Better Living, Inc. et al., 54 F.T.C. 648 (1957), aff'd, 259 F.2d 271, 3rd Cir. 1958). Crucially, the target audience is not expected to take the representation seriously; as with sarcasm and other non-literal utterances (see Chapter 9), there is no intent to deceive. Puffery constitutes a valid defense against charge of fraud or perjury. As attorney Diane Lockhart explains on the legal drama "The Good Fight" (CBS, S1E7), "Mere puffery is the legally acceptable practice of promoting through exaggeration". When Donald J. Trump (or his ghostwriter) declares in *The Art of the Deal,*

> People want to believe something is the biggest and the greatest and the most spectacular. I call it truthful hyperbole. It's an innocent form of exaggeration – and a very effective form of promotion. (Trump 1987: 58)

he is describing puffery. Instances include his touting of "Celebrity Apprentice" as the top-rated show on television in 2015 when it was the 67th ranked show by Nielsen figures, his building the 58-story Trump Tower in New York and repeatedly promoting it as 68 stories high (https://www.nytimes.com/2017/01/24/arts/television/for-trump-everything-is-a-rating.html), or his announcing on a variety of subjects before or during his presidency "I alone can fix it".

---

8 See Chapters 1 and 9 for other problems with Frankfurt's definition.

Puffing is a species of bullshit in Frankfurt's sense. The truth of p is of no concern to the speaker, whose goal is not to commit himself to the truth of p but to persuade the target to buy the product and/or the seller. But invoking puffery does not always succeed. In the 2015 case *Ferguson v. JONAH* [= Jews Offering New Alternatives to Homosexuality], New Jersey Superior Court No. L-5473-12, the defendant's ex-gay conversion "therapy" practice was found to constitute consumer fraud, notwithstanding JONAH's move to dismiss their own previous claims as mere non-actionable puffery (https://tinyurl.com/yybo4m43).

In the political sphere, Fox News lawyers defended their pundit Tucker Carlson against charges of slander by explaining to the court that "the 'general tenor' of Carlson's show should inform a viewer that he is not 'stating actual facts' about the topics he discusses and is instead engaging in 'exaggeration' and 'non-literal commentary'".[9] Similarly, attorney Sidney Powell argued (at least in part; see Kenyon & Saul's contribution to this volume for some complications) that Dominion's \$1.3 billion defamation lawsuit against her should be dismissed on the grounds that her false statements in court, alleging that their voting machines were part of an anti-Trump plot and linking them to Venezuela's late dictator Hector Chávez, were so patently absurd that "no reasonable person" would believe them (https://tinyurl.com/t8zrvhas). In each case, the defendant was essentially pleading puffery: if no reasonable person would believe the speaker is warranting the truth of his remarks, no assertion (or lie) is made (Stokke 2013), and thus no one would be deceived. Yet many acolytes – however unreasonably – have continued to take Carlson and Powell at their (implausible) word, with dire consequences.

It will be recalled that Grice distinguishes the first submaxim of quality, "Do not say what you believe to be false", from the second submaxim of quality, "Do not say that for which you lack evidence". Violating Quality-1 results in a lie, given a context in which the maxim is assumed to be in effect (Fallis 2009). Violating Quality-2 typically results in bullshit or puffery (including "truthful" hyperbole), especially if there is no reason to assume the target would be deceived.[10] If a reasonable person would be deceived, we are dealing with fraud and not mere

---

9 In some quarters, this is now known as the "Tucker Carlson defense": "It can't be slander if no reasonable person would believe it" (https://www.courierherald.com/opinion/the-tucker-carlson-defense/). An early practitioner of the Tucker strategy was the spokesperson for Sen. Jon Kyl (R-AZ) who defended Kyl's baseless 2011 assertion that "well over 90%" of Planned Parenthood's activity is devoted to performing abortions (when the true figure was shown to be more like 3%) by helpfully explaining that the senator's earlier remark "was not intended to be a factual statement" (http://knowyourmeme.com/memes/events/not-intended-to-be-a-factual-statement).

10 See Stokke 2019 on problems with the attempt to identify bullshit with violations of Quality-2 and Chapter 7 for a reconsideration of bullshit.

puffery. Perjury requires the intentional misrepresentation of material facts, excluding puffery and bullshit (cf. Lawrence 2014; Horn 2017a,b; Chapter 14). Bluffing overlaps with puffery, extending to true cases of fraudulent misrepresentation as well as less consequential instances that fit the profile of puffery. As in other cases, what the bluffer fears is being called. (For more on puffery, fraud, bullshit, and the law, see the discussion in Chapter 14.)

7 Bluffing and equivocation

We have seen that among humans bluffing is purposive; the bluffer intends that the target come to believe that the displayed misrepresentation is accurate. One class of such cases involves dual displays, one accurate and one not. This is the realm of equivocation:

> Oxford English Dictionary, s.v. EQUIVOCATION, 2a:
> The use of words or expressions that are susceptible of a double signification, **with a view to mislead**; *esp.* the expression of a **virtual falsehood** in the form of a proposition which (in order to satisfy the speaker's conscience) is **verbally true**.

Historically, the Jesuits exploited the possibility of equivocation to extricate their clients from an awkward position. In his Provincial Letter No. 9, Pascal (1656: 101–2; translation and emphasis mine) recruits a resourceful Jesuit monk to explain this method for dealing with "embarrassing cases . . . when one is **anxious to avoid lying while inducing a false belief**". As the (fictitious) monk puts it, "In such cases, our doctrine of equivocations has been found of admirable service . . . it is permitted to use ambiguous terms, leading people to understand them in another sense from that in which we understand them ourselves." One pre-Jesuit master of this technique was the 13th century Dominican father St. Raymond de Peñafort, who recommended deflecting the murderer at the door without lying by swearing of the intended victim *"Non est hic"*, which is designed to be understood as the virtual falsehood 'he is not here' but which, given the homonymy of the 3rd person singular form *est*, is "verbally true" on the alternate reading 'he does not eat here' (Cavaillé 2004: ¶10; Cardenas 1702 in Roussel de la Tour 1763: 464).[11]

In our own day, equivocation – especially among politicians – often exploits not homonymy but polysemy arising from technical definitions with which hearers are unlikely to be familiar. In a speech on 26 January 1998, Bill Clinton asserted "I did not have sexual relations with that woman, Monica

---

11 It may be worth noting that Raymond is the patron saint of lawyers.

Lewinsky", relying on the technical meaning of this term of art excluding oral stimulation (Tiersma 2003, 2005; see also Chapter 1). On whether the earlier statement "There is no improper relationship" between himself and Lewinsky was technically a true representation, Clinton famously observed "It depends on what the meaning of the word 'is' is" (testimony before grand jury, footnote 1128 in the Starr report), leaning on the polysemy between the present tense and timeless meanings of the copula. (Dowd 1998 offers a more cynical take.)

Similarly, when asked in March 2013 whether the National Security Agency "collect[s] any type of data at all on millions or hundreds of millions of Americans", the then National Intelligence Director James Clapper responded in what he later described as "the most truthful, or the least untruthful manner" by saying no, despite strong evidence to the contrary. He subsequently justified this denial on the grounds that what the NSA had been doing was not **collecting** data but merely **aggregating** it, given the equivocation on *collect*: "There are honest differences on the semantics . . . when someone says 'collection' to me, that has a specific meaning, which may have a different meaning to him" (James Clapper to Andrea Mitchell, 9 June 2013 interview, https://tinyurl.com/yya9pcjp).

Each instance of equivocation is also a case of bluffing in that the equivocator risks the embarrassment or legal peril of discovery by a more perspicacious audience ("What exactly do you mean by *est/sexual relationship/is/collect*?"). The same is true for the well-worn ploy of equivocation in reference. In one of the later books in his detective series, Michael Connelly's Harry Bosch is employed by the San Fernando Police Department, but runs a bluff with outsiders by introducing himself with studied underspecification as "Detective Bosch with the SFPD out in Calfornia":

> Bosch had taken to using the abbreviation SFPD when making calls outside the city because the chances were good that the receiver of the call would jump to the conclusion that Bosch was calling from the San Francisco Police Department and be more willing to help than if they knew he was calling from tiny San Fernando.
> – Michael Connelly (2016), *The Wrong Side of Goodbye*

Some instances of onomastic misdirection are more personal:

> [Chelsea is at a party, using "Beulah" as a *nom de fête* . . .]
> "I'm Beulah."
> "Well, that's just beautiful. Is it a family name?" she asked me.
> "Yes", I said. **Technically, it wasn't a lie.** Beulah had to be somebody's family name.[12]

---

12 Similarly, the Russian Jewish immigrant Abe Trillin, inspired by the "Dink Stover at Yale" books, christened his son Calvin Marshall Trillin so he would blend in with his future fellow Yalies. When asked about his moniker, the celebrated journalist/author/foodie explains, "It's

– Chelsea Handler (2005), *My Horizontal Life: A Collection of One-Night Stands*

Sister Marie came to greet the new girl, who was clearly troubled, as so many of these motherless children were . . . She asked the girl for her name.

"Lillie Perrin, Madame."

The girl's eyes were lowered, which led the mother superior to believe this was not the truth.

"Lillie is your given name?"

Lea had been given it, by own dear mother before leaving Berlin, so perhaps when she said yes **it was not truly a lie**.

– Alice Hoffman (2020), *The World That They Knew*, p. 158

To describe a claim as *technically* or *verbally* true, *not exactly false*, or *not truly a lie* (see Chapter 2) is to implicate that it misses the mark in other ways in terms of ethics or cooperation; recall the OED's unpacking of equivocation as virtual falsehood. Virtual falsehood, however, is still technical or minimal truth, as on David Lewis's formal definition of the practice:

Take a sentence σ which is assigned multiple interpretations by [Language] *L* on an occasion *o* of its utterance. One can be minimally truthful in *L* with respect to σ on *o* by taking any one of those interpretations and doing whatever one would have to do to be truthful in *L* if that interpretation were the only one assigned to σ on *o* by *L*. A trickster is being **truthful in this minimal way** if, knowing that Owen is going to the shore of the river, he says "Owen is going to the bank" during a conversation about Owen's lack of cash. (Lewis 1969: 193)

In each scenario the equivocator, whether actual – Clinton, Clapper, Chelsea – or constructed – Lea, Lewis's trickster – is bluffing, counting on not being pressed further. But equivocation isn't ipso facto bullshit: the goal (with the possible exception of Chelsea Handler's case) is not self-aggrandizement. Nor is equivocation puffery; it is distinct from vague claims or self-promotion. Like other bluffers, equivocators display a false face (along with a true one!) and are vulnerable to being called ("What do you mean by ____?" "Is that your family name?")

8 The ethics of bluffing

It will have been noticed that the equivocation bluffs of Section 7 elicit different ethical judgments based on the context; we tend to judge a politician's misdirection in the service of his goal of escaping legal scrutiny for previous unforced errors differently from a refugee from the Nazis seeking to hide her true identity or a man seeking to protect his friend from a murderer, while a raunchy uninhibited

---

an old family name. Not our family, but still an old family name" (https://tinyurl.com/27vsh2w5).

humorist adopting a pretense as a lark might fall somewhere in between. In terms of legal ethics, the question is whether the target of the bluff (or their counsel) is the one responsible for not pushing the matter; in the words of the unanimous Supreme Court judgment overturning Bronston's perjury conviction, "If a witness evades, it is the lawyer's responsibility to recognize the evasion and to **bring the witness back to the mark**, to flush out the whole truth with the tools of adversary examination". But what of the targets of the Jesuits' casuistry or of Lewis's trickster?[13] Aside from the superordinate question of whether it is always better (or less bad) to intentionally mislead than it is to lie (see Saul 2012, Timmermann & Viebahn 2021, and Chapters 2 and 4), just how unethical is it to bluff?

Beyond the forensic field of play, it is widely recognized that adversaries across a business or poker table are engaged in a contest in which strategic misrepresentation is part of the house rules. Important work on the ethics of bluffing has appeared in the business journals, beginning with Carr 1968 and extending to a series of reconsiderations (e.g. Sullivan 1984, Dees & Cramton 1991, Carson 1993, Allhoff 2003, Guidice, Alder & Phelan 2009, Carson 2010). For Carr, bluffing in business negotiations is much like bluffing in poker: accepted by general implicit convention as "part of the game" and as something everyone does in such games. But are morality and ethics really matters of community norms or local customs? For Dees & Cramton and for Carson, it's not **convention** that licenses bluffing but **self-defense**; one bluffs lest one be bluffed. We can distinguish attempting to deceive an adversary in a business negotiation about matters of empirical, material facts (such as the condition or history of items for sale, extent of injuries) or about one's own bargaining position (as when one interlocutor states disingenuously " . . . and that's our final offer"). Dees & Cramton (1991) and Carson (1993), citing ABA model rules for Professional Conduct, observe that bluffing in the former case is frowned on and may rise to the level of fraud, while attempting to mislead in the latter case is generally regarded as permissible (and even expected) given the context of rational mutual distrust.

The boardroom and used-car lot, like the courtroom, provide a natural home for the bluff, while actual lies remain beyond the pale; this is another locus for the literal truth defense, while allowing for or even encouraging false inferences is fair game:

---

13 The radical intentionalism of the Jesuits, who recommended equivocation and mental reservations as means of "telling the truth in a low key, and a lie in a loud one" (Pascal 1656: 102), continues to flourish.

Kellyanne Conway, the designer of "alternate facts", played the radical intentionalist card against critics of President Trump: "You always want to go by what's come out of his mouth rather than look at what's in his heart" (https://www.nytimes.com/2017/01/09/opinion/trump-trapped-in-his-lies-keeps-lying-sad.html).

> You regard me as an adversary, someone trying to get the best of you. Hence you expect me to give the literal truth in our discussion. In this respect, you can expect no more from me than our legal system expects from a hostile witness being cross-examined by an attorney in a trial: the witness may not lie, without committing perjury; but **as long as what the witness says is literally true, he is not responsible for false suggestions arising from his statements.** Similarly, . . . in a competitive negotiation, as long as one remains in the realm of settlement preferences, one is not responsible for more than the literal truth of what one says; one is permitted to defend oneself.   (Strudler 2005: 462–463)

As Solan (2012) has stressed, good lawyers master the distinction between outright lying and "deception by misdirection", to **"encourage others to draw false inferences about the underlying facts"**. But their skill at the practice of "truthful insincerity" contributes to the low esteem in which lawyers are often held and, by the same token, the disrepute in which the Bill Clintons and Jesuits of the world have been held by those – from Pascal's day to our own (cf. Dowd 1998) – who scorn any attempt to "dissemble" or "parse the truth", or (as some might put it) to engage in "paltering" (Schauer & Zeckhauser 2007).

Within international politics and global economics, bluffing is expected, and the decision on whether or not to call a possible bluff can be risky in the extreme. The imminent decisions by Greece and the U.K. on whether to leave the European Union were frequently characterized in terms of bluffing, as seen in this blogpost on "The Economics of Bluffing" (Buttonwood blog, *The Economist*, 28 May 2015 (https://tinyurl.com/yyt2785h):

> Will Greece default on its debts and leave the euro? Will Britain decide to leave the European Union? Politicians in the two countries have threatened, implicitly or explicitly, to take these drastic steps if their European colleagues do not offer them inducements to stay.
>
> **Many people regard these threats as a bluff.** They think that Greece does not really want to leave the euro, and that David Cameron, Britain's prime minister, does not want his country to exit the EU. When push comes to shove, Greece will do a deal and Mr Cameron will persuade British voters to stay in the EU in his planned referendum. But there are risks that neither outcome will turn out as planned. In both cases, political leaders are making a risky bet. [. . .]
>
> The other nations within the euro zone and the EU may decide to **call Greece or Britain's bluff.** In Britain, the electorate also has the right to exercise the option of exit – which they might use in the referendum to protest against government policies in general rather than voting on the merits of EU membership in particular.

With the wisdom of hindsight, we know that while the bluff by Greece was correctly called by the EU, the Brexit threat proved not to be a bluff at all.

While lacking the repercussions on the international political sphere, there are also high stakes for individuals when it comes to bluffs within the criminal justice system. Police interrogators regularly lie to or mislead suspects as to the evidence connected to a crime or the statements of putative confederates in order to procure confessions of guilt. This has led to the prevalence of false confessions, which according to the Innocence Project comprise up to 25–30% of confessions of convicted defendants who were later exonerated by DNA. The false evidence ploy, where suspects are confronted with apparently strong evidence that in fact does not exist, while legal in the U.S. (*Frazier v. Cupp* 394 U.S. 731 (1969)), is often deemed unfair. As such, it has been replaced in some jurisdictions by the technique of the bluff, in which interrogators – often armed with irrelevant or blank case files and dossiers – pretend to possess evidence that has not yet been examined. In theory, this practice should lead guilty suspects but not innocent ones to confess. In fact, as Perrillo & Kassin (2011) demonstrate (see also Kassin 2005 and his contribution to this volume), actual innocence works against suspects, who tend to reason that they have nothing to worry about, given their expectation of future exoneration. In lab experiments, bluffing increased the false confession rate by 60%, as much as the presentation of fake evidence.

Kassin and Perrillo point out that that misrepresentations presented by law enforcement personnel need not be false. A police officer imparting to a suspect that he has DNA evidence when that evidence has not actually been tested bluffs by feigning knowledge that p when he doesn't know that p (or that not-p). This too has its analogue at the poker table: contrary to what is often claimed (e.g. Perrillo 2019: 278), the bluffer need not "hold the knowledge of what is actually true". A player may execute what might be termed a blind bluff – betting as if she had a strong hand when she has covertly avoided inspecting her cards. Just so, in cases like that summarized by Perrillo (2019: 281), police interrogators – while "unaware of the actual strength of the evidence" – confidently treat this evidence as conclusive and damning.

As we saw in SWISS BANK and TAKING THE STAND in §3, the defense team can play the same card, feigning the possession of exculpatory or exonerating evidence that would in fact collapse if called. Across the animal kingdom, the strategic misrepresentation of strength or weakness by prey or predator are presumably beyond good and evil, and a similar assessment might be offered for human bluffing *in extremis*, such as Athanasius's stratagem or Raymond de Peñafort's equivocation, in which each future saint bluffed epistemic weakness. But such bluffs, whether or not ethically justified, crucially risk collapse upon being called by predator or persecutor, as do the bluffs of those engaged on either side of a business negotiation. On the other hand, when police interrogators act as if they possess, or are in the process of developing, definitive inculpatory evidence against a suspect, their bluffs are on particularly dubious ethical grounds, often resulting in false confessions leading to unjust convictions. Worse still, several studies have shown that "bluffing significantly increased the rate of false confessions, though it did not affect the rate of true confessions" (Perrillo 2019: 286). Furthermore, false confessions induced by the bluff tactic are unlikely to be detected by prosecutors, juries, or judges and are exceedingly rarely overturned. In other words, *Caveat reus* – Let the suspect beware![14]

9 All's fair in love and war: The ethics of bluffing reconsidered

When contexts for the practice of bluffing beyond the card table are considered (e.g. Perrillo 2019), the focus is standardly on the realms of the courtroom (or interrogation room) and the boardroom (or negotiating table), as we have seen. Ethical implications are considered in terms of tactics and strategies in a context defined by rational mutual distrust and self-defense. But there are other domains in which, as has long been recognized, bluffing and related deceptions are known to flourish and to be anticipated if not always accepted.

The dictum that all is fair in love and war can be traced back in its English incarnation to Aphra Behn's 1687 farce "The Emperor of the Moon": "All Advantages are fair in Love and War".[15] The apparent original of the observation is a bit older:

> [L]ove and war are the same thing, and as in war it is allowable and common to make use of wiles and stratagems to overcome the enemy, so in the contests and rivalries of love the tricks [*embustes*, 'deceptions'] and devices employed to attain the desired end are justifiable, provided they be not to the discredit or dishonour of the loved object.
>
> (Miguel de Cervantes (1620), *Don Quixote*, Part 2, Chapter 21;
> http://www.readprint.com/chapter-1702/
> Don-Quixote-Miguel-de-Cervantes)

A representative instance of the Cervantes/Behn half of the "all's fair" aphorism is the reprehensible – but not actionable – misbehavior of Willoughby toward Marianne, the "sense" of Austen's *Sense and Sensibility*, who upon interrogation by her sensible sister Elinor is forced to acknowledge that her ex-beau never actually lied to her:

---

14 In June 2021, Illinois became the first state to bar police from purveying "false information about evidence or leniency" to minors (https://tinyurl.com/2ajvyef2). Adult suspects remain fair game.
15 Thanks to Garson O'Toole, a.k.a. the Quote Investigator (https://quoteinvestigator.com/), for locating the Behn (1687) reference.

EDWARDIAN CAD

"But **he told you** that he loved you?"
"Yes – no – **never absolutely**. It was **every day implied, but never professedly de-clared**. Sometimes I thought it had been, but it never was."
(Jane Austen (1811), Sense and Sensibility, Chapter 29)

Alas, those who – like Marianne – suspect their suitor's actions and declarations amount to a bluff have reason to avoid the risk of calling it. Whether all is indeed fair in such cases depends on whether you're the bluffer or the bluffee.

The latter half of the "all's fair" dictum is legion, and this should be no surprise. As Clausewitz (1832: 86) remarks, one page before his famous definition of war as merely politics by other means, "In the whole range of human activities, war most closely resembles a game of cards". Perhaps the best-known military bluff postdating Odysseus's Trojan horse was the World War II exploit Operation Fortitude. This was General George S. Patton's "phony army" (also known variously as his phantom, fake, or ghost army), which leveraged fake intelligence and inflatable tanks, as in Figure 4, to successfully bluff the German military command into anticipating an Allied attack at Pas-de-Calais and overlooking the actual preparations for the amphibious D-Day landing on the beaches of Normandy. (See Beyer 2015 for more on the *trompe l'oeil* army.)

Figure 4: Inflatable dummy tank modeled after M4 Sherman.
(http://www.psywarrior.com/DeceptionH.html)

So **is** all fair in love and war – and in poker, in business and diplomatic negotiations, in the courtroom and the interrogation room? As we have seen, the context is critical in determining the moral and ethical parameters, including the risk incurred by the bluffer and the access of the bluffee to the means for calling the bluff. To return to our Jesuit colleagues from §7, the key principle licensing equivocation (as well as mental reservation; cf. Pascal 1656, Fauconnier 1979, Horn 2017a) is the difference between lying and hiding the truth *(cacher la vérité)*: "not with the express intention of deceiving others but only of allowing them to deceive themselves" (Fegeli 1750, in Roussel de la Tour 1763).[16] The core strategy traces back to the fathers of the church. Aquinas, after pointing out that it is not permitted to tell a lie, allows – following Augustine (*De Mendacio*) – that "it is lawful to hide the truth prudently [*veritatem occultare prudenter*]" (*Summa Theologiae*, Part IIb, Q110; https://tinyurl.com/4wnhbh3h). But the practical assessment of the landscape of deception stubbornly remains in play for moral philosophers – and the rest of us.

Canonical instances of bluffing, in particular at the poker table and in the realms of adversarial exchanges in business and law, count as intentional misleading that falls short of lying (and, as we have seen, differ from lies in not being restricted to linguistic acts). In poker, lying about specific cards one claims to hold (as opposed to merely acting through betting or facial expressions as if one's hand were stronger or weaker than it actually is) is not bluffing but "coffeehousing" or "table talk" and is usually frowned upon if not banned.

But elsewhere bluffs can also be lies. Jennifer Saul (p.c.) points out that if, on coming downstairs in my house, I confront a thief in flagrante and warn him that I have called the police (when I haven't actually had the opportunity to do so) and that they will be arriving in five minutes, what I say is a lie, but it is also a bluff – one that will be called by a skeptical thief who stays put.[17] Along the same lines, consider a scenario we will call PILATES, taken from a contemporary novel. A bit of background: Kit, the narrator, is a bakery manager having an affair with Matt, a carpenter renovating her shop. Whenever she wants to visit Matt she tells her husband David she's going to Pilates, which is a lie – but also a bluff. Kit counts on David not calling the bluff even though (as with SWISS BANK, GUERNICA, et al.) he has enough information to see through it or could pursue the matter with a bit of additional effort. But she counts on his not doing so. When David finally confronts her over her absences, Kit responds

---

16 Compare Kant's view, summarized by Macintyre (1995: 336–37), "that my duty is to assert only what is true and that the mistaken inferences that others may draw from what I say or what I do are, in some cases at least, not my responsibility, but theirs." Similarly, for the 19th century moral theologian Peter Scavini, as cited by Cardinal Newman (1864), equivocation is permissible because "then we do not deceive our neighbour, but allow him to deceive himself". See also Lackey 2013, Dynel 2018, and Chapter 1 on the ethical distinctions between active deception and withholding the whole truth.

17 The idle threats considered in §2 (scarecrows, fake "Armed Response" or "Beware of dog" signs) are bluffs but not lies because they do not involve assertions or promises.

with a whataboutism, while acknowledging to herself that she's not only lying to David but also running a bluff on him:[18]

PILATES

"But what *are* you doing? Just going to Pilates? Really? I have to be honest, it just doesn't seem true."

He was, I realized, daring me to say it. "You have a lot of nerve", I said instead. "To judge me for pursuing solitary interests when your job forces me to spend a lot of time alone. *You'd* rather work than be with me, not the other way around." The accusations came to me as though there were nothing else to say, nothing to admit or explain, when I still barely understood what Pilates was. **If David looked at our credit card statements, there'd be no charges for Pilates class. My lies were sloppy, relying on my sureness – unchallenged until now – that David didn't care enough to ask questions.**

– Liv Stratman (2020), *Cheat Day*, p. 246

Whether all is fair for Kit and David depends on who's doing the grading.

Why can lies function as bluffs in PILATES or in Saul's idle threat case but not at the poker table or between business negotiators or courtroom adversaries? The key is the role of social conventions in those spheres, allowing a bright line to be drawn in contexts defined by rational mutual distrust (competitive games, bargaining, adversarial legal proceedings and perjury law) whereby bluffs and other strategic misrepresentations are excused or even encouraged but explicit lies remain beyond the pale.

Bluffing involves a misrepresentation of one's physical or epistemic strength with the intention – or rather (to include those mantises and frogs) the goal – to deceive the target, while lying involves an insincere (believed-false) assertion or promise with or without the intention to deceive the addressee. Thus the two categories are both distinct and non-disjoint, practically and morally. Lying, from Augustine's and Kant's days to our own, is generally (although not invariably) judged as ethically wrong; bluffing presents a more complicated profile.

---

18 The PILATES scenario contrasts robustly with the case of bald-faced adultery-themed deceit I have cited elsewhere (Horn 2017a: 27) in which the "deceived" partner isn't actually deceived at all:

That he worked late she did not doubt, but she knew he did not sleep at his club, and he knew that she knew this . . . The regularity of his evening calls, however much she disbelieved them, was a comfort to them both . . . Even being lied to constantly, though hardly like love, was sustained attention; he must care about her to fabricate so elaborately.

– Ian McEwan (2001), *Atonement*, p. 139

Sullivan (1984: 4) depicts tactics like puffery and bluffing as occupying "an ethically gray area" and Guidice, Alder & Phelan (2009: 538) cite empirical findings that "suggest that bluffing is an ethically gray issue".[19] Gray areas emerge when there are black and white domains to surround them. For Aristotle (*Categories*, Chapter 10), the existence of shades of gray as intermediaries between white and black is a prime instance of polar contraries licensing an unexcluded middle. In the same way, the strategic bluff is recognized as neither right nor wrong, bespeaking the absence of any exhaustive opposition between simple moral contradictories. Whether we must admit fifty or more shades of gray will be left as an exercise for the reader.

References

Adler, Jonathan. 1997. Lying, deceiving, or falsely implicating. *Journal of Philosophy* 94: 435–52.

Allhoff, Fritz. 2003. Business bluffing reconsidered. *Journal of Business Ethics* 45: 283–289.

Baldwin, William. 1571. Treatice of morall philosophie: Contaynynge the sayinges of the wise, wherin you may see the woorthy and pithy sayings of philosophers, emperors, kynges, and oratours: of their liues, their aunswers, of what linage they came of, and of what countrey they were . . . London: In Fleetestrete within Temple barre, at the signe of the hand and starre, by Richard Tottyll.

Behn, Aphra. 1687. The emperor of the moon a farce: as it is acted by Their Majesties servants at the Queens Theatre. London: R. Holt. (Database: Early English Books Online.)

Beyer, Rick. 2015. Weapons of mass distraction. *Works that Work*, No. 6. https://worksthat work.com/6/ghost-army.

Brenkert, George. 2013. Bluffing. In H. LaFollette (ed), *International Encyclopedia of Ethics*, 587–9. Oxford: Blackwell.

de Brunhoff, Jean. 1934. *The Travels of Babar*. New York: Random House.

Carr, Albert. 1968. Is business bluffing ethical? *Harvard Business Review* 46: 143–153.

Carreyrou, John. 2018. *Bad Blood: Secrets and Lies in a Silicon Valley Startup*. New York: Knopf.

Carson, Thomas. 1993. Second thoughts about bluffing. *Business Ethics Quarterly* 3: 317–341.

Carson, Thomas. 2010. *Lying and Deception: Theory and Practice*. Oxford: Oxford University Press.

Cavaillé, Jean-Pierre. 2004. *Non est hic*: Le cas exemplaire de la protection du fugitif. http://dossiersgrihl.revues.org/300?lang=en.

Chisholm, Roderick and Thomas Feehan. 1977. The intent to deceive. *Journal of Philosophy* 74: 143–59.

von Clausewitz, Carl. 1832. *On War*. Translated by Michael Howard and Peter Paret. Princeton: Princeton University Press, 1989.

Colapinto, John. 2021. *This is the Voice*. New York: Simon & Schuster.

---

19 Other metaphors point to this intermediary status, e.g. in the title of Dees & Cramton (1991)'s paper on the ethics of bluffing in business negotiations, "Shrewd bargaining on the moral frontier".

Coleman, Linda and Paul Kay. 1981. Prototype semantics: The English word *lie*. *Language* 57: 26–44.

Darwin, Charles. 1886. *The Expression of the Emotions in Man and Animals*. New York: Appleton.

Dawkins, Richard and John Krebs. 1978. Animal signals: Information or manipulation. In J. Krebs and N. Davies (eds.), *Behavioural Ecology: An Evolutionary Approach*, 282–309. Oxford: Blackwell.

Dawkins, Richard and John Krebs. 1979. Arms races between and within species. *Proceedings of the Royal Society of London* 205: 489–511.

Dees, Gregory and Peter Crampton. 1991. Shrewd bargaining on the moral frontier: Toward a theory of morality in practice. *Business Ethics Quarterly* 1: 135–167.

Dowd, Maureen. 1998. The Wizard of Is. New York Times, 16 Sept. 1998. https://tinyurl.com/86ydm96k.

Dynel, Marta. 2020. To say the least: Where deceptively withholding information ends and lying begins. *Topics in Cognitive Science* 12: 555–582.

Fallis, Donald. 2009. What is lying? *Journal of Philosophy* 106: 29–56.

Fauconnier, Gilles. 1979. Comment contrôler la vérité. *Actes de la Recherche en Sciences Sociales* 25: 3–22.

France, Clemens. 1902. The gambling impulse. *American Journal of Psychology* 13: 364–407.

Frankfurt, Harry. 2005. *On Bullshit*. Princeton: Princeton University Press.

Friedman, Lawrence. 1971. Optimal bluffing strategies in poker. *Management Science* 17: B764–B771. https://doi.org/10.1287/mnsc.17.12.B764.

Grice, H. P. 1989. *Studies in the Way of Words*. Cambridge: Harvard University Press.

Guidice, Rebecca, G. Stoney Alder, and Steven Phelan. 2009. Competitive bluffing: an examination of a common practice and its relationship with performance. *Journal of Business Ethics* 87: 535–553.

Horn, Laurence. 1972. *On the Semantic Properties of Logical Operators in English*. UCLA dissertation. Distributed by Indiana University Linguistics Club, 1976.

Horn, Laurence. 1984. Toward a new taxonomy for pragmatic inference: Q-based and R-based implicature. In Deborah Schiffrin (ed.), *Meaning, Form, and Use in Context (GURT '84)*, 11–42. Washington: Georgetown University Press.

Horn, Laurence. 1989. *A Natural History of Negation*. Chicago: University of Chicago Press. (Reissue edition, Stanford: CSLI, 2001.)

Horn, Laurence. 2007. Neo-Gricean pragmatics: a Manichaean manifesto. In N. Burton-Roberts (ed.), *Pragmatics*, 158–83. Basingstoke: Palgrave.

Horn, Laurence. 2017a. Telling it slant: Toward a taxonomy of deception. In Dieter Stein and Janet Giltrow (eds.), *The Pragmatic Turn in Law*, 23–55. Berlin: De Gruyter.

Horn, Laurence. 2017b. What lies beyond: Untangling the web. In Rachel Giora and Michael Haugh (eds.), *Doing Pragmatics Interculturally*, 151–174. Berlin: De Gruyter.

Kassin, Saul. 2005. On the psychology of confessions: Does *innocence* put *innocents* at risk? *American Psychologist* 60: 215–228.

Lawrence, James. 2014. Lying, misrepresenting, puffing and bluffing: Legal, ethical and professional standards for negotiators and mediation advocates. *Ohio State Journal on Dispute Resolution* 29: 35–58.

Lewis, David. 1969. *Convention: A Philosophical Study*. Cambridge, MA: Harvard University Press.

MacIntyre, Alasdair. 1995. Truthfulness, lies, and moral philosophers: what can we learn from Mill and Kant? *The Tanner Lectures on Human Values* 16: 307–361.

Mahon, James. 2015. The definition of lying and deception. In E. Zalta (ed.), *Stanford Encyclopedia of Philosophy*. http://plato.stanford.edu/entries/lying-definition.

Markham, James. 1981. For Spain, 'Guernica' stirs memory and awe. *New York Times*, Nov. 2, 1981. http://tinyurl.com/omlwl8a.

Meibauer, Jörg. 2005. Lying and falsely implicating. *Journal of Pragmatics* 37: 1373–99.

Meibauer, Jörg. 2014. *Lying at the Semantics-Pragmatics Interface*. Berlin: de Gruyter.

Meibauer, Jörg. 2019. Bald-faced lies. In Meibauer (ed.), 252–263.

Meibauer, Jörg (ed.). 2019. *The Oxford Handbook of Lying*. Oxford: Oxford University Press.

Morton, Eugene. 1977. On the occurrence and significance of motivation-structural rules in some bird and mammal sounds. *American Naturalist* 3: 855–869.

Newman, John Henry. 1864. *Apologia pro vita sua*. http://www.newmanreader.org/works/apologia/.

Ohala, John. 1997. Sound symbolism. Proc. 4th Seoul International Conference on Linguistics [SICOL] 11–15 Aug 1997, 98–103. http://linguistics.berkeley.edu/~ohala/papers/SEOUL4-symbolism.pdf

Pascal, Blaise. 1656. *Les lettres provincielles*. Manchester: The University Press, 1920.

Perrillo, Jennifer. 2019. Bluffing. In J. Meibauer (ed.), 277–287.

Perrillo, Jennifer and Saul Kassin. 2011. Inside interrogation: The lie, the bluff, and false confession. *Law and Human Behavior* 35: 327–337.

de Reuck, Anthony and Julie Knight (eds.). 1966. *Conflict in Human Society*. Boston: Little, Brown.

Roussel de la Tour. 1762. *Extraits des assertions dangereuses et pernicieuses en tout genre, que les soi-disans jesuites ont, dans tous les tems & perseveramment, soutenues, enseignées & publiées dans leurs livres*. Paris: Pierre-Guillaume Simon.

Sartori, Anne. 2005. *Deterrence by Diplomacy*. Princeton: Princeton University Press.

Saul, Jennifer. 2012. *Lying, Misleading, and What is Said*. Oxford: Oxford University Press.

Schauer, Frederick and Richard Zeckhauser. 2007. Paltering. KSG Working Paper No. RWP07-006, Available at https://ssrn.com/abstract=832634 or http://dx.doi.org/10.2139/ssrn.832634.

Solan, Lawrence. 2012. Lawyers as insincere (but truthful) actors. *Journal of the Legal Profession* 36: 487–527.

Solan, Lawrence and Peter Tiersma. 2005. *Speaking of Crime: The Language of Criminal Justice*. Chicago: University of Chicago Press.

Sorensen, Roy. 2007. Bald-faced lies! Lying without the intent to deceive. *Pacific Philosophical Quarterly* 88: 251–264.

Steger, Rick and Roy Caldwell. 1983. Intraspecific deception by bluffing: a defense strategy of newly molted stomatopods (Arthropoda: Crustacea). *Science* 221: 558–560.

Stokke, Andreas. 2013. Lying and asserting. *Journal of Philosophy* 110: 33–60.

Stokke, Andreas. 2019. Bullshitting. In Meibauer (ed.), 264–277.

Strudler, Alan. 2005. Deception unraveled. *Journal of Philosophy* 102: 458–473.

Sullivan, Roger. 1984. A response to "Is business bluffing ethical?" *Journal of Business Ethics* 3: 1–18.

Tiersma, Peter. 2003. Did Clinton lie? Defining "sexual relations". 79 *Chi.-Kent L. Rev.* 927 (2004). Available at http://papers.ssrn.com/sol3/papers.cfm?abstract_id=470645.

Tiersma, Peter. 2005. The language of perjury (focusing on the Clinton impeachment). Available as download, http://www.languageandlaw.org/PERJURY.HTM.

Timmermann, Felix and Emanuel Viebahn. 2021. To lie or to mislead? *Philosophical Studies* 178: 1481–1501.

Trump, Donald, with Tony Schwartz. 1987. *The Art of the Deal*. New York: Random House.

Turri, John and Angelo Turri. 2015. The truth about lying. *Cognition* 138: 161–168.

Umbers, Kate, Jussi Lehtonen, and Johanna Mappes. 2015. Deimatic displays. *Current Biology* 25.2: R58.

Weissman, Benjamin and Marina Terkourafi. 2019. Are false implicatures lies? An experimental investigation. *Mind and Language* 34: 221–246.

Wickler, Wolfgang. 1968. *Mimicry in Plants and Animals*. New York: McGraw-Hill.

Wiegmann, Alex, Jana Samland, and Michael Waldmann. 2016. Lying despite telling the truth. *Cognition* 150: 37–42.

Elisabeth Camp
Just saying, just kidding: Liability for accountability-avoiding speech in ordinary conversation, politics and law

Abstract: Mobsters and others engaged in risky forms of social coordination and coercion often communicate by saying something that is overtly innocuous but transmits another message 'off record'. In both ordinary conversation and political discourse, insinuation and other forms of indirection, like joking, offer significant protection from liability. However, they do not confer blanket immunity: speakers can be held to account for an 'off record' message, if the only reasonable interpretations of their utterance involve a commitment to it. Legal liability for speech in the service of criminal behavior displays a similar profile of significant protection from indirection along with potential liability for reasonable interpretations. Specifically, in both ordinary and legal contexts, liability depends on how a reasonable speaker would expect a reasonable hearer to interpret their utterance in the context of utterance, rather than on the actual speaker's claimed communicative intentions.

> Never write if you can speak; never speak if you can nod; never nod if you can wink.[1]
> – attributed to Martin Lomasney, Boston ward boss

Indirect speech is often communicatively effective. And when it is, this is often not in spite of but precisely because it leaves its main point unstated. Among other things, relying on implicit interpretive assumptions enables speakers to communicate messages that are complex or imprecise without needing to articulate them explicitly. It also enables speakers to deny having meant messages that are risky.

---

[1] https://thewestendmuseum.org/the-life-legend-and-lessons-of-martin-lomasney-ward-boss-west-end-icon/.

Acknowledgement: Thanks to Larry Horn for the invitation to contribute, for substantive comments, and for longstanding inspiration. Thanks to David Cohen, Kim Roosevelt, and Rick Swedloff for conversation about law, and to Anhelina Mahdzyar for research into legal cases. Thanks to audiences at Oxford University, HaLO2 (Leibniz-Centre General Linguistics), the World Congress in Beijing, Colgate University, Columbia University, the University of Arkansas, Vassar College, Yale University, the Chapel Hill Colloquium, and the University of Pennsylvania for discussion at talks on which this essay is based.

Elisabeth Camp, Rutgers University, e-mail: elisabeth.camp@rutgers.edu

https://doi.org/10.1515/9783110733730-010

Indeed, such deniability is possible even when communication succeeds, so that all parties involved know that the speaker did mean what they deny having meant.

In §1, I offer a quick tour through some varieties of such accountability-avoiding speech, focusing especially on political discourse, and in particular on utterances by Donald Trump. In §2, I offer a brief theoretical characterization of insinuation and some of its accountability-avoiding cousins, and explain why they display the puzzling profile of deniability in the face of successful coordination. In ordinary and political discourse, indirection makes it more difficult to hold speakers accountable for what they meant. But it does not confer blanket immunity: speakers can sometimes be liable for messages they communicate indirectly. In §3, I argue that the same dynamic holds for legal liability, modulated to incorporate higher evidential standards and – especially in the United States – a concern for free speech.

1 Insinuation and its accountability-avoiding cousins

Agents who are interested in risky forms of social coordination and coercion often achieve those goals by saying something that is overtly innocuous but transmits another message 'off record'. Messages that are insinuated in this way can underwrite communication that is successful, at least in the sense that the speaker's target audience identifies the intended message, and often also in the sense that it produces the speaker's desired cognitive and practical effects. But even when communication succeeds, such a speaker preserves "plausible deniability" about their message, so that they can avoid accountability for the consequences that would have been entailed by a direct, explicit utterance of it. So, for instance, Henry II is famously reputed to have uttered something along the lines of

(1) Will no one rid me of this meddlesome priest?[2]

in 1170, as a veiled command for his knights to assassinate the Archbishop of Canterbury, Thomas à Becket. After Becket was brutally killed at the cathedral altar, Henry denied any intention to incite murder, and his defenders continue today to insist that the question was merely a rhetorical expression of exasperation.

---

[2] Although this is the most famous formulation, historical sources suggest that an alternate articulation may be more accurate: "What miserable drones and traitors have I nourished and brought up in my household, who let their lord be treated with such shameful contempt by a low-born cleric?" (Schama 2002, 142).

More recently, mobsters have taken up the mantle of intimidation by insinuation. Among early documented cases, *The Ludington (MI) Daily News* reported in 1926 that "Detroit Bandits Use Psychology in Bank Robbery. Pick Cashier Up on Street and Bring Him to Verge of Hysteria with Questions," where those questions included superficially innocuous inquiries about family life, such as

> (2) How are your children now? You think a lot of them, don't you? You have a nice little family, haven't you? Wouldn't it be a pity if anything happened to break it up?[3]

By the late 1960s, the trope "Nice X you've got here; it'd be a shame if something happened to it" had morphed from a locution of actual gangsters into a cultural meme. However, the technique of issuing directives via rhetorical questions and expressions of sentiment has persisted.

In particular, former President Donald Trump is often said to communicate "like a mobster." As Michael Cohen testified before the House Oversight Committee, Trump "doesn't give you questions. He doesn't give you orders. He speaks in a code."[4] One amply discussed example is Trump's utterance to then FBI Director James Comey of (3):

> (3) I hope you can see your way clear to letting this go, to letting Flynn go. He is a good guy. I hope you can let this go.

Asked during Senate Intelligence Committee hearings whether he took the President's utterance of (3) "as a directive," Comey agreed, adding that it "rings in my ears" as akin to Henry II's utterance of (1). In response to Comey's interpretative testimony, Senator James Risch and others denied that Trump had meant any such thing, insisting that he had merely expressed his personal feelings, something he'd also repeatedly done explicitly and in public.

While Trump's utterance of (3) was notably one-on-one, he has also regularly deployed insinuation in public speech. Thus, while campaigning in August 2016, Trump commented on Russia's 2014 annexation of Crimea,

> (4) This was taken during the administration of Barack *Hussein* Obama, OK?[5]

While it is not obvious precisely what motivation Trump is imputing to Obama with (4), his emphatic use of the President's middle name strongly suggests that Obama avoided a more forceful response to Russia's violation of international law because he was motivated by Islamist sympathies. While this imputation is wildly

---

3 https://www.barrypopik.com/index.php/new_york_city/entry/nice_place_you_got_here_ be_a_shame_if_anything_happened_to_it.

4 https://www.congress.gov/event/116th-congress/house-event/LC63709/text?s=1&r=64.

5 http://transcripts.cnn.com/TRANSCRIPTS/1608/11/ebo.01.html.

false, what Trump actually said is undeniably and uncontroversially true. Because any further interpretations are 'off record', the responsibility for identifying them is borne at least in part by the hearer, and can be disclaimed by Trump or his allies.

Examples (1) through (4) illustrate the characteristic structure of insinuation. A speaker S produces an utterance U of a sentence with an innocuous encoded message L, in order to communicate an unstated, risky message Q. Faced with the accusation of having meant Q, the speaker can demur that they were "just saying" the unobjectionable message L. When insinuation succeeds, all parties know, and know that the others know, that S did mean Q. Nonetheless, S's denial of having meant Q sticks: they avoid being held accountable for Q, and often thereby for its ensuing practical consequences.

This combination of effective transmission with plausible deniability makes insinuation a useful tool in the rogue's kit, or indeed for anyone navigating risky forms of social coordination. Sarcasm, figurative speech, and jokes offer distinct but related opportunities for accountability-avoiding communication. Thus, a speaker who says

(5) How about another small slice of pizza?

to someone who has conspicuously eaten more than their share (Kumon-Nakamura 1995, 4) can legitimately deny having asked or claimed anything at all, while still effectively rebuking the addressee for violating the norms of politeness that are evoked by their utterance (Camp 2012). Where the speakers in (1) through (4) do mean what they say, and can potentially be held to account for its (uninteresting) message L, a sarcastic speaker can disclaim having meant L, or indeed for having really claimed (or asked, or ordered) anything at all.

Jokes often function in a similar way to sarcasm. Consider the following exchange between President Richard Nixon and the formidable UPI reporter Helen Thomas at a 1973 press gaggle:

(6) NIXON: Helen, are you still wearing slacks? Do you prefer them actually? Every time 1 see girls in slacks it reminds me of China.

THOMAS: Chinese women are actually moving toward Western dress.

NIXON: This is not said in an uncomplimentary way, but slacks can do something for some people and some it can't. But I think you do very well. Turn around . . . . How does your husband like your wearing pants outfits?

THOMAS: He doesn't mind.

NIXON: Do they cost less than gowns?

THOMAS: No.

NIXON: Then change. [grinning, to roaring laughter][6]

The joking nature of Nixon's speech puts Thomas in a conversational bind. On the one hand, Nixon's speech clearly presupposes and enforces a set of potent gender norms. As such, it effectively conveys a message that would have been controversial if articulated explicitly, even in 1973. On the other hand, it would have been tactically unwise for Thomas to respond by directly challenging those norms or Nixon's embrace of them: doing so would simply have compounded the humiliation being enacted on her by revealing her to be a typically thin-skinned, literalistic female, incapable of "taking a joke."

With classic insinuation, the speaker can disavow having meant the risky message *Q* and fall back on the claim that they were "just saying" the innocuous encoded content *L*. With sarcasm and jokes, the speaker can likewise deny having meant *Q*; but they can also insist they were "just kidding" about *L*. Metaphor and hyperbole offer a slightly different profile of deniability. A speaker of metaphor or hyperbole cannot deny having claimed (or asked, or ordered) *something*, but they typically retain at least some wiggle room about just what that claim was (Camp 2006, 2017a). So, for example, erstwhile Trump advisor Steve Bannon was widely condemned for saying (7) on his podcast in November 2020:

> (7) Second term kicks off with firing [FBI Director] Wray, firing [NIAID Director] Fauci . . . No, I actually want to go a step farther but the president [Trump] is a kind-hearted man and a good man. I'd actually like to go back to the old times of Tudor England. I'd put their heads on pikes, right, I'd put them at the two corners of the White House as a warning to federal bureaucrats, you either get with the program or you're gone.

In response to those condemnations, Bannon's spokeswoman insisted that "Mr. Bannon did not, would not and has never called for violence of any kind," on the ground that his comments were "clearly meant metaphorically" and alluded "for rhetorical purposes" to a comment from the day before about Thomas More's trial for treason in Tudor England.[7]

Rudy Giuliani employed a similar disclaimer in response to accusations of having incited violence at the January 6, 2021 rally preceding the Capitol attack by uttering (8):

> (8) If they ran such a clean election, they'd have you come in and look at the paper ballots. Who hides evidence? Criminals hide evidence. Not honest people. Over the next 10 days, we get to see the machines that are crooked, the ballots that are fraudulent, and if

---

6 *The Evening Bulletin* [Philadelphia] (1973); cited in Goffman (1979), 1–2.

7 https://www.reuters.com/article/us-usa-election-facebook-idUSKBN27S35P.

we're wrong, we will be made fools of. But if we're right, a lot of them will go to jail. Let's have trial by combat.

Speaking to a reporter the following day, Giuliani claimed that his utterance was a reference to the HBO series *Game of Thrones*, and "the kind of trial that took place for Tyrion in that very famous documentary about fictitious medieval England."[8] Later, in a motion to dismiss a lawsuit against him for incitement to insurrection, Giuliani's lawyers declared,

> (9) The statement is clearly hyperbolic and not literal and, even if it were to be perceived literally, Giuliani is clearly referring to an event in the future after evidence of alleged Election fraud is collected. No one could reasonably perceive the "trial by combat" reference as one inciting the listeners to an immediate violent attack on the Capitol, which could have nothing to do with Giuliani's allegorical "trial by combat" over evidence of fraud in the Election.[9]

In both cases, representatives for the original speakers deny, plausibly enough, that the utterances were intended literally. Further, they use this denial to disclaim any responsibility for inciting violence – even though it is also highly plausible that those utterances contributed significantly to a violent public attempt to overthrow the duly elected President of the United States.

In paradigmatic cases of insinuation, the speaker says and asserts U's encoded message L, but this is merely a means for communicating their main message Q. In this sense, while they do commit to L, they don't really care about it. Something similar obtains with paradigmatic cases of sarcasm, jokes, and figurative speech, with the difference that the speaker avoids even minimal commitment to L. But often with accountability-avoiding speech, L's role is less instrumental and its status as meant more unsettled. For instance, while Nixon's final utterance in (6) is unquestionably a joke, it also communicates genuine endorsement of its literally encoded message, in an elegant illustration of what Al Franken (2004) calls "kidding on the square." That is, Nixon is *not* actually "just kidding" – he also really means L, even if he could use its status as a joke to get away with claiming that he doesn't.

Speakers also sometimes avoid accountability for L by claiming that their utterance was joking or non-literal when it is ambiguous or doubtful whether this is so. Thus, at the same August 2016 rally at which he uttered (4), Trump also said (10):

8 https://deadline.com/2021/01/rudy-giuliani-game-of-thrones-trial-by-combat-capitol-violence-1234673891/.

9 https://storage.courtlistener.com/recap/gov.uscourts.dcd.228356/gov.uscourts.dcd.228356.13.1_1.pdf.

(10) President Obama, he is the founder of ISIS. He is the founder in a true sense.

When invited to clarify whether he really only meant that Obama "created the vacuum" that helped ISIS take power, Trump repeatedly insisted that he'd meant what he actually said, saying "No, I meant he's the founder of ISIS. I do." Facing criticism for (10)'s obvious falsity, however, he pivoted to claiming that his utterance had been sarcastic, for instance tweeting:

(11) Ratings challenged @CNN reports so seriously that I call President Obama (and Clinton) 'the founder' of ISIS, & MVP. THEY DON'T GET SARCASM?[10]

A similar dynamic circulated around his July 2016 utterance of (12):

(12) Russia, if you're listening, I hope you're able to find the 30,000 emails that are missing. I think you will probably be rewarded mightily by our press. Let's see if that happens.

Although Russian operatives took (12) seriously enough to begin hacking Hillary Clinton's email, Trump demurred, saying "Of course I'm being sarcastic," while Giuliani and Newt Gingrich said that he was "telling a joke." Eventually, by 2019, Trump was claiming that the media had willfully distorted their representations of his original utterance:

(13) Remember this thing, 'Russia, if you're listening'?' Remember, it was a big thing, in front of 25,000 people. 'Russia if you're . . . ' It was all said in a joke. They cut it off right at the end so that you don't then see the laughter, the joke. And they said, 'He asked. He asked for help.'

Trump's claims in (13) about his utterance of (12) are verifiably false: (12) occurred during a small press conference and occasioned no laughter, only an inquiry about whether he seriously condoned foreign interference in US elections.[11] However, this falsity is not something the audience of (13) can immediately determine, nor is it something they are likely to be motivated to believe. Further, bracketing these claims about (12)'s objective context, Trump's claims about his original communicative intentions cannot be directly falsified, because they concern the black box of his inner psychology.

In (11) and (13), Trump disclaims responsibility for what he actually said in (10) and (12) by insisting that he was sarcastic or joking. A speaker can also take the opposite tack: of avoiding accountability for a message Q that results from

---

10 https://www.cnn.com/2016/08/12/politics/donald-trump-obama-clinton-isis-founder-sarcasm/index.html.

11 https://www.politifact.com/factchecks/2020/mar/02/donald-trump/donald-trump-rewrites-history-about-his-russia-if-/.

interpreting U as sarcastic, joking, or figurative, by insisting that they "just" meant U literally and sincerely. So, for instance, in his February 2019 State of the Union address, Trump uttered

(14) We must reject the politics of revenge, resistance and retribution and embrace the boundless potential of cooperation, compromise and the common good.

On the dais behind him, House Speaker Nancy Pelosi clapped alongside her Republican colleagues, but in a way that was conspicuously slow, with outstretched arms and a pointed gaze.[12] This is widely taken to be sarcastic, and hence to communicate something like scorn for Trump's hypocrisy in uttering (14). In defense, Pelosi denied being insincere:

(15) It wasn't sarcastic. Look at what I was applauding. I wanted him to know that was a very welcome message.

While Pelosi's claim to endorsing (14)'s encoded message might well be sincere, her denial of sarcasm, and her denial of having meant that Trump was contemptibly insincere in uttering (14), are dubious. Indeed, in a remarkable display of filial disloyalty, Pelosi's daughter Christine clapped back at her mother's clapback, while offering an elegant articulation of the sort of mutual knowledge that can be achieved through implicit meaning:

(16) Oh yes that clap took me back to the teen years. She knows. And she knows that you know. And frankly she's disappointed that you thought this would work. But here's a clap. #youtriedit.[13]

In some of these cases of joking, sarcasm, and figurative speech, the utterance U's encoded content L is risky, while in others it is innocuous. In some of these cases, the speaker claims that it is determinately true (or false) that they meant (only) L, where this claim about their meaning is probably or certainly a lie. In other cases, it is genuinely ambiguous or unsettled whether they meant L or were just kidding; indeed, they may not be entirely clear on the matter themselves. All of these variations in the relationship between L and Q, and in what a speaker is prepared to admit having committed to with U, make a difference to the specific contours of liability and responsibility that speakers undertake in conversation. These differences matter; and other articles have (and will) detail these variations in loving detail. For current purposes, though, what matters

---

12 https://www.youtube.com/watch?v=gaq4tV72tlk.

13 https://www.usatoday.com/story/news/politics/2019/02/07/nancy-pelosi-says-clapping-trump-state-union-wasnt-sarcastic/2806817002/.

is just that all these varieties of accountability-avoiding speech exemplify the more general structure characteristic of paradigmatic insinuation: they all afford some sort of interpretive deniability in virtue of some form of (putative) indirection, which enables speakers to disavow responsibility for something that they clearly did mean, even when all parties involved know that they did mean it.

This structure is *prima facie* puzzling. First, at a general level, insinuation is a form of strategic communication: speech that doesn't just superficially appear to, but actually does flout the principles of cooperation that have long been treated as fundamental to communication (Asher and Lascarides 2013, Camp 2018a). But what could determine the contents of non-encoded meaning in situations where cooperative principles have been abrogated? Second, it is unclear why the speaker should manage to get away with such a denial, given that, as Christine Pelosi says, everyone involved knows what's really going on. Can such denials really be taken seriously? And if so, what are their limits?

I have focused on cases of political discourse, especially recent cases by Trump and his allies, because they are familiar, clearly exemplify the relevant phenomenon, and hold obvious import for civic life. These cases push the boundaries of deniability (as well as other boundaries). In that sense they are atypical. Further, there are important systematic differences between speech directed at individuals and at general audiences; and political discourse operates under a distinctive profile of incentives and costs. Nonetheless, the communicative strategies employed in these political cases are fully continuous with those that govern accountability-avoiding speech in ordinary conversations; and they have provoked similar responses to those offered by ordinary hearers in response to those more ordinary conversational cases. Thus, in both their continuity with and difference from the ordinary, such political cases illuminate the structure and limits of accountability-avoiding communication in general.

2 How insinuation works

In this section, I'll use the term 'insinuation' in a broad sense, to refer to utterances with the basic structure outlined above: an utterance U of a sentence with literal meaning L communicates a distinct message, Q, implicitly and in a way that preserves deniability about having meant Q or L or both. In addition to illuminating the dimensions of variations in the relationships between L and Q outlined in §1, (1) through (14) also display other dimensions of variation. First, both L and Q can vary in *illocutionary force*: (1) and (2) encode questions and communicate directives; (3) encodes an assertion (or an expressive) and communicates a directive; and (4) and (10) encode assertions and communicate (false) information. Second, Q can vary in *determinacy*: Trump's 'ask' in (3) is quite specific, while the bandits' threat in (2) and Trump's innuendo in (4) are quite vague.

Third, U can vary in the *publicity* of its occasion of utterance and the *specificity* of its target audience: Trump's conversation with Comey was conspicuously private, while his criticism of Obama occurred at a political rally. Subsequent denials D of an insinuated meaning Q can also vary in context and audience, and in their relationship to U: they can occur in public or private, and at the time of the initial utterance or long after; they can be addressed to the original target or a third party; and they can be issued by the original speaker, the original target, or a third party, who may or may not have been present at U.

Finally, while most of the examples in §1 are nefarious in some way, insinuations need not be malicious, oppressive, or even manipulative (Camp 2018a). Speakers also regularly advert to insinuation for praiseworthy, potentially cooperative purposes: for instance, to express romantic interest or float the possibility of a promotion or raise; or to criticize a partner's performance of a household task or to back away from a burgeoning romance. In such cases, the primary motivations for indirectness are more likely to center around politeness, and sparing the speaker's and/or hearer's feelings and social status (Goffman 1979, Brown and Levinson 1987).

All of these differences affect what kinds of deniability are feasible. But the core phenomenon remains: insinuation affords speakers a significant measure of protection from accountability for an implicit message even when all parties know that the message has been successfully identified, and when they all know that it has achieved its intended cognitive and practical effects. This is the puzzling profile of successful coordination with plausible deniability that we need to explain.

Sometimes the gap between what is actually said and what is meant is absurdly narrow, as when evolutionary psychologists refer to "the Four F's" of "fighting, fleeing, feeding and mating," or when Trump said of Megyn Kelly's role as moderator during the Republican debate,

> (17) She gets out there and she starts asking me all sorts of ridiculous questions, and you could see there was blood coming out of her eyes, blood coming out of her . . . wherever.[14]

---

14 https://www.cnn.com/videos/us/2015/08/08/donald-trump-megyn-kelly-blood-lemon-intv -ctn.cnn.

And yet: even in these cases, refraining from actually tokening an expression makes a genuine difference, and can afford a measure of deniability. Thus, Trump resisted calls to apologize for (17) by pointing out that he didn't actually *say* the imputed content, and insisting that he never intended to evoke it and was bewildered why anyone would think he had:

(18) If I had said that, it would have been inappropriate . . . I didn't even finish the answer, because I wanted to get on the next point. If I finished it, I was going to say ears or nose, because that's just a common statement – blood is pouring out of your ears.[15]

(19) I cherish women . . . I said nothing wrong whatsoever. Do you think I would make a stupid statement like that? You almost have to be sick to put that together. Only a deviant would think anything else.[16]

As these and many of the other examples from §1 illustrate, meaning denials are often quite strained; as Pinker et al. (2008, 836) put it, "[a]ny 'deniability' in these cases is really not so plausible after all."[17] Indeed, in many cases, those denials amount to bald-faced lies, and repeated denials constitute a form of gaslighting. Nevertheless: in ordinary conversation and political discourse we regularly allow speakers to get away with them, and fail to hold them directly responsible for their insinuated message. We impose forceful sanctions on deceptions enacted through lying, for instance, that we do not for misleading, even when the misleading message is highly obvious and the practical and moral effects are arguably on a par (Saul 2012).

How is deniability in the face of manifestly successful communication possible, despite being implausible? The basic reason is that the process of identifying the implicit message Q on the basis of U requires appealing to interpretive presuppositions I that are implicit, nuanced, and local, in the sense of not being uncontroversially and ubiquitously shared. (Camp 2018a). These sorts of contextually-laden interpretive factors are endemic to what Fricker (2012, 89) aptly calls the "dodgy epistemics" of pragmatic interpretation. The competence of ordinary adult speakers to discern and attune to these factors, in an intuitive,

---

15 *Today*, August 10 2015.
16 *CNN State of the Union*, August 8 2015; "Donald J. Trump Statement on RedState Gathering," August 8 2015.
17 In this respect, insinuation tracks other instances of plausible deniability. The term originated in a National Security Council Paper issued by President Harry Truman in 1948 (NSC 10/ 2) defining covert operations as those "which are so planned and executed that any US Government responsibility for them is not evident to unauthorized persons and that if uncovered the US Government can plausibly disclaim any responsibility for them."

flexible way in real time, is essential to the flow of conversation; and our deployment of these interpersonal, social, and cultural competencies is so ubiquitous, fundamental, and automatic that we often fail to notice it. However, when they are diminished or absent, as in conversations with computers, children, or adults with markedly divergent neurological or cultural profiles, conversation becomes dramatically less efficient and more effortful.

A skillful insinuator exploits this general competency for attuning to implicit, nuanced, context-local interpretive factors to craft an utterance whose message is simultaneously intuitively comprehensible and also deniable. For the sake of achieving deniability, it is especially useful to exploit assumptions that are amorphous or interpretively perspectival, as in (4) or (6) (Camp 2017a). It is also useful to exploit features of U beyond the actual words uttered, such as uninformativeness through repetition, as in (2); the use of marked expressions rather than more common default alternatives, as in (10); unusual prosody, including intonation, focal stress, and/or pauses, as in (17); and facial expressions and bodily gestures, as in (14). Speaking "with a wink and a nod" by exploiting non-verbal and contrastive aspects of U signals to the hearer that there is an implicit message to be discerned and points toward a structure of relevant interpretive alternatives, but in a way that cannot easily be extracted from the original conversational context via testimonial report.

If this explains why an insinuated message Q is deniable, when does denial work? Fricker (2012) claims that a speaker "can be nailed as having stated that P; [but] never as having insinuated that P." "Given [the] complex epistemics" required to identify Q given L, she says, "it is not epistemically feasible to pin undeniable specific commitment onto a speaker: she can always wriggle out of it." In particular, Fricker argues that claims about insinuated meaning turn on claims about "private intentions," and that lies about such attitudes "may be suspected, but cannot be refuted" (2012, 87–9).

I disagree: while a competent insinuator retains at least some wiggle room as to what they meant by U, some denials are out of bounds, and some designedly indirect messages can be pinned on a speaker. More specifically, I think, a speaker retains "plausible deniability" when U admits of at least one alternative interpretation Q' which cannot be ruled out as a candidate meaning for U. For Q' to be an admissible alternative to Q, it must be *reasonable* to calculate Q' on the basis of the conversation thus far, the uttered sentence's encoded meaning L, and some alternative set of interpretive presuppositions I', such that having meant Q' would render U at least minimally cooperative as a contribution to the conversation (Camp 2018a).

In effect, the meaning denier "plays to a virtual audience" (Goffman 1967, cited by Lee and Pinker 2010, 7896), by pretending that U was performed in an alternative conversational context C' that would have generated Q'. Meaning denials are highly frustrating, because their hearers typically know that they have indeed identified I as the maximally relevant set of presuppositions, and they know that the speaker also knows this. But the implicit, nuanced, local nature of those presuppositions makes it difficult to prove that I is maximally relevant. In a fully cooperative conversation, the mutual obviousness of the fact that the presuppositions in I are operative would suffice to establish Q as U's meaning, and to render I' and Q' irrelevant. But the insinuator, and in turn the meaning denier, are not being fully cooperative: they act strategically, by refusing to overtly acknowledge what is in fact mutually obvious, at least among participants in the actual conversational context C.

In order for their communicative strategy to succeed, then, the insinuator must walk a fine line. On the one hand, they must produce a signal that is robust enough to induce confidence in their hearer that they meant Q – and often, to motivate their hearer to act on that basis. To accomplish this, they must rely on presuppositions I that are in fact maximally accessible, and whose maximal accessibility is mutually obvious. But on the other hand, they must also produce a signal that is veiled enough to preserve deniability. And to accomplish this, they need to construct U so that I' is also accessible, albeit not maximally so. Moreover, in many cases, the insinuator also expects and intends their audience to identify both of the presupposition sets I and I', and for it to be mutually obvious both that I is in fact more accessible than I', and that S is prepared to pretend that I' is more accessible than I. This is a delicate operation; but it is one that skilled insinuators manage to perform regularly. When they succeed, insinuations amount to a kind of conversational *jiu jitsu*, shifting a significant portion of the interpretive responsibility for Q away from the speaker and onto the hearer.

Faced with an insinuation, how should a hearer respond? Once the conversational hot potato of an insinuation has landed in their hands, they must do something with it; but all of their options are problematic. The most straightforward option is to allow the conversation to evolve on the basis of the assumption that the speaker meant Q. But doing so accommodates that assumption, so that it, and often Q itself, become part of the common ground (Stalnaker 1978, Lewis 1979). And this may not be something the hearer is happy to do.

In some cases, such as (4), (6) or (7), this may be because the hearer finds Q itself objectionable. However, they cannot straightforwardly object to Q, because the normal explicit mechanisms of rejection, like "That's not true" or "I refuse," will target the focal, at-issue content of L instead. They can actively block Q by redirecting the conversation with something like, "Hey wait a minute! Are you saying that Q? But that assumes I, which is false!" (von Fintel 2004, Langton 2018).

However, such a response risks playing into the insinuator's hands, in at least two ways. First, Q still ends up introduced into the explicit conversation, but now by the hearer and in a way that enables the insinuator to issue a demurral to having meant Q, as in "You said it, not me" or "Only a deviant would think that." Second and more fundamentally, any response that engages with Q, even without explicitly mentioning it, thereby confirms Q's accessibility, and in turn the reasonableness of entertaining its supporting presuppositions I. Especially in cases like (4), (7) or (14), where I evokes an intuitive interpretive perspective and/or visceral imagery, the mere fact of getting the hearer to entertain, and to be seen to entertain, these presuppositions can engender unwelcome cognitive and conversational complicity (Camp 2013, 2017a).

In other cases, such as (1) or (3), the hearer may be willing to go along with Q but want to confirm the speaker's commitment to it. But here too, normal forms of explicit agreement, like "I agree" or "Sure!", will target L's focal, at-issue content. More generally, any form of response that engages Q directly, such as "So the Archbishop should be eliminated?," transfers responsibility for Q onto the hearer. Worse, in cases where an insinuated proposal's success hinges on its being covert, a direct response purchases communicative clarity at the cost of practical success.

Given these risks, it is often strategically wiser for the hearer to avoid catching the conversational hot potato altogether. This can be done in various ways, with different risks and payoffs. One option, likely to be attractive to recalcitrant hearers, is to engage in *flat-footed pedantry* (Camp 2006, 2007, 2018a). Just as the meaning-denying insinuator can feign ignorance of the mutually obvious fact that they meant Q, so too can the hearer focus narrowly on the encoded content L, perhaps while requesting clarification about its conversational relevance. An alternative option, likely to be attractive to compliant hearers, is to engage in *reciprocal insinuation*. Just as the insinuating speaker manifests their communicative intentions and attempts to advance their goals without undertaking a risky explicit commitment, so too can the hearer respond in a way that makes their comprehension of Q manifest while preserving their ability to deny having done so. Repeatedly iterated reciprocal insinuation can end up constituting a "shadow conversation" of increasingly robust implicit coordination (Camp 2018a, 57) – or in more tragicomic cases, of increasingly sustained misinterpretation.

In this way, insinuations, meaning denials, and responses can constitute a complex dance of strategic interactions erected on an edifice of implicit, nuanced, context-local assumptions, which may be reliably coordinated, even mutually obvious, without ever being acknowledged as such. The discussion of (3) during Comey's testimony before the Senate Intelligence Committee provides an especially rich and elegant illustration of the resulting conversational opportunities and liabilities.[18] Here is one revealing exchange, between Comey and Senator James Risch:

(20) RISCH: I want to drill right down, as my time is limited, to the most recent dust-up regarding allegations that the president of the United States obstructed justice . . . You put this in quotes – words matter. You wrote down the words so we can all have the words in front of us now. There's 28 words there that are in quotes, and it says, quote, "I hope" – this is the president speaking – "I hope you can see your way clear to letting this go, to letting Flynn go. He is a good guy. I hope you can let this go."

COMEY: Correct.

RISCH: And you wrote them here, and you put them in quotes?

COMEY: Correct.

RISCH: Thank you for that. He did not direct you to let it go.

COMEY: Not in his words, no.

RISCH: He did not order you to let it go.

COMEY: Again, those words are not an order.

RISCH: He said, "I hope." Now, like me, you probably did hundreds of cases, maybe thousands of cases charging people with criminal offenses. And, of course, you have knowledge of the thousands of cases out there that – where people have been charged. Do you know of any case where a person has been charged for obstruction of justice or, for that matter, any other criminal offense, where this – they said, or thought, they hoped for an outcome?

COMEY: I don't know well enough to answer. And the reason I keep saying "his words" is I took it as a direction.

RISCH: Right.

COMEY: I mean, this is the President of the United States, with me alone, saying, "I hope" this. I took it as, this is what he wants me to do. Now I – I didn't obey that, but that's the way I took it.

RISCH: You – you may have taken it as a direction, but that's not what he said.

In questioning Comey, Risch insistently focuses on the gap between Trump's words and Comey's interpretation, and thereby insinuates – without saying outright – that Comey's interpretation is ill-founded. Other commentators offered alternative minimalist interpretations of the pivotal Trump-Comey exchange:

---

18 https://www.nytimes.com/2017/06/08/us/politics/senate-hearing-transcript.html?_r=2. My discussion here of Comey's testimony extends Camp (2017b).

Senator Roy Blunt paraphrased it as "So he said, "He's a good guy." You said, "He's a good guy." And that was it"; while Senator James Lankford commented, "If this seems to be something the president's trying to get you to drop it, this seems like a pretty light touch." Governor Chris Christie explained the conversation in terms of broader cultural differences: "What you're seeing is a president who is now very publicly learning about the way people react to what he considers to be normal New York City conversation."[19]

In response to these skeptical questions and reinterpretations, Comey in (20) justifies his interpretation by appealing to precisely the sorts of interpretive capacities, contextual factors, and hearer responses identified above. He invokes "a gut feeling," based on "a lot of conversations with humans over the years," as applied to "the circumstances – that I was alone, the subject matter, and the nature of the person that I was interacting with and my read of that person." He cites non-verbal gestures (and non-gestures) as features of the context: "I didn't move, speak, or change my facial expression in any way during the awkward silence that followed. We simply looked at each other in silence." And he reports his own insinuating response: "I remember saying, I agree [Flynn] is a good guy, as a way of saying, 'I'm not agreeing with what you asked me to do.'"

At the same time, Comey also consistently avoids putting himself on the testimonial hook for any further assessment of (3). He demurs that "it's not for me to say whether the conversation I had with the president was an effort to obstruct"; that it's not his job to "sit here and try and interpret the president's tweets." Instead, he "just says" how he in fact took (3) at the time, and offers at least the appearance of letting the facts 'speak for themselves' – where these 'facts' include that of a methodical, experienced hearer interpreting Trump's utterance as a directive to perform an action of at least questionable legality. He thereby provides his audience with the evidential ammunition they need to justify the conclusion that Trump obstructed justice, without assuming responsibility for that conclusion himself.

3 Insinuation under the law

In his questioning of Comey, Risch doesn't merely insinuate that Comey misinterpreted Trump; he also insinuates that insinuated messages are not subject to criminal liability. If so, this would provide concrete empirical support for Fricker's

---

19 https://www.cnbc.com/2017/06/08/chris-christie-trumps-pressure-on-comey-was-normal-new-york-city-conversation.html.

claim that a speaker "can be nailed as having stated that P; [but] never as having insinuated that P," because the "dodgy epistemics" of interpretation hinge on "private intentions," where lies about such attitudes "may be suspected, but cannot be refuted" (2012, 87–9).

Fricker's contention is significantly stronger than my central claim, which is that insinuation makes it harder to hold speakers to account for unstated messages but does not confer blanket immunity. For one thing, Fricker posits a universal, qualitative difference[20] between explicit statement and implicit implicature[20], which I reject. On the one hand, many of the same contextual interpretive factors operative in insinuation are also present in strategic communication exploiting lexically encoded context-sensitivity, as in indexicals like 'here' and 'now', or the scope for tense (Camp 2018a, 60): (recall Bill Clinton's infamous defense that he didn't lie in testifying that "There is no sexual relationship," because "It depends on what the meaning of 'is' is" (Saul 2000, 2012; see also Chapters 1, 8, 14). And on the other, it is not obvious that all indirect utterances are deniability-affording insinuations: in fully cooperative conversations, for instance, an indirect response to a question can entail commitments in a way that cannot be denied, and that arguably constitutes assertion (Soames 2008).

More importantly, even restricting attention to insinuations in the sense of deniable indirect messages, I want to reject both the claim that a speaker can never be nailed for meaning Q, and that meaning is constituted by inaccessible private intentions. In §2, I said that deniability is possible when it would be *reasonable* to attribute a minimally cooperative meaning Q' to U on the basis of L plus an accessible set of interpretive presuppositions I'. This affords a skillful insinuator some wiggle room. But by the same token, if the only reasonable interpretation of U is one on which it means Q, or if any of the interpretations in the range of reasonable meanings $\{Q_1, Q_2, Q_3 \ldots\}$ would also entail the same objectionable consequence as Q, then a speaker can be held responsible for it.

This is a high epistemic bar to clear. Moreover, in ordinary conversations and in political discussions, it is often not exactly clear what "holding responsible" amounts to, and so hearers often roll their eyes and let the interpretation of U go, even if they also hold the speaker to account for a broader pattern of behavior to which U contributes.

By contrast, the law is a system for levying specific consequences for specific actions. And especially in the United States, it relies on an argumentative, analytic process grounded in ordinary intuitions about responsibility. This makes it a valuable case study for analyzing the workings of speech as a form

---

20 Cf. also Lee and Pinker (2010) and Asher and Lascarides (2013).

of action, and specifically of insinuation. In this section, I argue that legal accountability for insinuation displays the same basic contours as ordinary and political discourse, but implemented with greater precision and a heightened standard. As in ordinary speech, insinuation affords speakers significant protection from liability, but not blanket immunity. The law recognizes that interpretation turns on context-local interpretive factors and on communicative intentions and responses. But those intentions and responses are not treated as ineluctably private; rather, they are filtered through the attitudes that a reasonable, suitably informed audience would attribute to the speaker, and/or that a reasonable speaker would anticipate such an audience to have, in that context.

Legal systems formalize liability for speech in different ways. A key initial distinction concerns what kinds of messages can be regulated by law. English common law and its descendants distinguish between speech that expresses an idea and speech that performs an action – between what J. L. Austin (1962) called *constatives* and *performatives* – and take the former to merit protection from legal regulation in a way the latter does not. However, it is also generally recognized that utterances that are declarative on their face can constitute performatives in virtue of the circumstances and consequences of their utterance. As J. S. Mill (1859/1985, 119) wrote,

> No one pretends that actions should be as free as opinions. On the contrary, even opinions lose their immunity when the circumstances in which they are expressed are such as to constitute their expression a positive instigation to some mischievous act. An opinion that corn dealers are starvers of the poor, or that private property is robbery, ought to be unmolested when simply circulated through the press, but may justly incur punishment when delivered orally to an excited mob assembled before the house of a corn dealer, or when handed about among the same mob in the form of a placard.[21]

In the United States, the First Amendment protects speech that expresses opinions, even when those opinions are highly noxious and are expressed in a manner that foreseeably causes, and is intended to cause, significant harm.[22] However, the mere fact of expressing an opinion still does not automatically confer protection, if the utterance's primary purpose is to perform an action that is subject to legal regulation. The difficult, and theoretically interesting, question is how to treat

---

21 The Corn Laws were repealed in the face of widespread opposition in 1846, not long before Mill completed *On Liberty* (Jaconelli 2018, 259).

22 For instance, in *Snyder v. Phelps* (2011), the Court held that Westboro Baptist Church members could not be held liable for invasion of privacy and emotional and physical distress produced by picketing the funeral of a military veteran while displaying signs and chanting messages like "Thank God for 9/11" and "Semper Fi Fags," because those activities occurred in a public venue and "addressed matters of public import," like the presence of homosexuals in the military.

utterances that *prima facie* do express ideas while simultaneously performing speech acts that do not warrant First Amendment protection.

The most thoroughly analyzed cases of this involve threats; other types of unprotected speech include obscenity, defamation, fraud, and "speech integral to criminal conduct" (*United States v. Stevens*, 2010). Very roughly, the Supreme Court has established three (rough, contested, and overlapping) categories of threats.[23] First, *fighting words* "by their very utterance inflict injury or tend to incite an immediate breach of the peace," where this category is fixed by appealing to "what men of common intelligence would understand would be words likely to cause an average addressee to fight."[24] Second, *incitement* is advocacy "directed to inciting or producing imminent lawless action and likely to incite or produce such action," where "mere advocacy" using violent rhetoric is protected unless it poses a "clear and present danger" of immediate, serious lawless action or violence.[25] Third, *true threats* are "statements where the speaker means to communicate a serious expression of an intent to commit an act of unlawful violence to a particular individual or group of individuals," where the speaker need not have intended to actually commit the violent act, only to "plac[e] the victim in fear of bodily harm or death."[26]

All three categories focus on the imminence, probability, and specificity of the danger created by a threatening utterance. This might seem to suggest that the Court is implicitly employing a causal model of threats. On this sort of model, threatening words would encode threats in such a way that the target audience's very hearing of those words either predictably causes in them a visceral psychological state that is inherently harmful, such as intense fear; or

---

23 For helpful discussion of threats and the First Amendment, see Karst (2006), Carnley (2014), and Pew (2015).

24 *Chaplinsky v. New Hampshire* (1942). The Court assumed without argument that "the appellations 'damned racketeer' and 'damned Fascist' are epithets likely to provoke the average person to retaliation."

25 *Brandenburg v. Ohio* (1969). The defendant, a member of the Ku Klux Klan, said at a rally, "If our President, our Congress, our Supreme Court, continues to suppress the white, Caucasian race, it's possible that there might have to be some revengeance [sic] taken." Echoing and amplifying Mill, the Court overturned conviction on the ground that the resulting risk was not imminent: "[As we have stated], 'the mere abstract teaching . . . of the moral propriety or even moral necessity for a resort to force and violence, is not the same as preparing a group for violent action and steeling it to such action.' A statute which fails to draw this distinction impermissibly intrudes upon the freedoms guaranteed by the First and Fourteenth Amendments. It sweeps within its condemnation speech which our Constitution has immunized from governmental control."

26 *Virginia v. Black* (2003).

else (or thereby) predictably causes them to act in a way that poses a "clear and present danger" – much as "natural tendency and reasonably probable effect" of someone's shouting fire in a theatre is to cause mayhem.[27] Conversely, one might then think, in cases where the connection between utterance and effect is indirect – say, when the utterance expresses an abstract proposition representing the speaker's opinions, hopes, or fears – then any downstream effect in the hearer must result in significant measure from their own powers and processes of reasoning, and hence should not be something for which the speaker is liable. And in this case, it would seem to follow that insinuating utterances like (1) through (3) would be immune from legal liability.

A causal model along these lines does appear to have dominated the Court's early thinking about incitement (Pew 2015), and it does currently govern the regulation of obscenity.[28] In other areas, however, and especially with respect to threats, the Court has employed a much more nuanced treatment of the relationship among utterance, meaning, and effect.

The clearest cases in support of Risch's contention that "words matter," in his narrow insinuated sense that speakers can only be held legally liable for what they actually utter, occur within the courtroom itself. Thus, the Court has held that a witness who engages in "lie by negative implication," by responding to a question in a way that is literally true but not maximally relevant, does not commit perjury, "so long as the witness speaks the literal truth."[29] However, the Court's grounds for this conclusion turned on how the circumstances of the courtroom, and specifically its formal, antagonistic structure, affect how it would be reasonable for interlocutors to act. That is, under the "pressures and tensions of interrogation," in contrast to "casual conversation," it is not surprising for witnesses to answer in ways that are not "entirely responsive"; given this, the burden falls on attorneys, who are highly trained and financially

---

27 *Schenck v. United States* (1919).

28 Thus, the Court has upheld the government's right to regulate even isolated tokenings of "indecent" terms, including mere mentionings (as in George Carlin's comedy routine "Filthy Words") and used as expletives (*FCC v. Pacifica Foundation*, 1978; also *FCC v. Fox Television Stations, Inc.*, 2009). Such speech is not protected because it is both "low-value" – expressing no substantive ideas or opinions – and inherently harmful. Because the damage inflicted is immediate and involuntary, "[t]o say that one may avoid further offense by turning off the radio . . . is like saying that the remedy for an assault is to run away after the first blow." Further, because broadcast media are so ubiquitous, the Court held that it would be unduly burdensome for audiences to protect themselves – and crucially, their children – from such "blows" in the absence of governmental regulation.

29 *Bronston v. United States* (1973). For discussion, see e.g. Solan and Tiersma (2005), Asher and Lascarides (2013), and Robbins (2019), as well as Chapters 1, 8, and 14.

compensated, to "recognize the evasion" and "bring the witness back to the mark," to clarify exactly what they are committing themselves to. The limitation of liability to literally encoded speech also extends to other courtroom roles. So, for instance, the Court held that a prosecutor who merely insinuated that the defendant had perjured himself, by saying things like

> (21) He gets to sit here and listen to the testimony of all the other witnesses before he testifies . . . That gives [him] a big advantage, doesn't it?

did not infringe on the defendant's right to due process.[30] Conversely, legal liability has also been held to apply to literal courtroom speech even when the speaker plausibly did not mean to commit to its encoded content; for instance, a judge was disqualified for an utterance he claimed was a joke.[31]

So courtroom speech provides some support for Risch's and Fricker's focus on explicit, literal content. By contrast, the standard governing legal liability for speech that occurs outside of the courtroom is much less literalistic, and much closer to Comey's characterization. First, a speaker S is typically *not* liable for U's encoded message L if a reasonable hearer would take S not to have meant U literally and sincerely in the context of utterance C. Thus, in the decision defining 'true threats', the Court overturned a draft protester's conviction for threatening the President in saying (22):

> (22) If they ever make me carry a rifle the first man I want to get in my sights is L.B.J.

on the ground that he was engaging in "political hyperbole." And they justified this analysis by citing context-specific features: that (22) is a conditional, that it occurred at a political rally, and that it elicited laughter.[32] Likewise, in a decision excluding defamation from First Amendment protection, the Court noted that

---

30 *Portuondo v. Agard* (2000).

31 *Brofman v. Florida Hearing Care Center., Inc.* (Florida District Appellate Court, 1997). Unfortunately, the decision is silent as to the specific joke uttered: "We see no need to reproduce the text of the joke here. While the joke was not particularly offensive to race or religion, it was not particularly funny either." In rendering its judgment, the Appellate Court invoked the 'reasonable hearer' standard: "While the trial judge may have meant the remark to be a joke, rather than a reflection on his belief as to the merits of the petitioner's complaint, the standard is the reasonable effect on the party seeking disqualification, not the subjective intent of the judge."

32 *Watts v. United States* (1969). Similarly, in *NAACP v. Claiborne Hardware* (1982), the Court held that when the NAACP Mississippi Field Secretary, Charles Evers, said at a rally "If we catch any of you going in any of them racist stores, we're gonna break your damn neck," he engaged in speech that was "impassioned" and "emotionally charged," but rhetorical, and so did not rise to the level of incitement – again citing the contextual circumstances and audience reception to justify their interpretation.

"imaginative expression" is integral to robust civic discourse, and that "statements that cannot 'reasonably [be] interpreted as stating actual facts' about an individual" should not be treated as defamatory.[33] More recently, the use of a ':P' emoticon to represent a face with its tongue sticking out, was judged to "make it patently clear" that the commenter was making a joke, and so that "a reasonable reader could not view the statement as defamatory."[34] At the same time, however, a speaker's mere claim to have meant U in a non-literal or non-serious way does not confer automatic immunity. Thus, as early as 1894, an appellate court held that "Where a person's conduct and conversation is such as to warrant a reasonably prudent man to believe that he is in earnest, he will not be permitted to say . . . that he was 'codding' and joking."[35] Rather, a legitimate claim of "just kidding," or of otherwise not meaning L by uttering U, requires the speaker to demonstrate that a reasonable person would have understood U as a joke, and that it was in fact understood in this way.[36]

Second, and conversely, a speaker *can* be held liable for meaning something other than L, when it can be demonstrated that a reasonable person in those circumstances would have interpreted U to have meant Q, and that the actual hearer did interpret U as meaning Q. More specifically, the Court has found that speech can constitute a "true threat" even when its *prima facie* encoded message would be protected, and even when the speaker disclaims having intended any implicit message. Thus, anti-abortion activists were held liable for having threatened doctors who provided abortions, on the ground that they had distributed posters with titles like "Nuremberg Files" and "Guilty of Crimes Against Humanity" listing those doctors' names and addresses, accompanied by the legend "Black font (working); Greyed-out Name (wounded); Strikethrough (fatality)." In holding the activists liable, the Court relied heavily on contextual factors to justify their interpretation of those speech acts. Among other such factors, one of the defendants had written a book, *A Time to Kill*, articulating Biblical justifications for murdering abortion providers; the previous three years had seen multiple murders of abortion providers; and "the poster format itself had acquired currency as a death threat."

---

33 *Milkovich v. Lorain Journal Co.* (1989). Earlier, in *Greenbelt Cooperative Publishing Association v. Bresler* (1970), the Court had held that a newspaper's quoted reports of a real estate developer's negotiations with public officials over a land sale as "blackmail" did not count as libel, because "even the most careless reader must have perceived that the word was no more than rhetorical hyperbole, a vigorous epithet used by those who considered [his] negotiating position extremely unreasonable."

34 *Ghanam v. Does* (Michigan Court of Appeals, 2014).

35 *McKinzie v. Stretch* (Illinois Appellate Court, 1894); *Ragansky v. United States* (1918).

36 E.g. *People v. Barron* (Illinois Appellate Court, 2004).

Although each one of the elements of the poster individually might count as protected speech, the Court concluded that given the totality of these circumstances, "No one putting [these names] on a 'wanted'-type poster . . . could possibly believe anything other than that each would be seriously worried about being next in line to be shot and killed."[37] Similarly, animal rights activists who maintained a website listing "Top Twenty Terror Tactics" along with information about employees for an animal testing company were held liable for making "true threats" despite having issued only veiled exhortations and tongue-in-cheek disclaimers like

(23) [We are] excited to see such an upswing in action against Huntingdon and their cohorts. From the unsolicited direct action to the phone calls, e-mails, faxes and protests. Keep up the good work!

(24) Now don't go getting any funny ideas! . . . We operate within the boundaries of the law, but recognize and support those who choose to operate outside the confines of the legal system.[38]

Given the state's vested interest in preventing violence and the fear of violence, it is not surprising that the Court would carve out an exception to First Amendment protection for veiled threats. Otherwise, the Court noted, insinuation would effectively constitute a literal "get out of jail free" card: "[R]igid adherence to the literal meaning of a communication without regard to its reasonable connotations derived from its ambience would render the statute powerless against the ingenuity of threateners who can instill in the victim's mind as clear an apprehension of impending injury by an implied menace as by a literal threat."[39]

Beyond threats, legal liability for indirect meaning also extends to other types of communication. For instance, in a case strikingly similar to Trump's utterance in (3), the National Labor Relations Board found that a supervisor's mere statement of personal feeling, in (25),

(25) Well I hope you won't continue to be an agitator or antagonize the people in the newsroom.

---

37 *Planned Parenthood* v. *American Coalition of Life Activists* (2003). Earlier, *Madsen* v. *Women's Health Clinic* (1994) held that "threats to patients or their families, however communicated, are proscribable under the First Amendment." So, for instance, *United States* v. *McMillan* (United States District Court, S D Mississippi, 1999) upheld the conviction of a protester who repeatedly stated "Where's a pipe-bomber when you need him?" every time he saw a particular Mississippi abortion provider. See Cohen and Connon (2015), esp. ch. 9, for discussion of threats to abortion providers and legal responses to those threats.
38 *United States* v. *Fullmer* (2009).
39 *United States* v. *Malik* (United States Court of Appeals, 2nd Circuit, 1994).

constituted unlawful coercion because it had a "chilling effect" and "interfered with [the employee's] exercise of rights" to advocate for improved working conditions and wages. Here too, the Board relied on the words and the context in justifying its finding, noting that (25) was uttered during a required meeting, alone in the supervisor's office, and in lieu of a requested meeting with the employee committee.[40] Liability for bribery follows the same pattern: establishing liability requires demonstrating that the parties achieved a "meeting of the minds" on a "clear and unambiguous" *quid pro quo* exchange of professional benefit for personal reward, rather than just a "vague expectation of future benefit." However, the parties' agreement on those terms may be entirely implicit; and in determining whether it was achieved, the Court may "consider both direct and circumstantial evidence, including the context in which a conversation took place."[41]

Thus far in this section, we've seen that legal liability for speech occurring outside the courtroom tracks ordinary standards for accountability. Speakers are liable for what their utterances mean, as distinct from what their uttered words mean. And utterance meaning is determined by a reasonable audience's interpretation and/or a reasonable speaker's expectation of a reasonable audience's interpretation given the particular circumstances of utterance, rather than by the speaker's private communicative intentions.[42] However, legal liability requires more than establishing just that the reasonable, contextually-grounded message conveyed by an utterance would be liable. It also requires demonstrating an actual harm. Moreover, it arguably requires demonstrating culpability or "*mens rea*": a speaker's knowledge that their action will produce that harm. In the remainder of this section, I offer a bit more detail about what the law considers to be a "reasonable interpreter," what contextual factors are relevant to their judgments, and what role is played by the actual addressee's responses and the actual speaker's intentions in establishing liability.

It might seem anodyne to hold, as the Court once did, that "written words or phrases take their character as threatening or harmless from the context in which they are used, measured by the common experience of the society in which they are published."[43] The problem is that employing "the common experience of

---

40 *KNTV, Inc. and American Federation of Television and Radio Artists, AFL-CIO.* Case 32-CA -12732 (1995). For discussion, see Liberman (2017).

41 *United States v. Tomblin* (1995). *United States v. Ring* (2013) likewise upheld conviction for bribery under the public sector honest-services fraud statute for merely implicit agreement, considering both direct and broader circumstantial evidence.

42 Different courts have focused to different degrees on the responses of a reasonable hearer and on the effects foreseen by a reasonable speaker. See Crane (2006), Karst (2006), Principe (2012), and Pew (2015) for discussion of speaker-centered and hearer-centered versions of the 'objective' reasonableness standard for threats.

43 *United States v. Prochaska* (1955).

society" as the measure of reasonable interpretation would ignore exactly the sorts of context-local factors that competent insinuators deploy to communicate with deniability. At the same time, at the opposite extreme, if we include *all* contextual factors within the relevant context of interpretation, including all of the speaker's and hearer's possibly idiosyncratic assumptions and preferences, then we undermine the notion of a reasonable interpretation as distinct from actual intention and reception. The question, then, is which context-local factors should be abstracted away from, and which should be admitted as interpretively relevant.

Generally, courts accept evidence concerning a broad range of "objective" facts about the circumstances of utterance. This may include publicly accessible information about other related events, as in the case of the "Nuremberg Files" posters. (In a similar vein, a protester's parking two Ryder trucks in front of two abortion clinics was deemed a true threat, given that Timothy McVeigh had recently used a Ryder truck to bomb a federal building in Oklahoma City.[44]) It may also include information about highly specific, non-public past interactions between speaker and addressee. For instance, a defendant's writing (26)

(26) You are my most desired goal, and I will Stop at nothing to reach you.

to a judge was held to constitute a true threat, given such contextual factors as his having met her only once, at a hearing; his having written her more than sixty letters over seven years, some including poems describing rape; his phoning her home and traveling to try to meet her; and his violating an agreement to avoid contacting her in exchange for dropping harassment charges.[45]

By contrast, "subjective" facts are typically not admissible in establishing what constitutes a reasonable interpretation, but instead only in demonstrating the occurrence of an actual harm or an actual intent. Thus, courts have consistently rejected the admissibility of expert testimony about the speaker's mental state and predicted actions. For instance, in determining whether a series of letters containing sentences like

(27) 17 little Angels Murdered by Beast Blythe and his 666 Molesters . . . . William Jefferson Blythe 3rd, Mr. buzzard's feast, WANTED For MURDER, DEAD OR ALIVE.

constituted threats against President Clinton, the Court held that testimony from Secret Service agents had rightly been excluded, because admitting such evidence would pose "a significant danger of misleading the jury into believing that

---

44 *United States v. Hart* (8th Circuit 2000); cited by Cohen and Connon (2015), 260.
45 *United States. v. Whitfield* (8th Circuit, 1994); discussed in Karst (2006).

it should judge [the] letters from the perspective of a highly trained Secret Service agent instead of from the perspective of an average, reasonable person."[46]

The implicit assumption governing these standards for admissibility appears to be that any relevant objective facts can be entered directly as evidence, and that admitting experts' interpretive judgments inappropriately infringes on the jurors' appointed role as "reasonable interpreters." However, the distinction between objective facts and subjective interpretation is not always clear. It becomes especially murky when determining how a reasonable audience would respond to U, especially in cases where the response itself constitutes the potentially liable harm, as with true threats. In many cases, as we've seen, the "totality of circumstances" relevant to determining whether it would be reasonable to feel fear upon hearing U is both quite rich and highly specific, and involves something more like an open-ended intuitive perspective than a fixed set of assumptions and preferences.

For these reasons, in their discussion of threats against abortion providers like the "Nuremberg Files" posters, Cohen and Cerrone (2015, 264) propose an alternative "reasonable abortion provider" standard, on the ground that only someone who shares the "collective memory" of violence against abortion providers is in an epistemic position to judge the reasonableness of experiencing fear in response to those acts. Their proposal extends a strand of theorizing about workplace discrimination to the criminal domain. The standards governing workplace discrimination differ significantly from those for threats, both because the liability is civil rather than criminal, and because the accusation is typically that an employing company knowingly permitted a pervasively hostile environment rather than that a specific employee made one or a few harassing utterances. Nonetheless, liability for speech in workplace discrimination manifests the same basic profile of responsibility as we have found for ordinary speech and criminal law. It is neither necessary nor sufficient that speakers utter words that are inherently offensive.[47] Rather, what matters is the "objective severity of harassment," as

---

46 More generally, "without additional assistance, the average layperson is qualified to determine what a 'reasonable person' would foresee under the circumstances," *United States v. Hanna* (1990).

47 Thus, "mere utterance of an . . . epithet which engenders offensive feelings in an employee does not sufficiently affect the conditions of employment to implicate Title VII," *Meritor Savings Bank v. Vinson* (1986). Rather, "[t]he real social impact of workplace behavior often depends on a constellation of surrounding circumstances, expectations, and relationships which are not fully captured by a simple recitation of the words used or the physical acts performed" *Oncale v. Sundowner Offshore Services, Inc.* (1998). And "mere jokes" have a social impact that constitutes harassment, even unaccompanied by more 'direct' forms of harassment such as slurs, graffiti, or images *Swinton v. Potomac Corporation* (9th Circuit, 2001). For discussion of racial jokes in workplace harassment, including application to the "reasonable person" standard, see Hughes (2015).

determined "from the perspective of a reasonable person in the plaintiff's position," where this "requires careful consideration of the social context in which particular behavior occurs and is experienced."[48]

As with the criminal cases the question is how to determine "the perspective of a reasonable person in the plaintiff's position." Early definitions of workplace harassment relied on the norms operative at the particular company being investigated, on the ground that in jobs where "humor and language are rough hewn and vulgar" and "[s]exual jokes, sexual conversations and girlie magazines abound," it is not reasonable to be offended by those things.[49] But this created a Catch-22: the very evidence required to demonstrate the existence of a pervasive pattern of hostile, abusive behavior also thereby undermined the reasonableness of being offended by it. Moreover, it is plausible that not just the male employees at that company, but many 'regular guys' more generally, would not feel offended by (or even notice) behaviors which it would be reasonable for a woman to feel fear in response to or to interpret as harassing, simply because they have not experienced being raped, groped, catcalled, repeatedly propositioned, and/or belittled, and hence lack the attunement to those behaviors that typifies many women's experiences and perspectives.

Given such a systematic gender-based perspectival disparity, some courts and theorists have advocated an alternative, "reasonable woman" standard for workplace harassment.[50] However, most courts and legal theorists have been reluctant to adopt a gender-based standard of reasonableness, either because they take it to presuppose an objectionable essentialism, on which women as such are different from and more delicate than men;[51] or because it sets up a slippery slope that ultimately transmutes the objective, universally accessible standard of reasonableness into a subjective, individualistic one.[52]

---

48 *Oncale v. Sundowner Offshore Services, Inc.* (1998).

49 *Rabidue v. Osceola Refining* (1986).

50 *Ellison v. Brady* (9th Circuit, 1991).

51 This echoes earlier rulings that it is not reasonable to regulate behavior that is rude but culturally ubiquitous. Thus, *Swentek v. USAIR, Inc.* (4th Circuit, 1987) argued that "The workplace is not a Victorian parlor, and the courts are not the arbiters of etiquette." Employing such a standard would attempt to protect women from everyday insults "as if they remained models of Victorian reticence," *DeAngelis v. El Paso Municipal Police Officers Association* (5th Circuit, 1995).

52 An alternative interpretation of the "reasonable woman" proposal retains a univocal standard grounded in the perspective of a reasonable person, but construes the proposal as an imperative for the person rendering the assessment to imagine those systematic gender-based experiences as part of the "totality of circumstances" that constitute the context, analogously to the way in which, in *Roberts v. State* (Louisiana Appellate Court, 1981), the court appealed

These challenges in defining what constitutes a reasonable interpretation of and response to a pervasive pattern of behavior or a specific action are both substantive and complex. They arise whenever we decide whether to hold someone accountable for their actions, whether legally or informally. They are especially endemic to the interpretation of speech, especially in cases where speakers have a vested interest in avoiding accountability, and especially when the accountability in question is legal. Nonetheless, they are challenges that interpreters, including judges and juries, must – and actually do – address.

The final component in establishing legal liability for speech returns from the utterance's "objective," "reasonable" meaning to the psychological states of the parties involved. Here again the law employs a heightened and precisified version of standards implicit in ordinary practice. In general, a speaker is potentially liable for meanings and responses that any reasonable person would have attributed to U. The vicissitudes of contextual interpretation entail that this typically covers a range of possible meanings $\{Q_1, Q_2, Q_3 \ldots \}$ and responses $\{R_1, R_2, R_3 \ldots \}$. In some cases of ordinary speech, it is sufficient for liability that a reasonable audience *would* have interpreted and responded to U in one of these ways, even if nobody actually did respond in those ways, and even if the speaker didn't intend anyone to so respond. For instance, we may justly criticize someone who uses a slur like 'midget' even if they are ignorant of its offensive connotations and it provokes no anguish or stereotype threat in its hearer (Camp 2016, 2018b); similarly for a misgendering use of a pronoun, or a culturally insensitive joke. However, in ordinary life, actual effects also matter, as do actual intentions: intentional infliction of harm is worse than damage caused via neglect, and both are worse than a lucky miss. Likewise in law, liability requires actual harm, and culpability requires knowing or at least anticipating that the harm will ensue.

On the side of the harm done, we've seen that it is not necessary that the audience actually suffer physical, financial, or other damage in order for an utterance to count as a liable threat, because the fear of such damage is itself inherently harmful and engenders further harmful effects. Indeed, for workplace discrimination, it suffices that the victim reasonably "subjectively perceives" the work environment as abusive, because this perception itself makes it harder for them to perform their duties, even if does not "seriously affect [their] psychological well-being" in other ways.[53]

---

to reasonable behavior for a blind person in deciding what counted as negligence. For discussion of the 'reasonable woman' standard and the balance between "objective" and "subjective" standards for reasonableness, see Treger (1993) and Peterson (1999).

53 The reasoning here about psychological effects echoes that for the Catch-22 about company-specific norms above. A requirement of actual trauma would turn psychological resilience into a

On the side of the speaker's awareness of that harm, we've seen that the dominant principle has been that it suffices to establish an utterance as a true threat that a reasonable person in the speaker's position would foresee that the audience would perceive the utterance as a serious expression of intent to harm.[54] However, it should be noted that a minority strand of more recent decisions apply the general criterion of *mens rea* for criminal culpability to speech, in order to require that a speaker specifically intend harm,[55] or at least that they "kn[e]w the threatening nature of [their] communication"[56] and produced it with "reckless disregard" for its likely effect. Thus, the specific relevance of actual speaker intent to liability for threatening speech is currently somewhat unsettled. However, none of the candidate criteria for establishing liability follow Fricker in accepting that a speaker's claims to communicative intent are definitive and unchallengeable. Rather, a speaker must provide publicly accessible evidence about their intent or lack thereof – for instance, by arguing that they were drunk at the time they made the utterance.[57] So, for instance, returning to Trump's utterance of (3), the fact that he cleared the room prior to speaking to Comey constitutes relevant circumstantial evidence that he knowingly intended to pressure Comey to drop the Flynn investigation and induce fear of retribution. While other sorts of evidence might also be relevant to establishing Trump's intent in uttering (3), Trump's claims about his private intentions do not automatically trump them, as Fricker maintains.

4 Conclusion

Insinuation enables speakers to effectively navigate risky waters of social coordination and coercion while preserving deniability about their communicative intentions: they can insist that they were "just saying" the message literally encoded by the sentence they uttered, while demurring about having intended their main message. Other forms of indirect speech, including sarcasm, jokes, hyperbole, and metaphor, offer distinct but related profiles of deniability: the speaker can insist that they were "just kidding" or "speaking rhetorically," including in cases where they really did mean what they said.

---

liability, because someone who managed to continue working at the company without being incapacitated would risk undermining their claim to harm (*Harris v. Forklift Systems, Inc.*, 1993).
54 *United States v. Hanna* (1991).
55 *Virginia v. Black* (2003).
56 *Elonis v. United States* (2015).
57 *United States v. Bagdasarian* (9th Circuit, 2011).

The puzzling question is why hearers accord speakers "plausible deniability" for those risky messages even when all parties involved know what the speaker really meant, on any independently plausible sense of 'know'. The reason is that indirect speech trades on the "dodgy epistemics" of interpretation, and specifically on assumptions that are implicit, nuanced, and context-local. This affords a crafty insinuator significant wiggle room. Especially in the context of political discourse, and especially during the candidacy and presidency of Donald Trump, it can indeed seem as if such speakers can "never be nailed" – that they can get away with even the most outlandish denials and alternative interpretations.

However, at least in the cases of ordinary conversation and of law, insinuation and related forms of accountability-avoiding speech do not confer blanket immunity. Speakers can be held liable for their unstated messages when any reasonable, suitably informed audience would have interpreted them as meaning something within the range of liable messages. "Words matter," as Senator Risch says, because they constitute the initial common denominator from which interpretation proceeds (Camp 2006, 2016). But an utterance's meaning also depends, as Comey says, on "the circumstances, the subject matter and the person" making the utterance.

The law is a system designed to hold actors to account by imposing specific, enforceable consequences when certain substantive and evidential thresholds have been met. In private interactions and politics, it is typically much less clear what "holding to account" amounts to, or how it should be levied. Further, at least in private interactions, the decision whether to pursue liability, and how far, involves performing a complex calculus on strategic costs and benefits. Should a hearer accommodate an objectionable insinuating message, push back in equally insinuating terms, or opt out of the conversation, and perhaps the relationship, altogether?

The same issues beset political discourse, which also brings along its own distinctive set of motivations, challenges and rewards. Who is the target audience? What difference does it make how non-target audiences respond? What sorts of consequences can effectively be levied on speakers who willfully attempt to avoid accountability; and what longer-term effects will those consequences have? Further, much of our discourse, in ordinary life and especially in politics, is not centered on disseminating information, but is rather directed at arousing emotion and exercising power (Stanley 2015). To the extent that political utterances function not merely, or not at all, to convey specific messages, forms of direct and indirect response aimed at effecting coordination on a common ground of information and plans risk being irrelevant and ineffectual.

In particular, to the extent that political utterances function primarily to construct and enforce partisan tribal identities, the question of what assumptions and preferences should be relevant for determining "the perspective of a reasonable person in the [hearer's] position" becomes much more unsettled, even unstable. So does the question of how to respond to utterances from across the perspectival divide, and what institutional structures and conversational techniques might foster genuine civic coordination across social groups. Unfortunately, these are pressing questions for another day.

References

Asher, Nicholas and Alex Lascaride. 2013. Strategic Conversation. *Semantics & Pragmatics* 6. 1–62.

Austin, J. L. 1962. *How to Do Things with Words*. Oxford: Oxford University Press.

Brown, Penelope and Stephen Levinson. 1987. *Politeness*. Cambridge: Cambridge University Press.

Camp, Elisabeth. 2006. Contextualism, Metaphor, and What is Said. *Mind and Language* 21:3. 280–309.

Camp, Elisabeth. 2007. Prudent Semantics Meets Wanton Speech Act Pluralism, *Context-Sensitivity and Semantic Minimalism: New Essays on Semantics and Pragmatics*, ed. G. Preyer and G. Peter. Oxford University Press, 2007, 194–213.

Camp, Elisabeth. 2012. Sarcasm, Pretense, and the Semantics/Pragmatics Distinction. *Noûs* 46:4. 587–634.

Camp, Elisabeth. 2013. Slurring Perspectives. *Analytic Philosophy* 54:3, 330–349.

Camp, Elisabeth. 2016. Conventions' Revenge: Davidson, Derangement, and Dormativity. *Inquiry* 59:1. 113–138.

Camp, Elisabeth. 2017a. Why Metaphors Make Good Insults: Perspectives, Presupposition, and Pragmatics. *Philosophical Studies* 174:1. 47–64.

Camp, Elisabeth. 2017b. Sounds of Silence, or Are We Responsible for What We Avoid Saying? *Blog of the APA* (online), 6/20/2017.

Camp, Elisabeth. 2018a. Insinuation, Common Ground, and the Conversational Record, *New Work on Speech Acts*, edited by Daniel Fogal, Daniel Harris, and Matt Moss. Oxford: Oxford University Press, 40–66.

Camp, Elisabeth. 2018b. Slurs as Dual-Act Expressions, in *Bad Words*, ed. D. Sosa. Oxford University Press, 2018, 29–59.

Carnley, Kara. 2014. *Harm and the First Amendment: Evolving Standards For "Proving" Speech-Based Injuries In U.S. Supreme Court Opinions*. PhD dissertation, University of Florida.

Cohen, David and Krysten Connon. 2015. *Living in the Crosshairs: The Untold Stories of Anti-Abortion Terrorism*. Oxford: Oxford University Press.

Crane, Paul. 2006. 'True Threats' and the Issue of Intent. *Virginia Law Review* 92. 1225–1277.

Franken, Al. 2004. *Lies (and the Lying Liars Who Tell Them)*. New York: Dutton/Penguin Press.

Fricker, Elizabeth. 2012. Stating and Insinuating, *Proceedings of the Aristotelian Society*, S86:1. 61–94.

Goffman, Erving. 1967. On Face Work, in *Interaction Ritual: Essays in Face-to-Face Behavior*. Chicago: Aldine, 5–45.

Goffman, Erving. 1979. Footing. *Semiotica* 25: 1/2. 1–29.

Hughes, Melissa. 2015. Through the Looking Glass: Racial Jokes, Social Context, and the Reasonable Person in Hostile Work Environment Analysis. *Southern California Law Review* 76. 1437–1482.

Jaconelli, Joseph. 2018. Incitement: A Study in Language Crime. *Criminal Law and Philosophy* 12. 245–265.

Karst, Kenneth. 2006. Threats and Meanings: How the Facts Govern First Amendment Doctrine. *Stanford Law Review* 58:5. 1337–1412.

Kumon-Nakamura, Sachi, Sam Glucksberg, and Mary Brown. 1995. How About Another Piece of Pie? The Allusional Pretense Theory of Discourse Irony. *Journal of Experimental Psychology: General* 124:1. 3–21.

Lee, James and Steven Pinker. 2010. Rationales for indirect speech: the theory of the strategic speaker. *Psychological Review* 117. 785–807.

Langton, Rae. 2018. Blocking as Counter-Speech, *New Work on Speech Acts*, edited by Daniel Fogal, Daniel Harris, and Matt Moss. Oxford: Oxford University Press, 144–164.

Lewis, David. 1979. Scorekeeping in a language game, *Journal of Philosophical Logic* 8. 339–359.

Liberman, Mark. 2017. Coercive Hopes. *Language Log*, 6/10/2017.

Mill, John Stuart. 1859/1985. *On Liberty*. London: Penguin Classics.

Peterson, Linda. 1999. The Reasonableness of the Reasonable Woman Standard. *Public Affairs Quarterly* 13:2. 141–158.

Pew, Bradley. 2015. How to Incite Crime with Words: Clarifying Brandenburg's Incitement Test With Speech Act Theory. *Brigham Young University Law Review* 2015:4. 1087–1114.

Pinker, Steven, Martin Nowak and James Lee. 2008. The Logic of Indirect Speech. *Proceedings of the National Academy of Sciences* 105(3). 833–838.

Principe, Craig. 2012. What Were They Thinking?: Competing Culpability Standards For Punishing Threats Made to the President. *American University Criminal Law Brief* 7:2. 39–54.

Robbins, Ira. 2019. Perjury by Omission. *Washington University Law Review* 97. 265–294.

Saul, Jennifer. 2000. Did Clinton Say Something False? *Analysis* 60:267. 219–295.

Saul, Jennifer. 2012. *Lying, Misleading, and the Role of What is Said*. Oxford: Oxford University Press.

Schama, Simon. 2002. *A History of Britain: At the Edge of the World?: 3000 BC–AD 1603*. London: BBC Books.

Soames, Scott. 2008. The Gap Between Meaning and Assertion: Why What We Literally Say Often Differs from What Our Words Literally Mean, *Philosophical Papers* Vol. 1: Natural Language: What It Means and How We Use It. Princeton: Princeton University Press, 278–297.

Solan, Lawrence and Peter Tiersma. 2005. *Speaking of Crime: The Language of Criminal Justice*. Chicago, IL: University of Chicago Press.

Stalnaker, Robert. 1978. Assertion. *Syntax and Semantics* 9. 315–332. New York Academic Press.

Stanley, Jason. 2015. *How Propaganda Works*. Princeton: Princeton University Press.

Treger, Tracy. 1993. The Reasonable Woman? Unreasonable!! Ellison v. Brady. *Whittier Law Review* 14:3. 675–694.

von Fintel, Kai. 2004. Would you Believe it? The King of France is Back! (Presuppositions and Truth-Value Intuitions), in *Descriptions and Beyond*, ed. M. Reimer and A. Bezuidenhout. Oxford: Oxford University Press, 315–341.

IV Crossing the perjury threshold: Deceit and falsehood in the courtroom

Roger W. Shuy
Perjury cases and the linguist

Abstract: Other chapters in this book deal with research and theory about lying, ethics, deception, bluffing, puffing, trickery, deceit, and morality in public and private life. This chapter describes the U.S. statutes concerning perjury and provides examples of how linguistic analyses were used with the language evidence in four representative perjury cases.

Linguists retained to examine the language evidence in perjury cases must first accommodate their analyses to the boundaries of perjury law. For this reason, United States statutes for perjury are first discussed. This is followed by a description of the important roles played by prosecutors, judges, and jurors who evaluate the language evidence. Finally, four perjury cases describe the way linguistic analysis has been used when a linguist was retained by either the defense or the prosecution.

1 U.S. perjury law

At common law, perjury has been described as "the willful giving, under oath, in a judicial proceeding or course of justice, of false testimony material to the issue or the point of inquiry" (J. Bishop, 2 *Commentaries on the Criminal Law* § 860, 1858). The "course of justice" refers to a proceeding, such as a creditor's examination of a debtor or during a grand jury inquiry.

Other general understandings of perjury are specified in state and federal statutes and in the Court's explanations of previous cases. The federal statutory definition in 18 United States Code Section §1621 declares:

> Whoever – having taken an oath before a competent tribunal, officer, or person, in any case in which a law of the United States authorizes an oath to be administered, that he will testify, declare, depose, or certify truly, or that any written testimony, declaration, deposition, or certificate by him subscribed, is true, willfully and contrary to such oath states or subscribes any material matter which he does not believe to be true; or in any declaration, certificate, verification, or statement under penalty of perjury as permitted under section 1746 of title 28, United States Code, willfully subscribes as true any material matter which he does not believe to be true: is guilty of perjury and shall, except as otherwise expressly provided by law, be fined under this title or imprisoned not more than five years, or both. This section is applicable whether the statement or subscription is made within or without the United States.

Roger W. Shuy, Georgetown University, e-mail: rshuy@montana.com

https://doi.org/10.1515/9783110733730-011

Linguists will notice that Section §1621 says, "does not *believe* to be true" rather than "*knows* not to be true" or "*knows* to be false." To *believe* something is true or false is not the same as to *know* it is true or false, a semantic distinction that was apparently not conveyed in this US statute. Searle (1979, 12) observes that *belief* is a determinable rather than a determinate and asks, "When does the report of a man's belief entail that the belief is true?" (159). Green (1989, 3) writes, "belief is what makes the difference (obviously) between a lie and a mistake. When people say something false that they believe to be false, they are lying, but if they say something that is false but they happen to believe it to be true, they are merely mistaken." The United Kingdom's Perjury Act of 1911 includes both of the words, "know," and "believe," in its definition of perjury: "If any person lawfully sworn as a witness or as an interpreter in a judicial proceeding willfully makes a statement material in that proceeding, which he knows to be false or does not believe to be true, he shall be guilty of perjury . . ."

It seems that both the US and UK perjury statutes permit people's *belief* that something is true or false as sufficient to protect them from being charged with perjury. The question is how we can determine their intentions. If concrete evidence of this can be found in other communications by the witness, this evidence can be used to support their intended meanings. Lacking such evidence, Nunberg (1978, 94–97) suggests two context conditions that aid interpretation: (1) when a community's normal belief about the meaning of that word is consistently used, and (2) when that word refers to things that are not normally represented by it. It would appear that legal communities in both the US and UK rely on statutes that consistently and normally state that when *witnesses* state that they *believe* something is either true or false such statements are at least somewhat equivalent to *knowing* that thet are either true or false. This also is consistent with *Black's Law Dictionary*'s definition of *believe* as 1. "to feel certain about the truth of; to accept as true; 2. to think or suppose," and *knowing* is "1. having or showing awareness or understanding: well informed. 2. deliberate; conscious."

1.1 False statements about material matters under oath

Under §1621(1), the following is required to prove that persons have committed perjury: a speaker under oath before a judicial proceeding makes an oral statement relating to a material matter that the speaker knows to be false.

Perjury is not just any misstatement made while under oath. The Supreme Court of the United States has explained:

A witness testifying under oath or affirmation violates this section if she gives false testimony concerning a material matter with the willful intent to provide false testimony, rather than the result of confusion, mistake, or faulty memory.

(*United States v. Dunnigan*, 507 U.S. 87, 94 (1993))

Here a "material matter" refers to false statements that have the potential of influencing the outcome of a trial or other official proceedings. But some false statements are not material to trial outcomes. *False statements* are different from simply making a mistake, because the law requires that any perjurious statements be made intentionally. These are considered different from evasion or half-truth because the perjury must be about some matter that the witness believes to be false and while doing so creates a misleading impression through indirection. While this may be dishonest, the law does not always consider this perjury.

1.2 Possible errors

The possibilities of errors in judging whether or not perjury has taken place appear in several contexts that are not relevant to the application of linguistic analysis. For example, there is a requirement that the witness be under oath. On rare occasions there have been times in which a witness testified without first being sworn. Another type of error happens when the person giving the oath is not qualified to do so. This arises more frequently in out-of-court proceedings, as in depositions and in those "courses of justice" referred to in the common law definitions of perjury. Obviously, these are issues for attorneys to address rather than linguists.

1.3 Perjury vs. false statements

There is a difference between perjury and false statements, because perjury connotes corruption and recalcitrance, while false swearing, usually made in a written statement while not under oath, connotes mere falsehood without moral judgment (Garner 1995). This difference is recognized in some jurisdictions but not in others.

In the U.S. federal jurisdictions, some false statements given while not under oath can be as damning as perjury (18 U.S.C. § 1001). In fact, if persons falsely deny that they committed a crime during their interviews with federal officers and give an exculpatory "no" or a mere "I didn't do it," they can be imprisoned for five years and be fined $10,000 (*Brogan v. United States*, 552 U.S. 398, 1998). For example, this was the 2004 stock trading crime for which the famous television personality, Martha Stewart, was convicted.

1.4 Linguistics and perjury

Linguistic analysis fits into this picture because the determination of perjury rests on the language used by both the questioner and the person who responds, which language in turn has to be assessed by the jury or by the judge in bench trials. Their task is how to determine whether perjury actually took place, whether it was misunderstood or misconstrued, or whether a speaker or writer simply made a harmless mistake.

Virtually all of the language processes that involve exchanges between witnesses, prosecutors, and judges can provide data for linguists to analyze, including instructions, directives, questions, answers, interruptions, and clarifications. In addition, certain witnesses, particularly those with limited intelligence or those whose command of the language is underdeveloped or poor, provide additional opportunities for linguistic analysis. Some persons might not be able to competently understand the Miranda oath or the questions asked them. As noted above, some slips, errors, or untrue statements made by a witness under oath that are not considered *material* are not subject to charges of perjury even if witnesses know that what they said is false. For example, witnesses knowingly might represent their position in a company as greater or lesser than it actually is. Although this is not true, it has no bearing on the substance (materiality) of the case.

A common opportunity for linguistic analysis relates to questions that are insufficiently clear to witnesses. That is, witnesses may believe they are giving true answers to what they understood the question to be. False answers to misunderstood questions are not considered perjury. But because there is no available window into the minds of witnesses, it can be difficult to determine whether persons intended to give false answers to questions they misunderstood. Without accessibility into human minds, the actual language that speakers use becomes the best tool for understanding what they appear to be trying to say and how their listeners may appear to misunderstand or misconstrue it.

Alert witnesses will request clarification to unclear questions at the time they are asked. Fair-minded questioners will then try to clarify. When neither takes place, witnesses can be insulated from perjury charges when their own attorneys object to the other attorney's unclear questions. Poorly phrased questions and answers can be discovered and repaired at trials where lawyers are present to object and to ensure that the exchanges were unclear or ambiguous. It is not unusual, therefore, for an attorney to object to a judge that the other lawyer's questions are at fault.

The situation is different, however, during grand jury proceedings where only the prosecutor, the witness, and the grand jury are present. No defense lawyer is pemitted to be there to monitor the clarity of questions and answers because, as a general rule, the U.S. legal system prohibits witnesses, including targets, from having legal counsel with them. This means that a grand jury inquiry is fraught with peril for witnesses when the language exchanges are not clear, leaving it to the members of the grand jury to discover or infer unclarity as they decide whether or not to indict the witnesses.

In grand jury testimony witnesses face several requirements that are counter to the conventional and predictable speech events in their everyday life. For example, some witnesses find it difficult to challenge questions that they believe to be inadequately worded, causing them to guess or to respond to questions that seem to exceed the questioner's authority. The only role of witnesses is to testify about things that they know (*Blair v. United States,* 250 U.S. 282 (1919). Witnesses are not permitted to remain silent before a grand jury unless they are subject to valid Fifth Amendment claims, and the prosecutor is not obligated to advise witnesses about the dangers of testifying falsely (*United States v. Mandujano,* 425 U.S. 564, 581–82 (1976).

2 Roles of the participants in perjury cases

The following summarizes the perjury litigation roles of participants that can be relevant for linguistic analysis. Even though perjury charges for the most part arise from testimony during a criminal procedure or before a grand jury, perjury charges also can arise as a result of false testimony in civil cases, whether at trial or during depositions.

2.1 Role of the prosecutor

To sustain a perjury charge, prosecutors have to demonstrate their use of careful and clear questioning and be able to show that the defendants *were aware of the meaning* of the questions and the potential falsity of their answers. For grand jury testimony to support a perjury charge, the meaning of the prosecutor's question must be clear, and participants should not simply have to try to assume its particular meaning. In order for defendants to be properly charged with perjury, the indictment must evidence a meeting of the minds of prosecutors and defendants during their verbal exchanges.

If a prosecutor's question is excessively vague or fundamentally ambiguous, the answer to that question should not form the basis of a prosecution for perjury or false statements. Therefore, a prosecutor should ask clear and precise questions in order for responses to those questions to support a perjury charge, although depending on the context sometimes it is deemed sufficient that the question's requisite precision can be inferred. Nevertheless, it is generally agreed that an indictment charging the defendant with perjury should set forth the particular falsehood with clarity, along with the government's factual basis for the charge.

2.2 The roles of the triers of the facts

A jury at a perjury prosecution should not have to speculate about the questioner's meaning, about the defendant's meaning, and about the grand jury's conclusion that gave rise to the charge.

When words or phrases of common usage form the predicate of a perjury charge but are arguably susceptible to more than one construction, it is the job of the triers of the facts to determine whether the witnesses and their examiners held a shared understanding.

Common sense meanings of the words used should be considered in light of the knowledge of the witness at the time of the testimony in order to determine whether a challenged statement is sufficiently specific to support perjury charges.

In determining whether statements are clear enough to support a perjury conviction, it is necessary to consider the defendants' own uses and perceptions of the challenged terms in order to determine whether they understood the meaning of the questions posed.

Perjury charges cannot be sustained when statements of the accused are lifted out of their immediate contexts, thus assigning meanings wholly different from that which their contexts clearly demonstrate. For example, in one of the four cases cited below, the social context of the normal duties of a parish priest was ignored by the prosecutor who theorized that the priest was a close personal friend of the person to whom he provided these conventional priestly functions. The matter of whether a question or answer is fatally ambiguous and thus may not constitute evidence of perjury, should consider the social context of the interaction.

In summary, jurors must determine whether the prosecutors' questions were asked in a clear and unambiguous way and had a factual basis for comparison with known facts in order to determine that the witness's answers were untrue in the social context in which witnesses could clearly understand what was being asked of them. The jury should not speculate about what is meant by both prosecutors and witnesses and they also have to determine whether the language used by prosecutors to witnesses was specific enough for a perjury charge to be supported. Furthermore, jurors have to view the language evidence in the overall context of the questions and answers.

2.3 Proving perjury

Perjury is an area of law that invites linguistic analysis. So how can perjury be proved?

2.3.1 Intelligence gathering and analysis

When prosecutors decide whether or not to charge a crime, they typically carry out an intelligence analysis of the evidence, which in perjury cases means comparing the known or strongly suspected factual evidence with the language used by the defendant. I have suggested elsewhere that an effective and accurate intelligence analysis is central to all law case investigations (Shuy 1990, 125–26). To be effective, such analysis has to precede indictments for perjury.

When the charge of perjury is based on testimony under oath, the obvious way for the government to proceed is to analyze the language used by suspects and compare it with other evidence in the case. When requested, linguists are available to help prosecutors perform their intelligence analyses. Although this happens only rarely, one of the four perjury cases described below shows how linguists helped one district attorney decide to withdraw his prosecutor's charges before the case went to trial.

Intelligence analysis is often thought to be the sole territory of law enforcement organizations, but it is also used by governments as they deal with international affairs and by commercial businesses as they deal with competitors. Intelligence analysis is central to the legal context for discovering such things as false leads, incomplete information, and ambiguity as the legal process makes reasonable, accurate interpretations from its intelligence gathering. Andrews and Peterson (1990) provide a detailed description of the relevance of intelligence analysis for police interrogations and for eliciting confessions, much of which can be traced to the seminal work of Godfrey and Harris (1971). Effective intelligence analysis leads directly to accurate procedures during law cases such as perjury.

One definition of intelligence analysis is: "that activity whereby meaning, actual or suggested . . . is derived through organizing and systematically examining diverse information" (Harris 1976, 30). An important primary step in good intelligence analysis is for the analyst to develop more than one hypothesis:

"The analyst must formulate alternative hypotheses . . . to probe allegations and suggestions of criminal activity rather than to build an evidentiary case" (Harris 1976, 34). This suggests that analysts must be able to go against their own presuppositions, recognize the unlikely, and provide connections that are not immediately obvious to others.

In perjury cases, this means that in their role of intelligence analysts prosecutors should not only consider hypotheses of defendants' guilt, but also envision hypotheses of innocence. According to a former intelligence analyst at the CIA's Directorate of Operations, the best use of information is to challenge the assumption that the analyst likes best (Heuer 1999). Failure to do this can lead to wastes of public resources, as I have noted elsewhere in such cases as the *United States v. John DeLorean*, the automobile manufacturer who was acquitted because the prosecutor's single hypothesis of guilt apparently prevented him from exploring any alternative hypotheses (Shuy 1993, 73). This is similar to a syndrome of some law enforcement agencies that has been called "detective myopia" or "tunnel vision." For example, when interviewed about the FBI anthrax investigation of Dr. Stephen J. Hatfill for spreading anthrax, former Los Angeles Police Chief Daryl F. Gates said that he believed that the agents became so focused on Hatfill that they lost their objectivity (Freed 2010, 52). Dr. Hatfill was never indicted, however, because the intelligence gathered was discovered to be myopic and inaccurate.

Some of the attributes of good intelligence analysis include the ability to be impersonal and neutral, to be receptive to new data, to view the evidence in its social and linguistic contexts, to analyze inductively, and to be free of bias. In short, intelligence analysis follows the familiar and standard scientific principles of systematically collecting relevant and representative data, evaluating what is collected, analyzing all of it with the proper tools, and arriving at an informed conclusion that is accurate and generalizable. This is illustrated in a scientist's proposed research in the form of a research grant, where the proposal is soundly reviewed before the research even begins. If something is lacking in the proposal, the project is stopped at that point. If the proposal is supported by peer evaluations, the project may be supported and the research begins. After the research is completed, the scientists' conclusions are then carefully scrutinized and evaluated by other experts in that field.

In a similar way, criminal law cases also begin with an unresolved problem or issue that provides these same opportunities to systematically gather representative evidence data, a step called intelligence gathering, usually done by law enforcement agencies. The next step, intelligence analysis, is carried out by prosecutors who employ the most useful and appropriate analytical tools in order to reach accurate and proper conclusions. Different from the scientific process, however, is that the ultimate review and final evaluation of this intelligence gathering and analysis is ultimately carried out by triers of the facts – juries or judges.

When scientists, law enforcement, or lawyers fail in their intelligence gathering and intelligence analysis efforts, it is often likely that they haven't first gathered accurate or representative data systematically, that they failed to use the proper analytical tools and procedures, or, based on their data gathering and analysis, that they failed to reach an accurate and proper conclusion.

In science, the penalty for bad intelligence gathering and analysis leads to professional criticism by the scientist's peers, which in turn leads to embarrassment and possible loss of professional reputation. In contrast, the penalty for bad intelligence gathering and analysis of indicted defendants can (and probably should) lead to an acquittal when the jury or judge recognizes that the intelligence gathering and analyses are faulty or insufficient. Absent such judgment, defendants can be convicted when triers of the facts fail to recognize the shortcomings of that intelligence gathering and analysis processes. From that point on, things become difficult, because when these triers of the facts make wrong decisions, appealing the verdict is complex and time consuming. In contrast, good intelligence gathering and intelligence analysis can lead to the proper conviction of defendants when the evidence and analyses are accurate, convincing, and proper. By the same token, improper intelligence gathering and intelligence analysis can lead at best to wasted resources and at worst to false convictions.

3 Linguistic tools in perjury cases

The major analytical tools of linguistics include phonetics, phonology, lexicon, morphology, syntax, semantics, pragmatics, speech acts, speech events, language variation, and discourse analysis. Evidence in each type of law case can call on at least some of these tools. Analyses of speech events and speech acts, for example, are particularly relevant in bribery, solicitation, and perjury cases where it is essential to discover the participants' mutual understandings about the speech event that they understand they are in. The language that they use in their topics and responses is the best evidence of their schemas about what they say. Their conversational strategies reflect the participants' efforts to promote their own positions and goals. Their syntax, lexicon, and phonetics are critical for addressing potential ambiguity in their language.

During the past forty years linguists have used these tools to become more and more active in the legal arena in civil cases involving trademarks (Shuy 2002), product warning labels (Dumas (1990), and discrimination, defamation, plagiarism, and deceptive trade practices (Shuy 2008). At the same time, linguistic analyses have contributed to criminal cases involving charges of bribery (Shuy 2013), murder (Shuy 2014), perjury (Shuy 2011), sex crimes (Matoesian 1993; Cotterill 2007; Shuy 2012), domestic terrorism (Shuy 2021), and analyses of interviews conducted by law enforcement (Fraser 2003; Heydon 2005; Haworth 2018, 2021; Shuy 2005, 2011, 2017).

As linguists examine evidence in perjury cases, it is prudent to begin with the broadest language element, the speech event, because it contextualizes that which follows. The participants reveal their schemas about that speech event in the topics they introduce and their responses to the topics introduced by the other participants. Their contributions to the exchange provide evidence of their lexicon, syntax, and semantics. The exchanges throughout the body of evidence provide opportunities for the participants to use conversational strategies such as intentional ambiguity, interruptions, overlaps, changes of subjects before the other persons have a chance to respond, and camouflaged illegality.

Linguists have found that it is useful to analyze these elements sequentially: from speech event to schemas to agendas (revealed by topics) to speech acts to syntax to lexicon to phonology in what has been called an inverted pyramid (Shuy 2013). This analytical sequence begins with the largest elements of the language evidence in order to assure that the smaller elements that most often are the focus of an investigation are not separated from their context.

Perjury lawyers not familiar with linguistic analysis tend to focus on the smallest units of language– sentences and words. Many prosecutors believe that these provide the "smoking gun" evidence to prove their cases. The four case examples that follow demonstrate how it is not always these smallest units of language that should be their main focus.

4 The perjury case of Steven Suyat

One case in which linguistic analysis was used was in the defense of Steven Suyat, who was charged with perjury in Honolulu (Shuy 2005; 2017), where the linguistic analysis of the speech event, schema, syntax, semantics, and conversational strategies provided the context that the prosecutor had overlooked.

4.1 Background

Suyat was a second-generation Filipino born and raised on the backwater Hawaiian island of Molokai. Many of its 12,000 inhabitants worked in the cane fields or factories where Hawaiian Creole and Hawaiian Pidgin English were spoken. After working seven years as carpenter, Suyat was promoted to business agent in the local carpenters' union. When that union picketed a site of a local construction company that did not use union workers, the builder filed unfair labor practices with the National Labor Relations Board. He claimed that the picketing had gone beyond merely providing information and illegally had tried to recruit members for the union.

Although this dispute was settled before litigation took place, 18 months later, based on subsequent covertly recorded conversations made by the contractor, the U.S. District Attorney's office came to believe that two of the union's business agents had filed false reports. These tapes revealed knowledge that these two business agents had not merely picketed for information purposes, but rather to pressure the non-union carpenters to join their union. Subsequently these two business agents were indicted and convicted.

4.2 Suyat's grand Jury appearance

Although Steven Suyat was not a party to those tape recordings, he was subpoenaed to testify as a witness at the trial of his two colleagues. After he testified, the prosecutor was not satisfied that Suyat had told the truth and subsequently brought him before a Grand Jury, charging him with giving false testimony during the previous trial of his two union colleagues. The perjury charges against him grew out of his Grand Jury testimony in a case for which he was not even a suspect. He understood his role in that speech event was to be a witness for the prosecution. His schema, revealed by his responses, demonstrated that he was trying to be as helpful as possible.

4.3 Suyat's testimony

Suyat tried to be cooperative when the prosecutor asked him, "One of the jobs of the business agent is to organize non-union contractors, is that right?" Suyat answered, "no," for he knew very well that his union organized workers but not contractors. Even a dictionary would support this understanding. To him, the prosecutor had made an error that required clarification. But his answer of "no" became one of the perjury counts against him.

Two more counts of perjury followed when the prosecutor asked Suyat if the job of each of his two fellow business agents was to organize contractors. Suyat said "no" again, still apparently believing that the prosecutor had erred in his description of the jobs of union business agents.

So why didn't Suyat simply correct the prosecutor's misuse of "organize contractors" here? His schema about this speech event did not allow him to correct a powerful prosecutor who knew far more about courtroom speech events than he did. If Suyat had responded, "I don't understand your question," this could have only seemed uncooperative. At any rate, the prosecutor took full advantage of Suyat's answers by quickly changing the subject before there could be any clarification about the meaning of "organize contractors." Elsewhere I have described this practice as the "hit and run" conversational strategy, which is commonly used by police interviewers and prosecutors (Shuy 2005).

The prosecutor did not stop here. An interesting question is why the prosecutor even asked his next question because it had little or nothing to do with the purported reason why he called Suyat to testify before the grand jury. Nevertheless, he asked Suyat to define "scab," to which Suyat responded, "I have no recollection." Suyat surely must have known the meaning of a word that refers to non-union workers who are hired as replacements for union members on strike.

One likely reason why Suyat didn't clearly indicate that he knew what "scab" means relates **to his participation in a** speech event that he had never experienced before. He was a poorly educated union carpenter who suddenly found himself in the unfamiliar and terrifying context of the courtroom where he knew that he had to do the best he could when he responded to questions posed by a very well-educated and powerful prosecutor. Since he could not be represented by a lawyer at his grand jury hearing, he had nobody there to advise him. Suyat was at sea about how to define this word. He searched his memory for a correct dictionary definition and came up blank. He knew that any answer he could give must be correct and decided it better to say "I have no recollection" rather than to give a faulty definition.

The prosecutor's next question referred to Suyat's own appointment book: "So you don't remember what you meant by it when you put it down here?" Suyat began to answer by saying "yeah, but –" and was cut off immediately by the prosecutor's, "Thank you. I have no further questions, your honor," still another example of the "hit and run" conversational strategy that prosecutors sometimes use to block an answer before it can be made.

Suyat's "yeah but –" did not indicate that he didn't *know* what a scab meant; only that he couldn't provide a definition that he felt would be satisfactory in this social context. His answer produced still another count in his indictment for perjury.

This also illustrates how the prosecutor did not seem to understand English language rules of concord (Huddleston and Pullum 2002, 847–48). The prosecutor asked his question negatively, "So you *don't* remember," indicating that agreement to this would be conveyed by a negative response from Suyat, such as "no," or "I don't remember," either of which would have indicated that he agreed that he didn't remember. But Suyat's response did not convey that he didn't remember using this word. His "yeah, but –", strongly suggests that he was starting to disagree with the prosecutor's question but the prosecutor interrupted him before he could complete his sentence, another example of the conversational strategy of blocking the witness before he could say what he was about to say.

Suyat's experience briefly illustrates the linguist's analysis of speech events, schemas, syntax, semantics, and conversational strategies in the language evidence in this perjury case. This was one of those cases, however, in which the judge did not permit expert witnesses to testify. The defense lawyer did the best he could with this linguistic analysis but was not successful in warding off Suyat's conviction of perjury.

5 The perjury case of Father Joseph Sica

Unlike the Suyat case in which the speech event was not mutually understood, a well-educated parish priest named Father Joseph Sica fully understood the speech event of a grand jury hearing. The linguistic analysis in this case called on the tools of semantics, social context, and the role of Father Sica's faint memory of events that had happened some twenty years previously.

5.1 Background

This all began when the priest was called before a grand jury as a witness during an investigation of Scranton businessman Louis DeNaples, who was trying to open a gambling casino at his resort in the Pocono Mountains area. The government separately had indicted DeNaples for perjury based on his testimony during the trial of William D'Elia, who at that time was the alleged head of the crime syndicate formerly led by Russell Bufalino, who had been convicted of murder and had recently died while in prison. The prosecutor claimed that he was interested in Father Sica's testimony only to the extent that he could reveal something incriminating about DeNaples' connection with D'Elia and Bufalino during the period when DeNaples was trying to open his gambling casino.

Unlike Suyat, Father Sica was highly educated, held advanced degrees in both philosophy and theology, and had published several books and articles. He had created various programs for the poor, homeless, and needy that had endeared him to the Scranton community. When he was called to testify before the grand jury, he was quite aware that his role at that hearing was to truthfully convey all he knew that might aid the prosecutor.

5.2 Sica's grand jury testimony

The prosecutor's stated purpose for Sica's grand jury appearance was to glean information that would associate DeNaples with William D'Elia, who allegedly headed the crime syndicate formerly run by the now deceased Bufalino. This effort failed when Sica testified that he had no memory of ever socializing with D'Elia or of talking with DeNaples about him. Receiving no useful information that related to his goal, the prosecutor then suspected that Sica was lying about his relationship those two men.

During the priest's grand jury testimony, the prosecutor apparently believed that Sica was trying to protect his good friend, DeNaples. There was no doubt about the priest's close relationship with DeNaples, for he mentioned multiple times that they had grown up together in the same neighborhood, had often visited and dined together, and had remained very close friends. When asked if he knew anything about DeNaples' plans for a gambling casino or his possible relationship with Bufalino's crime syndicate, the priest had nothing helpful to report. After Sica openly admitted that he and DeNaples had continued to be very close friends, the prosecutor turned his focus to the priest's past relationship with Bufalino.

Meanwhile, the prosecutor had discovered information from Sica's radio interview that took place after the grand jury hearing, and amended his original complaint, probably believing that he had strengthened it because during that interview the priest mentioned that Bufalino had given him a gift at his ordination ceremony some twenty years earlier. Unbeknownst to the priest, the prosecutor also had discovered Sica's past correspondence with Bufalino along with a photo of them having dinner together. The prosecutor apparently believed that these items were crucial to his intelligence analysis and were the very smoking guns he needed to indict Sica for perjury at trial.

At that point the priest retained a defense attorney to represent him. Before the trial started, however, the prosecutor's superior, the district attorney, made his own intelligence analysis by reviewing the evidence, which caused him to fear that if this case was brought to trial it might result in a failed prosecution that would embarrass his office. He then asked linguists Robert Leonard and me to analyze the language found in both the prosecutor's evidence and prosecutor's grand jury examination.

5.3 Linguistic analysis of the evidence

The linguists examined the priest's grand jury testimony, particularly his use of the expression, "personal relationship." During that hearing the prosecutor asked when Sica last had any contact with Bufalino, to which the priest answered, "about twenty years ago . . . when I was a deacon in Sayre." Sica then explained that he met Mr. and Mrs. Bufalino when he was a chaplain at a hospital after Bufalino had come there to visit a patient. The prosecutor apparently believed that the priest was lying and continued to probe, next asking, "Did you have any personal relationship beyond that?" to which the priest answered, "no."

5.3.1 Semantic analysis of "close personal relationship"

At that point the prosecutor confronted Sica with four items that he considered clear evidence that Sica was lying. One item was a photo of the priest sitting with others at a table with Bufalino during what appeared to be a celebration. A second item was a note in which Sica told Bufalino that he had written a character witness for him to the governor just before that crime leader's sentencing. The third item was a Sica's thank-you note to Bufalino for an unspecified gift. The fourth item was a tape recording of the interview Sica had given to a local radio station shortly after he was indicted. In it the priest said that on a few occasions when he would go home to Scranton to visit his parents, Bufalino would invite him to dinner but the priest never complied. Sica admitted that Bufalino did give him a gift of a hundred dollars at his ordination service. He also mentioned that Mrs. Bufalino had asked him to give her husband a "blessing and prayer" immediately before he was sent to prison.

Based on the prosecutor's intelligence analysis, Sica's indictment concluded: "The evidence demonstrates that Sica had a substantial relationship with Russell Bufalino. The testimony to the contrary is false."

During his grand jury hearing the prosecutor asked Sica five times whether he had a personal relationship with Bufalino and each time the priest responded that he did not. Clearly there was no meeting of the minds about the meaning of a "personal relationship." Next, the prosecutor upped the ante by asking whether this was a "close personal relationship." Sica answered, "not close and personal."

Interestingly, neither the prosecutor nor the priest had tried to define "close personal relationship." The prosecutor inferred his definition from the four pieces of evidence he had collected. Although Sica did not try to define the expression in specific words, he did so quite nicely in his description of his "close personal relationship" with DeNaples. When asked if he knew DeNaples, Sica said "extremely well," adding that he had known him for fifty years, adding, "He and my dad were very close . . . so we knew Mr. and Mrs. DeNaples as our family." He added that after their parents died, "I became much closer to Louis and Betty and to all his children. If I would go home on weekends, I would stop to see him or he'd be at my dad's house." When asked how frequent this contact was, Sica replied, "I would say consistent and constant contact with him," adding that they shared meals "frequently and consistently . . . I considered myself a very close friend." This constituted Sica's implicit definition of a "very close relationship." It wasn't a dictionary definition, but it was compellingly clear. So here we have the prosecutor's inferred but undefined meaning of "close personal relationship" in contrast with Sica's effort to implicitly explain that meaning with concrete examples.

5.3.2 Sociolinguistic context analysis

The prosecutor also failed to view the evidence that he collected in the social context in which it took place. In ceremonial events such as ordination of a parish priest, it is common practice for photos to be taken, for gifts to be given to the honoree, and for the new priest to follow the conventional politeness rules of sending thank-you notes to those who gave gifts. It is clear that Sica's offering of prayers and blessings when Bufalino was about to go to prison was a professional service of a priest rather than a friendly social visit. The prosecutor's intelligence analysis had leaped to conclusions that this priestly obligation was evidence of a close personal relationship.

5.3.3 Sica's contextual memory of past events

Although linguists are not commonly considered experts about memory, they can examine the social context in which memory is called upon. The prosecutor believed that Sica lied about his lack of a close personal relationship with Bufalino. This belief was bolstered by Sica's subsequent radio interview in which he mentioned his

$100 gift from Bufalino and recalled the time that he prayed over Bufalino before he was sent to prison. When the prosecutor learned of this from the radio interview, he amended his indictment to include these undisputed facts.

During Sica's grand jury testimony he was specifically asked to recall the relevant social context of what had happened many years earlier. To the priest, these events were unrelated to the context of the prosecutor's questions about close personal relationships. Many gifts were given at his ordination ceremony and many thank you notes followed. Father Sica had administered priestly comfort to hundreds of people, many of whom he did not even know, much less have any personal relationships. They were duties and activities common to the context of what parish priests do.

At the forthcoming trial, the prosecutor was prepared to ask Sica to explain the $100 gift and his visit with Bufalino, inferring that the priest had deliberately avoided these topics during his grand jury testimony. If the prosecutor had simply asked Sica about these events during his grand jury testimony, the priest could have explained how they came about. Instead, he followed the common prosecutorial strategy of saving them to use as a surprise. But Sica had openly recalled them during that radio interview after his grand jury testimony. It would appear that Sica's purported "selective memory" during his grand jury testimony was not as selective as the prosecutor believed.

5.3.4 Review of the analysis

The linguists prepared a written report explaining the relevance of semantics, sociolinguistic context, and contextually influenced memory. The district attorney reviewed this report and recognized that the indictment was seriously flawed. He subsequently dropped all charges against Father Sica, saying, "Perjury cases are very difficult in that language is the weapon we need to prove. We engaged expert linguists, something a little different but it is a perjury case, where precision of questioning is important." (Shuy 2011).

6 False statements by a man who applied for a big game hunting license

Like most states, Montana's rules concerning residency relate to such things as taxation, educational benefits, and licenses for driving, fishing, and hunting. Applicants who own two homes, one of which is in the state, can qualify for hunting licenses although they face more restrictions than full-time Montana residents. Hunters who are full-time residents can be as protective about these restrictions as the state's fish, game and hunting officials. This happened when a local hunter reported a man that he believed was violating state hunting regulations.

6.1 Background

This report led to an investigation of the man suspected of taking deer and elk illegally. The investigators discovered that in recent years this man had been purchasing non-resident licenses but since he had recently bought a home in the state, he now believed that he had satisfied the Montana requirement of having a residence in the state.

Based on this accusation by a local hunter, state officials obtained a search warrant of the suspect's Montana cabin and confiscated some evidence including photographs and various game mounts on the walls, after which they indicted this man on four counts, one of which was a felony charge for possessing and unlawfully taking 2,000 pounds of bull elk and buck mule deer along with a misdemeanor charge that he had made a materially false statement on his hunting license application form. On this form he claimed to be a legal resident of the state while the state officials believed he was not.

Before the case reached the court, the prosecutor dropped the felony change that the man had taken illegal amounts of elk and buck mule deer because he was unable to find sufficient evidence of this. At the same time, however, he increased the charge of making false statements from a misdemeanor to a felony. The defendant might have agreed to a misdemeanor charge simply to get rid of the case, but it was very important to him to avoid the felony charge because he was currently serving on boards of several corporations that had strict rules prohibiting felons from serving on them. Therefore, the man retained a lawyer to oppose the felony charge. Since this lawyer realized that this case dealt primarily with language, she retained a linguist to analyze the language of the licensing procedure.

6.2 Ambiguity and vagueness in the licensing process

The defense lawyer first considered filing a motion to dismiss the felony charge, based on her belief that the Montana statute was void on its face about residency requirements. She believed that it was unclear to ordinary readers exactly what the statute prohibited and that it encouraged arbitrary and discriminatory actions by law enforcement. She also believed that even the two initial steps required of applicants provided ample evidence that the entire application process was flawed.

6.3 Ambiguity and vagueness in the conservation form

Obtaining a Montana hunting license is a two-step process. The first step is to fill out the "conservation form," which applicants can pick up at many grocery and hardware stores or at gun and fishing tackle shops. The following is the entirety of that form:

> I hereby declare that I have been a LEGAL resident of the state of Montana, as defined by MCA 87-2-102 for at least 180 consecutive days. All statements on this form are correct and true. I understand that if I subscribe to or make false statements on this form, I am subject to criminal prosecution.

Most ordinary readers simply sign it in the same cavalier that way that they sign other bureaucratic forms which they often don't even bother to read. One problem is that for applicants to follow the state's requirements to be considered a legal resident of Montana, they either have to already know what MCA 87-2-102 says or they have to look it up, neither of which is very likely. The second sentence curiously states, "all statements on this form are correct and true." It does not specify that the applicant's *answers* are correct and true and there is no way that the reader could know whether or not what the state specifies on this form is accurate. Finally, "at least 180 consecutive days" does not indicate whether these 180 consecutive days must be during the current year, whether they can carry over from a previous year, or whether they can be accumulated during some period in the past.

6.4 The ambiguity of the application form

After applicants submit the conservation form, the state sends them the hunting license application form to fill out and sign. The first sentence clumsily tries to clarify the "180 consecutive days" requirement: "I hereby declare that I have been a legal resident for at least 180 days____years____months immediately prior to making application for this license," and concludes with a warning that all statements on this form are true and correct, referencing Montana statute MCA-87-2-102, but unlike the conservation form, clarifying that the requirement of truth and accuracy pertains to the applicant's answers rather than to the veracity of the statements printed on the form itself.

The first problem is how the applicant should fill in the blank spaces between "days, years, months." A second problem concerns the meaning of "immediately prior to," which was directly copied from the state's statute. *Black's Law Dictionary* defines "immediately" as:

(1) within a reasonable time having due regard to the nature of the circumstances of the case, and

(2) not deferred by any period of time.

This application form could have been made clearer by replacing its vague "immediately prior to" with a specific time period such as "180 consecutive days during the twelve months preceding the date of this application," if that is indeed what the form intended.

Further confusion in the application form is caused by the fact that it is directed to multiple audiences, not just to conventional hunters but also to hunters who qualify for a 30-day military exemption and those who qualify for the special exemption given the Montana Job Corps. The application form is silent about the details relating to these exemptions. Although packaging the form used for three types of applicants onto one document may save space, a multiple-purpose form only increases the potential for ambiguity and vagueness.

The application form's reference to Montana statute MCA-87-2-102 repeats the problem found on the earlier conservation form. Both mention the *source* of the information while omitting to say what that information is. It is not likely that most applicants have committed this statute to memory.

6.5 The ambiguity and vagueness in the statute

After pointing out the ambiguity in the application form, the linguist examined the wording of the statute (MCA 13-1-112). It informs applicants that they must meet five behavioral requirements: (1) they must have "physically resided in Montana as the person's principal or primary home or place of abode for 180 consecutive days immediately before making application;" (2) they must have filed Montana state income tax returns; (3) any vehicles owned and operating for hunting in Montana must have state licenses; (4) applicants must be registered to vote only in Montana; and (5) they must not have applied for hunting licenses in any other state or country. All of these can be fact-checked by the state. So far, so good.

The statute also specifies that applicants are considered "residents" if they have resided in the state as "that person's principal or primary home or place of abode for not less than 120 days a year." *Black's Law Dictionary* defines "residence" as "something more than physical presence and something less than a domicile," concluding that these words "have no precise legal meaning." Montana courts had previously ruled that "residence," "abode," and "domicile" are variables that mean "practically the same thing." Other Montana statutes, however, including those concerning state taxation and motor vehicle registration, differentiate these terms, saying, "the word, 'residence,' may not be confused with the word, 'domicile.'" Montana's statute on elections uses "residence" 13 times and never mentions "domicile" or "abode" at all.

The defendant's lawyer at first thought she might address the statute's ambiguity of "resident" and "residence," but she eventually believed that in this case any criticism of the statute might be a hill too steep to climb and might create unwanted negative reactions. At any rate, most ordinary Montana readers seem to have a sense that a "residence" is a place where you live.

Differing from the conservation and application form, the statute says that applicants must have lived in their principal or primary "home" in Montana for *180 consecutive days* immediately before applying for a hunting license. As noted above, applicants must also *reside* in their principal or primary Montana *home* for no less that 120 days a year, file state income tax returns, license their vehicles used and operated in the state, not possess or apply for hunting licenses in any other state or country, and register to vote only in Montana. It leaves ambiguous which 120 days per year are required or whether they need to be consecutive.

It would have helped applicants if they had been informed about these five behavioral requirements when they filled out their application form. In the long run, however, the defense lawyer concluded that attacking the state statute would be less effective at trial than exposing the ambiguities found in the license application process.

6.6 Survey of how Montana residents understood the issue

Even though there is ambiguity in the conservation form, the application form, and the statute itself, the ultimate issue is what Montana residents understand these to convey. This is where linguistic analysis of ambiguity and vagueness can be supported by that which applicants actually report that they understand. To discover this, a survey was made of residents in the rural county where the trial was about to be held. It included questions relating to the meanings of the previously discussed language found in the conservation and application forms as well as the wording of the statute. The questions dealt with the confusing statements about the 120-day and 180-day residence requirements, the vague meaning of "immediately prior to," "all statements on this form are true and correct," and the intended meaning of "legal." The results of this survey showed that jurors relied on the perlocutionary effect of these forms regardless of the illocutionary force of the statements (Searle 1969, 1980).

The results of this survey strongly favored the defense. The prosecutor apparently had believed that since rural Montanans were protective of their hunting opportunities, they would find it easy to convict a wealthy outsider with two or more homes in different states. Based on the linguistic analysis of ambiguity and vagueness found in the conservation form, application form, and statute, the survey demonstrated that the jury focused not only on the rights of the defendant, but also on their own potential rights if they ever should find themselves in the same situation. At trial the jury found the defendant not guilty of the felony of making false statements.

7 False statements made while getting a bank loan

The perjury evidence was based on spoken language in Steven Suyat's case, both spoken and written in Father Sica's case, and only on written language in the Montana hunter's case. Both written and spoken language served as the evidence in the investigation of businessman, Larry Kopp, who was charged with both perjury and fraud.

7.1 Background

Larry Kopp and his brother Marvin Kopp had created a real estate development of various corporations and partnerships. After Marvin died, the business fell on hard times and the brothers' nephew, David Kopp, became the managing partner while Larry continued to be marginally involved.

As the company began to founder, Larry negotiated a 13 million-dollar second mortgage with Ensign Bank, most of which was to help support a financially troubled shopping mall that the partnership already owned. When the bank holding the initial mortgage learned of this second mortgage, it was disturbed enough to prepare default procedures on their initial loan. In an effort to prevent this from happening, the partnership's office manager, Stuart Sherer, fraudulently forged signatures on the leases of the mall's current tenants falsely inflating the partnership's rental income enough to make it appear to support its claim that the business was sound.

Ensign Bank then relied on this false information and provided Kopp with the requested second mortgage. The business continued to founder, however, and a year later it defaulted on this second mortgage from Ensign. This bank received the deed to the mall and subsequently sold it to a new buyer. Meanwhile, federal authorities discovered the fraud involving the allegedly forged signatures of the mall's tenants. Sherer admitted his complicity and said that he forged these tennants' signatures at the direction of Larry Kopp. The authorities then offered Sherer unspecified considerations in return for tape-recording his conversations with Larry Kopp that would confirm his alleged role in the false statements about the mall's rental income. Four such conversations provided the purported evidence against Larry.

One useful way to analyze continuous conversations is to perform a topic analysis showing who introduced which topics and what the other person's responses to them were. There is no way to reach into persons' minds to determine their intentions, but their topics and responses give the best clues to such intentions and agendas. This analysis showed that Larry Kopp introduced no topics that related to the illegal fraud and when Sherer tried to introduce them, Kopp gave no evidence that he understood these vague hints and suggestions about illegality.

Their first conversation took place during a long meeting between Larry and Sherer that yielded nothing incriminating. Larry introduced topics suggesting they could avoid a lawsuit by selling properties to repay their $13 million loan, but he was not aware that Sherer had already admitted making the false statements about leases and Larry expressed no concerns about his own possible criminal involvement. Sherer then introduced the topic that the government subsequently relied on to charge Larry with making a false statement: "Probably when you asked me to make up the phony leases, we shoulda got David more intimately involved." To Sherer's topic, Larry responded to only the main clause but not to the damaging dependent clause saying, "Probably it would have been easier if he actually was."

Sherer's sentence contains two elements that affected Larry's response. First is the recency effect. When given an uninterrupted series of statements or questions, listeners tend to respond to the most recent or last one in the series (Shuy 2005, 32,152). Larry did this here. Second, Sherer unpredictably produced lower stress and intonation on the dependant clause that conveyed the damaging information and greater stress and intonation on the main clause that was relatively benign, creating what can be considered grammatical camouflage. Larry responded only to Sherer's main clause saying, "We're gonna blame it all on him." This response was consistent with what Larry had said several times earlier– that his brother David was responsible for the company's financial problems because of his sloppy management: The government misinterpreted Larry's pronoun, "it," here to mean the phony leases, but Larry's "it" was consistent with his accusations of David throughout the conversation, i.e. his nephew's negligent business acumen.

After the investigators reviewed this conversation, they apparently were not satisfied that Sherer had uncovered any convincing evidence of Larry's guilt, so they arranged a second effort, a surreptitiously recorded phone tap from David to Larry. This call contained topics that were mostly about bids for the sale of the mall, their insurance, and their taxes, none of which yielded anything useful for the prosecution of Larry Kopp.

The government then sent Sherer back to record Larry a second time, still hoping to get something incriminating. When investigators keep going back, it's a clear sign that they didn't get what they wanted in their earlier efforts. Sherer again asked for Larry's advice about what to say to the investigators, while Larry remained firm that the company's financial stress was entirely his nephew David's fault and that he and Sherer had nothing to do with the fraudulent lease statements, having learned about them only after the fact. Both men introduced topics about various leases in the shopping mall, the offers they had received for it, and how they might stall Ensign from foreclosing on the loan, but nothing in this meeting demonstrated evidence that Larry was implicated in the fraud scheme.

A few days later, the government sent Sherer to talk with Larry a third time. Again, their topics were about various leases at the mall and when Sherer finally brought up the topic of how to prevent Ensign from foreclosing their loan, Larry's response was, "I believe they've kind of given up on that tack . . . I believe that they've convinced themselves that this property is really worth what they got in there . . . they were anxious to see if they could steal it away from me, because if they get any more, then they could keep the money." Nothing incriminating here.

Sherer's final effort was to ask, "Are we gonna stick with the same tack that David is the one who did everything?" after which he added, "otherwise we both go to jail for a long time and I don't want any part of that." The government took Larry's response to be an admission of his guilt: "No, we're in cahoots when it comes to that, obviously . . . my gut feeling is that it doesn't enter into the picture. I will agree with you that if we sell it, we come out okay."

Apparently, the government believed they had caught Larry here, but they overlooked the meaning conveyed by Sherer's ambiguous inclusive pronoun, "we," in "otherwise we both go to jail," which Larry interpreted to mean Sherer and David, not Sherer and himself. Larry's "we" in his response, "we're in cahoots when it comes to that," was equally ambiguous about what Sherer and Larry were in "cahoots" about. One possible meaning is that the two men were in agreement about not going to jail, which was far from incriminating. The government's view was that to be "cahoots with" admitted an illegal partnership. Dictionaries define "cahoots" only as "in agreement or partnership with," with no indication of illegality. After this exchange, Larry continued to affirm that the fraudulent statement was entirely David's doing.

Based on these four conversations, the government indicted Larry Kopp on several counts of false statements and fraud. The linguistic analysis demonstrated the separate agendas of the speakers revealed by their topics and responses, the ambiguities of pronoun references, the role of prosodic stress, and the recency effect. The judge did not allow the linguist to testify in this case. The defense attorney tried to include these linguistic findings but was not able to do so successfully and Larry Kopp was convicted.

8 Summary of the use of linguistics in perjury cases

Since people under oath use written or spoken language about material issues to commit perjury or false statements, this is a fruitful area for linguists to be helpful to the legal process. Sometimes statements constitute obvious perjury but on other occasions the language evidence is subject to misinterpretation that linguistic analyses can help identify. Contributions of linguistic experts often address the lack of clarity, grammatical referencing, ambiguity, vagueness, interrupted responses, unclear questions, and the cultural and sociolinguistic contexts in which the exchanges between representatives of law interact with their subjects.

Steven Suyat's perjury charges growing out of his courtroom testimony were challenged by a linguistic analysis that revealed the lack of the participants' required mutual understanding of the speech event. It demonstrated the participants' contrasting schemas about the language used and was aided by analysis of the speakers' syntax and semantics.

The linguistic analysis of the perjury charges growing out of Father Sica's grand jury testimony addressed both semantics and the culturally based discourse features of the sociolinguistic context. The district attorney believed that his prosecutor had not done an acceptable intelligence analysis of the evidence and called on linguists to show him how his own prosecutor's indictment was misguided.

The felony charge of the Montana hunter believed to have produced false statements was countered by showing the semantic ambiguity and vagueness of the hunting license application forms that were supported by a sociolinguistic survey demonstrating the confusion that the government's language actually conveyed to rural Montanans.

The indictment of Larry Kopp not only accused him of making false written statements based on covertly tape-recorded spoken language evidence gathered during Kopp's conversations with another company employee who had already pleaded guilty. Kopp's misunderstanding of Sherer's attempts to disguise the speech events and schemas did not produce language evidence of his guilt. Topic analysis and evidence of pronoun referencing in these conversations revealed the agendas not only of Sherer, but also of Larry Kopp, whose language evidence did not support his involvement in either perjury or fraud.

These four cases illustrate some of the ways linguistic analysis can offer guidance to the courts when both written and spoken forms of perjury charges are brought. The examples show that linguistic analysis can be helpful to both those who were indicted as well as to the prosecution.

References

Andrews, Paul & Marilyn Peterson. 1990. *Criminal Intelligence Analysis*. Loomis, CA: Palmer Publications.

Bishop, J. 1858. 2 *Commentaries on the Criminal Law* § 860.

Cotterill, Janet. 2007. *The Language of Sexual Crime*. New York: Palgrave.

Dumas, Bethany. 1990. Language and cognition in products liability. In Judith Levi & Anne G. Walker (eds.), *Language in the Judicial Process*, 203–244. New York: Plenum.

Fraser, Helen. 2003. Issues in transcription: Factors reflecting the reliability of transcripts as evidence in legal cases. *Forensic Linguistics* 10: 201–226.

Freed, David. 2010. The Wrong Man. *The Atlantic* (May 2010), 46–56.

Garner, Bryan. 1995. *A Dictionary of Modern Legal Usage*. New York: Oxford University Press.

Godfrey. Drexel E. & Don R. Harris. 1971. *Basic Elements of Intelligence*. Washington, DC: US Government Printing Office.

Green, Georgia. 1989. *Pragmatics and Natural Language Understanding*. Hillsdale, NJ: Lawence Erlbaum.

Harris, Don R. 1976. *Basic Elements of Intelligence*. Washington, DC: Law Enforcement Assistance Administration.

Haworth, Kate. 2018. Tapes, transcripts and trials: The routine contamination of police interview evidence. *The International Journal of Evidence and Proof* 22 (4): 428–450.

Haworth, Kate. 2021. Police interviews in the judicial process. In Malcolm Coulthard, Alison May & Rui Sousa-Silva (eds.), *The Routledge Handbook of Forensic Linguistics*, 169–194. London: Routledge.

Heuer, Richards J., Jr. 1999. *Psychology of Intelligence Analysis*. Washington DC: Center for the Study of Intelligence, Central Intelligence Agency.

Heydon, Georgina. 2005. *The Language of Police Interviewing: A critical analysis*. Basingstoke: Palgrave Macmillan.

Huddleston, Rodney & Geoffrey K. Pullum. 2002. *The Cambridge Grammar of the English Language*. Cambridge: Cambridge University Press.

Matoesian, Gregory. 1993. *Reproducing Rape*. Chicago: University of Chicago Press.

Nunberg, Geoffrey. 1978. *The Pragmatics of Reference*. Bloomington, IN: Indiana University Linguistics Club.

Searle, John R. 1969. *Speech Acts*. Cambridge: Cambridge University Press.

Searle, John R. 1979. *Expressing Meaning: Studies in the Theory of Speech Acts*. Cambridge: Cambridge University Press.

Searle, John R., Ferenc Kiefer & Manfred Bierwisch. 1980. *Speech Act Theory and Pragmatics*. Boston: D. Reidel Publishing.

Shuy, Roger W. 1990. The analysis of tape-recorded conversations. In P.P. Andrews & M.B. Peterson (eds.), *Criminal Intelligence Analysis*, 117–48. Loomis, CA: Palmer.

Shuy, Roger W. *Language Crimes*. 1993. Cambridge, MA: Blackwell.

Shuy, Roger W. 1998. *The Language of Confession, Interrogation, and Deception*. London: Sage.

Shuy, Roger W. 2002. *Linguistic Battles in Trademark Disputes*. New York: Palgrave Macmillan.

Shuy, Roger W. 2005. *Creating Language Crimes*. New York: Oxford University Press.

Shuy, Roger W. 2008. *Fighting Over Words*. New York: Oxford University Press.

Shuy, Roger W. 2010. *The Language of Defamation Cases*. New York: Oxford University Press.

Shuy, Roger W. 2011. *The Language of Perjury Cases*. New York: Oxford University Press.

Shuy, Roger W. 2012. *The Language of Sexual Abuse Cases*. New York: Oxford University Press.

Shuy, Roger W. 2013. *The Language of Bribery Cases*. New York: Oxford University Press.

Shuy, Roger W. 2014. *The Language of Murder Cases*. New York: Oxford University Press.

Shuy, Roger W. 2017. *Deceptive Ambiguity by Police and Prosecutors*. Oxford U Press.

Shuy, Roger W. 2021. Linguistics in Terrorism Cases. In Coulthard, Malcolm, Alison May & Rui Sousa-Silva (eds.). *The Routledge Handbook of Forensic Linguistics*. London.

Saul Kassin

Trickery and deceit: How the pragmatics of interrogation leads innocent people to confess – and factfinders to believe their confessions

Abstract: Beginning with the nonconfrontational suspect interview aimed at assessing truth and deception, through the guilt-presumptive process of interrogation aimed at eliciting an admission of guilt, to the construction of a post-admission narrative confession, the psychology of confessions requires an understanding of language. In particular, this chapter will focus on confrontational approaches to interrogation that deploy trickery and deceit to communicate promises and threats covertly, through pragmatic implication, thereby circumventing the laws of evidence that ban confessions elicited through psychological coercion. Research to be described shows that these covert forms of communication are "effective" at leading people to infer promises and threats, leading suspects to confess, and misleading juries to infer guilt. Research also describes why Miranda does not serve as an effective safeguard and hence why all interrogations should be video recorded in their entirety.

In his classic treatise on evidence, John Henry Wigmore (1904/1985) described confession, even after it is recanted, as the most potent evidence presentable in court. Other legal scholars have since echoed this opinion (e.g., Kamisar et al., 1994) and a good deal of empirical research has confirmed it (e.g., Kassin, 2012, 2017; Scherr et al., 2020). Alongside this long-standing assessment as to the persuasive power of confession evidence, however, is the realization that confessions are fallible – sometimes reported secondhand by police or informants, raising questions as to authenticity; and at other times induced from a suspect through the processes of interrogation, raising questions about voluntariness, coercion, and reliability.

In recent years, a disturbing number of cases have surfaced across the United States involving innocent people who had confessed and were convicted at trial, only later to be exonerated by DNA or by other means, i.e., the discovery that no crime was committed, as when the presumed murder victim is found alive; evidence indicating that it was physically impossible for the confessor to have committed the crime, as when he or she was demonstrably elsewhere at the

Saul Kassin, John Jay College of Criminal Justice – City University of New York
e-mail: skassin@williams.edu

https://doi.org/10.1515/9783110733730-012

time; or when the real perpetrator, having no connection to the defendant, is apprehended. Although the precise incidence rate is not known, research indicates that false confessions and admissions are present in nearly 30% of all DNA exonerations reported by the Innocence Project and in nearly 15% of all wrongful convictions archived by the National Registry of Exonerations.

The stories behind the numbers are a varied lot. They involve men and women; children and adults; people of all racial and ethnic backgrounds, throughout history, and from countries all over the world. In some cases, a voluntary confession is initiated by the innocent suspect; in other cases the confession comes about through the processes of interrogation, leading innocent people to confess as an act of compliance or even, at times, to internalize the belief in their own guilt (Kassin & Wrightsman, 1985). Sometimes innocent individuals implicate not only themselves but others who are innocent as well. In a number of cases (e.g., the Central Park jogger case, the West Memphis Three, the Birmingham Six, and the Norfolk Four), multiple false confessions were taken for the same crime (for overviews, see Drizin & Leo, 2004; Gudjonsson, 2018). In rare instances, serial false confessions to multiple crimes are taken from the same person, such as Sture Ragnar Bergwall of Sweden, also known as Thomas Quick, who confessed to over thirty murders before being fully exonerated and released (Josefsson, 2015; Råstam, 2013).

The problem of false confessions is fourfold: (1) police routinely use psychologically unsound tactics to determine whether a suspect is telling the truth or lying, leading them to make judgments with confidence that are not diagnostic; (2) presuming the suspect's guilt, police use lawful forms of trickery and deceit that can lead innocent suspects to confess in detail to crimes they did not commit (especially suspects who are young, cognitively limited, psychologically disordered, or otherwise compliant or suggestible); (3) confronted with these narrative confessions, prosecutors, judges, juries, and others uncritically infer guilt from false confessions even when they are recanted, even when they lack corroboration – and even at times when they are flat out contradicted by other evidence; and (4) contradicting the belief that Miranda rights to silence and to counsel serve to safeguard the accused from these outcomes, research shows that Miranda has failed and that other reforms are needed.

What follows is a brief overview that highlights the venues of miscommunication by which both suspects and later factfinders are duped in part by the language of interrogations and confessions.

1 Historical overview

Beginning with *Brown v. Mississippi* (1936), the United States Supreme Court banned the use of physical violence and other third-degree methods of interrogation, e.g., simulating drowning by holding a suspect's head in water; prolonged confinement; deprivations of sleep, food, and other needs; extreme sensory discomfort (for a review, see Leo, 2008). To meet the challenges wrought by this ban, police had to develop psychological approaches to interrogation that would get crime suspects to confess. Although various such approaches were developed, the new paradigm was most popularly captured by the Reid technique, developed in the 1940s by criminologist Fred Inbau and Chicago Police Officer John Reid. As taught by the Chicago firm of John Reid & Associates, their approach was ultimately codified in *Criminal Interrogations and Confessions* (Inbau & Reid, 1962; for the most recent edition, see Inbau et al., 2013).

Essentially, the Reid technique consists of a three-step process. First, examiners conduct a pre-interrogation interview, the goal of which is to gather information and determine if a suspect is telling the truth or lying. Second, suspects who are identified as deceptive are subjected to an accusatorial nine-step interrogation aimed at eliciting an admission of guilt. Third, if the suspect makes a partial or full admission, the examiner seeks to convert that statement into a full and detailed narrative confession. This confession can be handwritten by the detective or suspect or audio or video recorded. In each phase, the psychologically-minded interrogator uses carefully scripted and worded language.

2 The pre-interview aimed at deception detection

Sometimes police reasonably determine that a suspect is deceptive and worthy of accusation on the basis of reports from witnesses or other forms of extrinsic evidence. Often, however, that judgment is based on a hunch. Originating from their initial use of the polygraph in the 1940's, Inbau et al. (2013) advise investigators to look for *behavioral* indicators of truth and deception in the form of purported verbal cues (e.g. long pauses, qualified or rehearsed responses), non-verbal cues (e.g. gaze aversion, frozen posture, slouching) and behavioral attitudes (e.g. unconcerned, anxious, guarded). They also recommend the use of various "behavior provoking questions" designed to evoke responses from suspects that are presumed diagnostic of guilt and innocence (e.g. 'What do you think should happen to the person who did this crime?').

Inbau et al.'s (2013) claim that they can train investigators to judge truth and deception at high levels of accuracy is not supported by research on the diagnostic value of the so-called behavioral symptoms (Kassin & Fong, 1999) or the behavior-provoking questions (Vrij et al., 2006). To the contrary, their claimed accuracy rates substantially exceed human lie-detection performance found in labs all over the world. Indeed, research has shown that people in general perform at no better than chance; that police training increases confidence while yielding only modest if any improvement in accuracy; that the behavioral symptoms cited by Reid and Associates are not diagnostic of deception; and that police in general exhibit a response bias toward seeing deception (for reviews of this vast literature, see Luke, 2019; Meissner & Kassin, 2004; Vrij, Hartwig, & Granhag, 2019).

The confidence that trained investigators have in their ability to make accurate judgments of truth and deception is dangerous on the ground, not a mere abstraction. I testified in a U.S. Army case several years ago, in Fort Drum, New York, where a CID investigator was asked why he targeted a young recruit for interrogation in a rape case. He proceeded to describe how he "knew" that this recruit was not telling the truth: "He tried to remain calm but you could tell he was nervous and every time we tried to ask him a question his eyes would roam and he would not make direct contact, and at times he would act pretty sporadic. He started to cry at one time. We actually called it off because his breathing was kind of impaired. There had to be something wrong." There was no evidence to support the prosecution so this particular defendant was speedily acquitted at trial. But the story of how he became a suspect because of a clinical hunch based on his demeanor repeats itself every day.

2.1 Hypothetical "bait questions"

One common interview ploy recommended by Inbau et al. (2013) is to ask hypothetical "bait questions" aimed at arousing behavioral cues to deception. Although lacking such evidence, a detective might ask, "Is there any reason we would find surveillance footage of you in the bank?" or "Is there any reason we should find your fingerprints on the knife?" As these examples illustrate, the bait question does not equate to the false evidence ploy, a devastating tactic that allows police to lie outright and that substantially increases the risk of false confessions (Kassin, 2021; for a survey of expert opinions on this issue, see Kassin et al., 2018). Rather, the bait question is designed merely to ask the suspect whether such evidence is plausible. Proponents of the bait question claim that a suspect's verbal and nonverbal reaction to the question will indicate whether their denials are true or false.

Alongside the relatively untested question of whether the bait question discriminates between deceptive-guilty and truthful-innocent suspects, and whether it serves to confuse them, lingers a secondary question. As more and more states require the video recording of interrogations, what effect might suggestive bait questions have on juries? Could exposure to bait questions lead jurors to incorporate the hypothetical into their memory of the evidence? Dating back to early experiments in which subjects were misled by post-event information into recalling the visual presence of a stop or yield sign in a slide show (Loftus, Miller, & Burns, 1978), a voluminous body of research has shown that misinformation in general, from various sources, can produce false memories (for an overview, see Loftus, 2005). Hence, there is reason to believe that the bait question may have this same effect if presented in court.

To test this hypothesis, Luke et al. (2017) conducted four experiments in which they presented subjects with a summary of a criminal case involving a liquor store robbery, after which they watched a video recorded interrogation of the defendant. Throughout the interrogation, the detective asked a number of bait questions. Some of these questions made casual reference to evidence from the case; others cited nonexistent evidence. For example, the summary cited fingerprints matching the suspects' on the murder weapon, but the interrogator can be heard asking, 'Is there any reason we would find your skin cells on the murder weapon?' When tested for their recall of evidence collected by police, subjects often mistakenly identified as evidence items that were referenced only in the misleading bait questions.

In followup studies, Crozier et al. (2020) found that bait questions not only altered memory but increased perceptions of the defendant's guilt in the process; even just one reference to nonexistent evidence had this effect. In fact, subjects who recalled a bait question as evidence were more likely to perceive the defendant to be guilty. In short, this tactic can mislead not only suspects but later their factfinders in court.

3 The accusatory interrogation aimed at confession

Following the nonaccusatorial interview, the suspect is either sent home or interrogated. The hunches-masked-as-science judgments of truth and deception thus become a potentially dangerous choice point in an investigation. It also means that the process of interrogation that follows is by its very nature guilt-presumptive and aimed at confession. As formally defined, "A guilt-presumptive process is a theory-driven social interaction led by an authority figure who holds a strong a priori belief about the target and who measures success by his or her ability to extract a confession" (Kassin, 2005). The consequence of this mindset is that a detective – now tunnel-visioned by a strongly held belief – will wittingly or unwittingly ask leading and provocative questions, reject the suspect's denials, and ratchet up the pressure, in turn making the suspect more anxious and the detective more determined to get a confession. This chain of events can create a feedback loop known as a self-fulfilling prophecy.

3.1 Consequences of a guilt-presumptive mindset

Social and cognitive psychology research suggests that a guilt-presumptive mindset may shape the questions a detective asks, how those questions are phrased, and hence the answers they elicit. In a classic study that illustrates the process, Mark Snyder and William Swann (1978) had pairs of subjects who were strangers to one another take part in a getting-acquainted interview. Within each pair, one subject was supposed to interview the other. But first, that subject was falsely led to believe that his or her partner was either introverted or extroverted (actually, participants were assigned to these conditions on a random basis) and told to select questions from a prepared list. Results shows that those who thought they were talking to an introvert chose mostly introvert-oriented questions ("Have you ever felt left out of some social group?"), whereas those who thought they were talking to an extrovert asked extrovert-oriented questions ("How do you liven up a party?"). Subjects thus unwittingly sought evidence that would confirm their expectations. By asking loaded questions, in fact, the interviewers actually garnered support for their beliefs. Thus, neutral observers who later listened to the tapes were also left with the mistaken impression that the interviewees really were as introverted or extroverted as the interviewers had assumed.

In a study modeled after this classic research, Kassin, Goldstein, and Savitsky (2003) had some subjects but not others commit a mock theft of $100, after which all were questioned by participant interrogators who by random assignment were led to believe that most subjects were guilty or innocent. Interrogators who presumed guilt asked more presumptuous questions (e.g., "How did you find the key that was hidden behind the VCR?" vs. "Do you know anything about a key that was hidden behind the VCR?"). They also conducted more coercive interrogations, and tried harder by all accounts to get their suspect to confess. In turn, this more aggressive style made the suspects sound defensive, which led observers who later listened to the tapes to judge them as guilty, even when they were not. To make matters worse, the most aggressive confession-driven

interrogations occurred when guilt-presumptive interrogators questioned innocent suspects whose vehement denials frustrated their efforts. Follow-up research has confirmed this chain of events in suspect interviews (Hill, Memon &
McGeorge, 2008; Lidén, Gräns & Juslin, 2018; Narchet, Meissner & Russano,
2011).

3.2 Minimization themes – promises by pragmatic implication

The guilt-presumptive structure of an American-style interrogation lays a foundation for the specific tactics that follow, a nine-step process designed to overcome a suspect's resistance, essentially by increasing the stress associated with
denial and decreasing the stress associated with confession.

Using the Reid technique, the interrogator begins with a "positive confrontation" by accusing the suspect with unyielding certainty. The interrogator then
develops psychological "themes" that justify or excuse the crime; interrupts all
statements of denial; overcomes the suspect's factual, moral, and emotional objections to the charges; ensures that the increasingly passive suspect does not
tune out; shows sympathy and understanding and urges the suspect to tell the
truth; offers the suspect face-saving alternative explanations that minimize the
moral seriousness of the crime; gets the suspect to recount the details; and con-
verts that statement into a full confession.

Recognizing the obvious risk to innocent people, U.S. courts over the years
have ruled to exclude confessions extracted from a suspect not only by direct
threats of harm or punishment but also by promises of leniency or immunity
from prosecution *(Bram v. United States*, 1898). But purveyors of the Reid technique use minimization tactics through which promises unspoken may be implied. Precipitated by expressions of sympathy and understanding, a detective
might downplay the moral seriousness of an offense by suggesting to a beleaguered suspect that his or her actions were spontaneous, accidental, provoked,
seduced, peer-pressured, alcohol- or drug-induced, caused by stress or raging
hormones, or otherwise justifiable by external factors. At times, detectives not
only "minimize" the crime but "normalize" what they allege the suspect to
have done by suggesting that they would have behaved in the same way.

The minimization theme used to excuse the crime depends on the crime
committed. For a sex crime, "Joe, no woman should be on the street alone at
night looking as sexy as she did. Even here today, she's got on a low-cut dress
that makes visible damn near all of her breasts . . . it's too much of a temptation
for any normal man." For a theft in the workplace, "Man, how in the world can
anybody with a family get along with the kind of money they're paying you? . . .

Anyone else confronted with this situation would do the same thing. Joe, your company is at fault." Different minimizing themes are scripted and worded for auto theft, blackmail, arson, child sex abuse, piracy, fraud, and numerous other crimes (Inbau et al., 2013; Senese, 2016).

On the surface, minimizing remarks may appear innocuous. No explicit promises are made regarding legal consequences. However, cases in which minimization tactics seduced innocent but beleaguered people to confess are everywhere to be found in the archives of wrongful convictions (Ofshe & Leo, 1997). In North Charleston, South Carolina, the detective told Wesley Myers, suspected in the murder of his girlfriend, "It's so easy to kill somebody when you're angry . . . Sometimes you don't know how strong you are." He even went so far as to plant this seed: "If you're guilty and it's an accident . . ." (Loudenberg, 2017). In Manitowoc County, Wisconsin, Brendan Dassey's *Making A Murderer* interrogators let him off the moral hook in no uncertain terms. After befriending Dassey and feigning sympathy, one detective said, "It's not your fault, remember that. You've done nothing wrong" (Demos & Ricciardi, 2015).

Explicit promises may not have been made in these cases, but research on the cognitive psychology of pragmatic implication suggests that leniency is communicated nevertheless. When people read text or hear someone speak, they tend to process information "between the lines," fill in the gaps, and recall not what was stated in words but what was pragmatically implied. In one study, for example, many subjects who read that "The burglar goes to the house" later mistakenly recalled that the burglar actually broke into the house. Those told that "the karate champion hit the cinder block," recalled that "the karate champion broke the cement block" (Harris & Monaco, 1978). In another study, those who heard that "The flimsy shelf weakened under the weight of the books" often recalled that the shelf actually broke (Chan & McDermott, 2006). In short, pragmatic inferences can change the meaning of a communication, leading us to infer something that is neither explicitly stated nor necessarily implied.

This research is directly applicable to the suspect being lulled by minimizing themes. In three studies aimed at examining the effects, Kassin and McNall (1991) tested the hypothesis that the minimizing remarks depicting a crime as spontaneous, accidental, provoked, pressured by others, or otherwise excusable would lead people to infer leniency in punishment. Basically, they had subjects read the interrogation of an actual murder suspect. The transcript was edited to produce three versions: One in which the detective made an explicit promise of leniency in exchange for confession, a second in which he made minimizing remarks by blaming the victim, and a third in which neither statement was made. Each subject read one version and then estimated the sentence they thought would be imposed on that suspect. The result: Compared to the no-techniques control group, subjects who read the minimization transcript had lower sentencing expectations – as if an explicit promise had been made.

This basic effect has been repeatedly replicated. Horgan et al. (2012) found that minimization tactics, even in the absence of a direct promise, induce accused laboratory subjects to view the potential consequences of confessing as less severe. In six experiments, Luke and Alceste (2020) presented MTurk subjects with an interrogation transcript in which the suspect was promised leniency outright, subjected to minimization themes, or merely questioned about the evidence. Across studies, they found that moral minimization led subjects to view the crime as less severe, which in turn reduced sentencing expectations – without leading them to believe that the interrogator had made a direct promise of leniency. They also found that communicating the importance of honesty also led subjects to draw the pragmatic inference that suspects will receive more lenient treatment if they confess. Presented with this pitch for honesty, nearly half of all subjects erroneously reported that the interrogator made a direct promise of leniency. Still other researchers went on to find not only that minimization communicates leniency but that it facilitates the innocent person's decision to confess. In fact, minimizing remarks had the same magnitude of effect on innocent suspects as an outright offer of leniency (Russano et al., 2005).

Inbau et al. (2013) have argued that minimization does not inherently communicate leniency but rather that a suspect infers the promise through a process of motivated reasoning, or "wishful thinking" (p. 213). This defense in light of the research makes little sense. It does not account for the fact that the aforementioned results are derived from and consistent with the basic *non*forensic literature on pragmatic implications. Nor does it account for the fact that minimization themes yield the inference of leniency even among neutral observers in the laboratory who are unmotivated to form the inference and have nothing to gain by doing so. In short, as U.S. courts do not typically accept confessions extracted by explicit promises of leniency because of the risk to innocents, promises unspoken via minimization essentially circumvent the law's intent.

4 The narrative confession – effects on factfinders

In *Miranda v. Arizona* (1966), the U.S. Supreme Court warned that even without using third-degree tactics, the psychological interrogation ploys advocated by the Reid technique and others compromise individual liberty and "[trade] on the weakness of individuals." This comment was accompanied by single footnote signaling the risk of a false confession. That footnote cited a then-recent New York case involving a nineteen-year-old African American man named George Whitmore.

On August 28, 1963, two young professional women in Manhattan were killed. Several months later, with these high-profile "career-girl murders" still unsolved, police questioned Whitmore, who was said to have produced an exquisitely detailed 61-page confession to both murders, a third murder, and a rape. The NYPD announced that Whitmore's confessions contained details that only the murderer could have known. Ultimately, it turned out that Whitmore had a solid if not ironic alibi: On the day of these crimes, he was on the south Jersey shore watching Reverend Martin Luther King's historic "I have a dream" speech from the Lincoln Memorial. After spending nearly three years in jail and a decade on bond, Whitmore was exonerated (English, 2011; Shapiro, 1969).

4.1 Contamination – details only the perpetrator could have known

What shocks the mind about Whitmore's false confession is that it contained 61 pages of surprisingly rich, specific, accurate chronological details about the career girl murders, many of which were not known to the public – the kinds of facts, we were told, that "only the perpetrator could have known." What is even more shocking, actually, is how often this happens and how blinding it is to factfinders even when the confessor claims innocence.

In 1987, Barry Laughman confessed to sexually assaulting and killing his elderly neighbor in Adams County, Pennsylvania. His confession contained vivid and accurate details about the positioning of the victim's body: that she was injured from a blow to the head, that her bra was pulled up to her neck, and that she suffocated on pills stuffed into her mouth. The detective testified that Laughman recounted these details on his own and he was convicted. In 2004, after 16 years in prison, Laughman was exonerated by DNA. In 2021, that DNA identified the likely perpetrator.

In 1996, Doug Warney confessed to stabbing a man in Rochester, NY. His confession detailed what the victim was wearing, what he was cooking for dinner, and how Warney used a paper towel to clean his own blood after cutting himself. Once again, this narrative described the crime; once again, the detective testified, incorrectly, that the defendant was the source of these details. In 2006, however, after 19 years in prison, Warney was DNA exonerated and the real perpetrator was identified.

For years, research has shown that confession evidence is potent in court. People do not adequately discount confessions – even when the confession was coerced (e.g., Kassin & Sukel, 1997; Wallace & Kassin, 2012), even when the confessor was a juvenile (e.g., Redlich et al., 2008), even when the confession was contradicted by DNA or other evidence (e.g., Appleby & Kassin, 2016), and even when it was reported secondhand by an informant who was incentivized to lie (e.g., Neuschatz, Lawson, Swanner, Meissner & Neuschatz, 2008).

Part of the reason that confessions are so persuasive in court is that they often contain these kinds of narrative details. Garrett (2010) examined the first 38 false confessions from the Innocence Project's DNA exoneration case files and found that 36 of these confessions contained facts about the crime that were both accurate and not in the public domain. In a followup analysis, Garrett (2015) expanded the database to 66 cases and found that 62 of these statements, or 94%, contained accurate and vivid facts that were not in the public domain: details that only the perpetrator (and police) could have known. Somehow in these cases, the police purposefully or inadvertently informed the innocent confessor about the facts of the crime through a process of "contamination."

False confession narratives contain other credibility cues as well. Appleby et al. (2013) analyzed the contents of 20 known false confessions and found that every single one contained visual and auditory details about the crime and the crime scene. They all referenced the victim and described the victim's behavior before, during, and afterward. Overall, 95% of the statements referenced co-perpetrators, witnesses, and other actors; 80% recounted what the victim allegedly said; 55% described the victim's alleged mental or emotional state ("She was scared, she could hear me coming"); and 40% expressed sorrow and remorse or heartfelt apologies for the crime they did not commit. "This was my first rape," said sixteen-year-old Korey Wise of the Central Park Five, who was innocent, "and it's going to be my last."

With police trained to withhold nonpublic crime facts from suspects so they can determine whether confessors can corroborate their admissions with demonstrations of guilty knowledge (Inbau et al., 2013), evidence of wholesale contamination is troubling. For starters, it means that both true and false confessions contain crime details and other credibility cues, making the two sets of statements difficult for factfinders later to distinguish. As to how it happens, there are two possible mechanisms: Often nonpublic crime details find their way into innocent people's confessions through purposeful acts of police misconduct (National Registry of Exonerations, 2020). At other times, however, the contamination process may be inadvertent (for a first-hand law enforcement account of how this can occur, see Trainum, 2014).

Research on conversational disclosure provides some insight into how contamination can occur even though police are advised against leaking privileged crime details. One benevolent possibility is that investigators reveal these details because interrogation has the characteristics of a conversation, and disclosure is a natural part of conversation (Shuy, 1998). In particular, linguistic scholars have noted that people tend to tailor their speech to match the intended recipient's knowledge (e.g., Grice, 1975; Stalnaker, 2002). In light of the guilt-presumptive nature of interrogations, as described earlier, perhaps police overshare nonpublic facts on the basis of their assumption that the suspect already possesses that common-ground knowledge.

To examine this possibility in a sample of laypeople who are not paid to close cases, Alceste et al. (2020) recruited 59 pairs of subjects. Half played the role of a suspect. Some were directed to break into a room in the lab, find a key hidden in a cup, use the key to open a filing cabinet drawer, which contained the combination to a briefcase behind a partition, which contained a folder, which contained an envelope, which contained a one-hundred-dollar bill. Others were not similarly instructed and knew nothing about the incident, spending time instead on an unrelated innocent task. All suspects were instructed not to confess to any crime if questioned.

Paired with each suspect, a second subject was instructed to play the role of an investigator. These investigators were taken through the various steps of the crime scene after which they interviewed a suspect in an adjacent room via Skype audio-only connection. These investigators were not informed as to whether their suspect was guilty or innocent. Half were incentivized with an offer of bonus cash to get the suspect to confess; the other half were not.

Results showed that mock investigators unwittingly communicated crime facts to both guilty and innocent participants at the same high rate, demonstrating that contamination occurred naturally in the information-gathering interview. They overshared information even when suspects steadfastly denied the charge. Two key findings in particular were noteworthy. First, investigators leaked crime details to suspects regardless of whether or not they were incentivized to get a confession. Overall, the questions they created contained an average of 3.94 out of nine key crime details (e.g., 78% disclosed the presence of a briefcase). Second, many of these leaked details found their way into all suspects' statements – including those who were innocent and had no firsthand guilty knowledge.

Through motives that are purposeful or inadvertent, it is clear that police – despite training that cautions them not to do so – will often communicate crime details while interrogating suspects. With false confessions containing the same details and content cues expected of true confessions, perhaps they can be distinguished by less overt linguistic markers. To examine this possibility, Rizzelli et al. (2021) analyzed the language of confessions from a corpus-based perspective, which allows for the discovery of systematic patterns of features across a large number of texts (Biber, 2015).

Using a psychological language analysis program called Linguistic Inquiry and Word Count (LIWC; see Pennebaker et al., 2015), they established a baseline of linguistic features in 96 presumed-true confessions drawn from a national sample of fully adjudicated state and local case files. Then they similarly analyzed 37 confessions proven false and compiled by the Innocence Project and other sources. All cases from both sources involved confessions to murder and/or sexual assault. A comparison of randomly selected confessions from these two sources revealed that they could be significantly distinguished by three linguistic categories: Compared to presumed-true confessions, proven false confessions contained fewer personal pronouns, especially "I," more impersonal pronouns (e.g., "it" and "that"), and fewer conjunctions (e.g., "and" and "then"). Together, these linguistic predictors discriminated the two sets of confessions at an accuracy rate of between 74% and 83%.

These results are preliminary and not easy to interpret. Indeed, one problem with false confessions is that the detective is the author/coauthor of these statements. Clearly, more research is needed using broader samples of confessions. From a practical perspective, however, it is clear that the similarities in content make it difficult for factfinders not privy to the process to trace the source of that information in deciding whether the confession can be trusted. This is yet another argument, perhaps the most important argument, for mandating that all interviews and interrogations be video recorded in their entirety – no if's, and's or but's (for a summary of arguments, see Kassin & Thompson, 2019; Sullivan et al., 2008).

5 Miranda: Safeguard for the accused?

Invariably, the first question people ask in response to a wrongful conviction caused by a police-induced false confession is, why didn't the exoneree exercise his rights to silence and a lawyer? Everyone thinks they know the drill: *"You have the right to remain silent. Anything you say can and will be used against you in a court of law. You have the right to have an attorney present. If you cannot afford an attorney, one will be appointed to you by the court."*

In 1966, the U.S. Supreme Court ruled that police must apprise all suspects in custody of their constitutional right to remain silent and have a lawyer present. Calling the process of interrogation "inherently compelling," Chief Justice Earl Warren's Court created a remedy: Any statement taken from a suspect without a knowing, intelligent, and voluntary waiver would be considered unlawful and barred from trial. At the time, this ruling set off an uproar among police, prosecutors, and "law and order" politicians. Almost from the start, however, research showed that a vast majority of suspects waived their rights. Fifty-plus years later, it is now clear that Miranda does not work; for views on how U.S. courts have eroded *Miranda* protections, see Kamisar (2017); Weisselberg (2008); and White (2001).

Setting aside ideological, political, and legal forces that have eroded the intended scope and power of these rights, research has shown that the language of Miranda warnings poses a problem, raising questions as to whether the waiver of these rights is knowing and intelligent. One misconception is that the Supreme Court composed the familiar warnings word for word. In fact, the Court declined to prescribe the language, leaving that to law enforcement agencies themselves.

Not surprisingly, the Court's decision not to offer a uniform warning unleashed a great deal of variability across the country. In comparisons of warnings from hundreds of jurisdictions, Rogers et al. (2007, 2008) found that the differences were often substantial (e.g., while the average warning is 53 words in length, they range from 21 to 231 words). These researchers also went on to measure the complexity of the various warnings using the Flesch-Kincaid grade estimate of readability and found that while the average Miranda warning required only a sixth-grade reading level, they ranged from a low of second grade to a post-graduate level. Substantively, 20% of all warnings examined failed to include a critical fifth admonition derived from the Supreme Court's opinion – namely, that suspects can choose to invoke their rights *at any time*.

These results have substantive implications. Over the years, psychologists have devised standardized instruments to assess Miranda comprehension. Early on, research using these tests showed that juveniles – particularly those younger than 16 – do not fully appreciate their rights and how to implement them (Grisso, 1981, 1998; Goldstein et al., 2012). These findings are so consistent that one has to wonder what the Seventh Circuit was thinking when it reinstated the overturned conviction of Brendan Dassey, "co-star" of *Making a Murderer*. Dassey was 16 years old and had a tested IQ of 73, making him even more vulnerable. In the film, he described himself as stupid. "They got into my head," he confided to his mother. Yet in defense of its decision to reinstate Dassey's conviction, this court argued that Dassey waived his Miranda rights – that he "nodded in agreement," which constituted prima facie proof that his ensuing statement was voluntary. As noted elsewhere, this analysis is only half right. "Dassey may have nodded but that gesture more likely conveyed the acquiescence to authority of a low-IQ teen – not a knowing and intelligent waiver of his constitutional rights" (Kassin, 2018).

Although cognitively capable adults fare better in their comprehension, many adults are still ill-informed about certain aspects of their rights. One third of American adults, for example, harbor the erroneous belief that "Once you give up the right to silence, it is permanent" (Rogers et al., 2010). The picture is even grimmer when situationally relevant factors are taken into consideration. Early on, Leo (1996) observed in over five hundred hours of interrogations a host of tactical approaches that police used to get suspects to waive their rights. In the course of these observations, he noted that several detectives invoked the metaphor of a confidence game, poker bluff, or skilled salesmanship to describe their approach, gaining the suspect's trust and then suggesting that they cannot help, which they would like to do, unless the suspect waives Miranda. Lab experiments using physiological and self-report data have also demonstrated that people experience a great deal of stress when accused of wrongdoing – and that this stress can substantially undermine Miranda comprehension (Scherr & Madon, 2012, 2013).

Well-intentioned as it was, Miranda has not served the protective functions that the U.S. Supreme Court intended. To more meaningfully protect the accused, one obvious possibility is to stop the practice of requiring suspects to self-invoke their constitutional rights to silence and to counsel. By making these rights the presumptive starting point of every interrogation, suspects would no longer have to break their silence in order to invoke their right to silence and the presence of a defense attorney might curb the use of excessive interrogation tactics (see Smalarz et al., 2016).

6 Looking ahead

The most obvious and necessary means of reform is to ban the tactics of the Reid technique that put innocent people at risk to confess and make it difficult for factfinders to make diagnostic judgments. Another approach is to think outside the box and adopt a whole new paradigm. Following recent developments in the UK and elsewhere in Europe, a new approach would reconceptualize "interrogation" as a process of information gathering, not confrontation aimed at confession (these alternatives fall outside the scope of this chapter; see Shepherd & Griffiths, 2013).

At the request of the American Psychology-Law Society, a division of the APA, a group of scholars wrote a Scientific Review or "White Paper" on false confessions. Joining me were Professors Steven Drizin, Thomas Grisso, Gisli Gudjonsson, Richard Leo, and Allison Redlich (Kassin, Drizin, Grisso, Gudjonsson, Leo & Redlich, 2010). After thoroughly reviewing the relevant literature, we landed on this most important remedy: "Without equivocation, our most essential recommendation is to lift the veil of secrecy from the interrogation process in favor of the principle of transparency. Specifically, all custodial interviews and interrogations of felony suspects should be videotaped in their entirety and with a camera angle that focuses equally on the suspect and interrogator."

This proposal was hardly new. And fortunately, approximately 30 states now require the recording of interrogations, not just confessions, for all or most crimes (Sullivan, 2019). These laws, however, provide too many loopholes. First, it is typically only "custodial interrogations" that trigger the need to record, not "pre-custodial" interviews during which judgments of truth and deception are made, information is communicated, and bait questions are asked. This distinction is highly problematic, however. It is too easy for police to claim that a suspect was not in custody and for judges to affirm that opinion during a pretrial suppression hearing, especially in light of research showing that "custody" is not a construct that elicits high levels of agreement within the legal community (Alceste & Kassin, 2021). Second, states that mandate recording contain too many exceptions that excuse the failure to do so rather than render the confession inadmissible – e.g., the suspect refused; the equipment was not available or malfunctioned; "inadvertence". Hence, more reform is needed on this most important means of preventing false confessions and their aftermath.

References

Alceste, Fabiana, Kristyn Jones & Saul Kassin. (2020). Facts only the perpetrator could have known? A study of contamination in mock crime interrogations. *Law and Human Behavior, 44*, 128–142.

Alceste, Fabiana & Saul Kassin. (2021). Perceptions of custody: Similarities and disparities among police, judges, social psychologists, and laypeople. *Law and Human Behavior, 45*, 197–214.

Appleby, Sara, Lisa Hasel & Saul Kassin. (2013). Police-induced confessions: An empirical analysis of their content and impact. *Psychology, Crime and Law, 19*, 111–128.

Appleby, Sara & Saul Kassin. (2016). When self-report trumps science: Effects of confessions, DNA, and prosecutorial theories on perceptions of guilt. *Psychology, Public Policy, and Law, 22*, 127–140.

Biber, Douglas. (2015). Corpus-based and corpus-driven analyses of language variation and use. In B. Heine & N. Heiko (Eds.), *The Oxford Handbook of Linguistic Analysis* (2nd edition). Oxford: Oxford University Press.

Bram v. United States, 168 U.S. 532 (1897).

Brown v. Mississippi, 297 U.S. 278 (1936).

Chan, Jason & Kathleen McDermott. (2006). Remembering pragmatic inferences. *Applied Cognitive Psychology*, *20*, 633–639.

Chomsky, Noam. (1959). Review of Skinner's *Verbal Behavior. Language*, *35*, 26–58.

Crozier, William, Timothy Luke & Deryn Strange. (2020). Taking the bait: Interrogation questions about hypothetical evidence may inflate perceptions of guilt. *Psychology, Crime & Law, 26 (9)*.Online publication, https://www.tandfonline.com/doi/full/10.1080/1068316X.2020.1742340.

Demos, Moira & Laura Ricciardi, directors. (2015). *Making a murderer* [A Netflix original documentary]. Los Angeles, CA: Synthesis Films.

Drizin, Steven & Richard Leo. (2004). The problem of false confessions in the post-DNA world. *North Carolina Law Review*, *82*, 891–1007.

English, T. J. (2011). *The savage city: Race, murder, and a generation on the edge*. New York: Harper Collins.

Garrett, Brandon. (2010). The substance of false confessions. *Stanford Law Review*, 1051–1118.

Garrett, Brandon. (2015). Contaminated confessions revisited. *Virginia Law Review*, *101*, 395–454.

Goldstein, Naomi, Heather Zelle & Thomas Grisso. (2012). *The Miranda rights comprehension instruments*. Sarasota, FL: Professional Resource Press.

Grice, H. P. (1975). Logic and conversation. In P. Cole & J. Morgan (Eds.), *Syntax and semantics: Vol. 3. Speech Acts* (pp. 41–58). New York: Academic Press.

Grisso, Thomas. (1981). *Juveniles' waiver of rights: Legal and psychological competence*. New York, NY: Plenum Press.

Grisso, Thomas. (1998). *Instruments for assessing understanding and appreciation of Miranda rights*. Sarasota, FL: Professional Resources Press.

Gudjonsson, Gísli. (2018). *The psychology of false confessions: Forty years of science and practice*. London: John Wiley & Sons.

Harris, Richard. (1974). Memory and comprehension of implications and inferences of complex sentences. *Journal of Verbal Learning and Verbal Behavior*, *13*, 626–637.

Harris, Richard & Gregory Monaco. (1978). Psychology of pragmatic implication: Information processing between the lines. *Journal of Experimental Psychology: General*, *107*, 1–22.

Heider, Fritz. (1958). *The psychology of interpersonal relations*. New York: John Wiley & Sons.

Hill, Carole, Amina Memon, & Peter McGeorge (2008). The role of confirmation bias in suspect interviews: A systematic evaluation. *Legal and Criminological Psychology*, *13*, 357–371.

Hilton, Denis. (1995). The social context of reasoning: Conversational inference and rational judgment. *Psychological Bulletin*, *118*, 248–271.

Horgan, Allyson, Melissa Russano, Christian Meissner & Jacqueline Evans. (2012). Minimization and maximization techniques: Assessing the perceived consequences of confessing and confession diagnosticity. *Psychology, Crime & Law*, *18*, 65–78.

Inbau, Fred & John Reid. (1962). *Criminal interrogation and confessions*. Baltimore, MD: Williams & Wilkins.

Inbau, Fred, John Reid, Joseph Buckley & Brian Jayne. (2013). *Criminal interrogation and confessions* (5th ed.). Sudbury, MA: Jones and Bartlett.

Josefsson, Dan. (2015). The strange case of Thomas Quick: The Swedish serial killer and the psychoanalyst who created him. London, United Kingdom: Portobello Books.

Kamisar, Yale. (2017). The Miranda case fifty years later. *Boston University Law Review, 97,* 1293.

Kamisar, Yale, Wayne LaFave & Jerrold Israel. (1994). *Modern criminal procedure* (8th ed.). St. Paul, MN: West Group.

Kassin, Saul. (2005). On the psychology of confessions: Does *innocence* put *innocents* at risk? *American Psychologist, 60,* 215–228.

Kassin, Saul. (2012). Why confessions trump innocence. *American Psychologist, 67,* 431–445.

Kassin, Saul. (2017). False confessions: How can psychology so basic be so counterintuitive? *American Psychologist, 72,* 951–964.

Kassin, Saul. (2018). Why SCOTUS should examine the case of "Making a Murderer"'s Brendan Dassey. *APA Online* June 12, 2018.

Kassin, Saul. (2021). It's time for police to stop lying to suspects. *The New York Times,* January 30, p. A23.

Kassin, Saul, Steven Drizin, Thomas Grisso, Gísli Gudjonsson, Richard Leo & Allison Redlich. (2010). Police-induced confessions: Risk factors and recommendations. *Law and Human Behavior, 34,* 3–38.

Kassin, Saul & Christina Fong. (1999). 'I'm innocent!' Effects of training on judgments of truth and deception in the interrogation room. *Law and Human Behavior, 23,* 499–516.

Kassin, Saul, Christine Goldstein & Kenneth Savitsky. (2003). Behavioral confirmation in the interrogation room: On the dangers of presuming guilt. *Law and Human Behavior, 27,* 187–203.

Kassin, Saul & Karlyn McNall. (1991). Police interrogations and confessions: Communicating promises and threats by pragmatic implication. *Law and Human Behavior, 15,* 233–251.

Kassin, Saul, Allison Redlich, Fabiana Alceste & Timothy Luke. (2018). On the general acceptance of confessions research: Opinions of the scientific community. *American Psychologist, 73,* 63–80.

Kassin, Saul & Holly Sukel. (1997). Coerced confessions and the jury: An experimental test of the "harmless error" rule. *Law and Human Behavior, 21,* 27–46.

Kassin, Saul & David Thompson. (2019). Videotape all police interrogations – Justice demands it. *The New York Times* Op-Ed, August 1, 2019.

Kassin, Saul & Lawrence Wrightsman. (1985). Confession evidence. In Saul Kassin & Lawrence Wrightsman (eds.), *The psychology of evidence and trial procedure* (pp. 67–94). Beverly Hills, CA: Sage.

Leo, Richard. (1996). Miranda's revenge: Police interrogation as a confidence game. *Law & Society Review, 30,* 259–288.

Leo, Richard. (2008). *Police interrogation and American justice.* Cambridge, MA: Harvard University Press.

Lidén, Moa, Minna Gräns & Peter Juslin. (2018). The presumption of guilt in suspect interrogations: Apprehension as a trigger of confirmation bias and debiasing techniques. *Law and Human Behavior, 42,* 336–354.

Loftus, Elizabeth. (2005). Planting misinformation in the human mind: A 30-year investigation of the malleability of memory. *Learning & Memory, 12,* 361–366.

Loftus, Elizabeth, David Miller & Helen Burns. (1978). Semantic integration of verbal information into a visual memory. *Journal of Experimental Psychology: Human Learning and Memory, 4,* 19–31.

Loudenberg, Kelly, director. (2017). *The Confession Tapes: A Public Apology.* Netflix S1, E3. https://www.netflix.com/title/80161702

Luke, Timothy. (2019). Lessons from Pinocchio: Cues to deception may be highly exaggerated. *Perspectives on Psychological Science, 14,* 646–671.

Luke, Timothy & Fabiana Alceste. (2020). The mechanisms of minimization: How interrogation tactics suggest lenient sentencing through pragmatic implication. *Law and Human Behavior, 44,* 266–285.

Luke, Timothy, William Crozier & Deryn Strange. (2017). Memory errors in police interviews: The bait question as a source of misinformation. *Journal of Applied Research in Memory and Cognition, 6,* 260–273.

Meissner, Christian & Saul Kassin. (2004). You're guilty, so just confess!" Cognitive and behavioral confirmation biases in the interrogation room. In G. Daniel Lassiter (ed.), *Interrogations, confessions, and entrapment,* 85–106. New York: Kluwer Academic.

Narchet, Fadia, Chris Meissner, & Melissa Russano (2011). Modeling the influence of investigator bias on the elicitation of true and false confessions. *Law and Human Behavior, 35,* 452–465.

National Registry of Exonerations. (2020). *Government misconduct and convicting the innocent: The role of prosecutors, police and other law enforcement.* September 1, 2020, pp. 1–218.

Neuschatz, J. S., Lawson, D. S., Swanner, J. K., Meissner, C. A., & Neuschatz, J. S. (2008). The effects of accomplice witnesses and jailhouse informants on jury decision making. *Law and Human Behavior, 32,* 137–149.

Ofshe, Richard & Richard Leo. (1997). The decision to confess falsely: rational choice and irrational action. *Denver University Law Review, 74,* 979–1122.

Pennebaker, James, Ryan Boyd, Kayla Jordan & Kate Blackburn. (2015). *The development and psychometric properties of LIWC2015.* Austin: University of Texas at Austin.

Råstam, Hannes. (2013). *Thomas Quick: The making of a serial killer.* Edinburgh, United Kingdom: Canongate Books.

Redlich, Allison, Jodi Quas & Simona Ghetti. (2008). Perceptions of children during a police interrogation: Guilt, confessions, and interview fairness. *Psychology, Crime & Law, 14,* 201–223.

Rizzelli, Lucrezia, Saul Kassin & Tammy Gales. (2021). The language of criminal confessions: A corpus analysis of confessions presumed true vs. proven false. *Wrongful Conviction Law Review, 2,* 205–225.

Rogers, Richard, Kimberly Harrison, Daniel Shuman, Kenneth Sewell & Lisa Hazelwood. (2007). An analysis of Miranda warnings and waivers: Comprehension and coverage. *Law and Human Behavior, 31,* 177–192.

Rogers, Richard, Lisa Hazelwood, Kimberly Harrison, Kenneth Sewell & Daniel Shuman. (2008). The language of Miranda in American jurisdictions: A replication and further analysis. *Law and Human Behavior, 32,* 124–136.

Rogers, Richard, Jill Rogstad, Nathan Gillard, Eric Drogin, Hayley Blackwood & Daniel Shuman. (2010). "Everyone knows their Miranda rights": Implicit assumptions and countervailing evidence. *Psychology, Public Policy, and Law, 16,* 300–318.

Russano, Melissa, Christian Meissner, Fadia Narchet & Saul Kassin. (2005). Investigating true and false confessions within a novel experimental paradigm. *Psychological Science, 16,* 481–486.

Scherr, Kyle & Stephanie Madon. (2012). You have the right to understand: The deleterious effect of stress on Miranda comprehension. *Law and Human Behavior, 36*, 275–282.

Scherr, Kyle & Stephanie Madon. (2013). "Go ahead and sign": An experimental examination of Miranda waivers and comprehension. *Law and Human Behavior, 37*, 208–218.

Scherr, Kyle, Allison Redlich & Saul Kassin. (2020). Cumulative disadvantage: A psychological framework for understanding how innocence can lead to confession, wrongful conviction, and beyond. *Perspectives on Psychological Science, 15*, 353–383.

Senese, Louis. (2016). *Anatomy of interrogation themes* (2nd edition). Chicago: John Reid & Associates.

Shapiro, Fred. (1969). *Whitmore*. Indianapolis: Bobbs Merrill.

Shepherd, Eric & Andrew Griffiths (2013). *Investigative interviewing: The conversational management approach* (2nd edition). Oxford, UK: Oxford University Press.

Shuy, Roger. (1998). The language of confession, interrogation, and deception. Thousand Oaks, CA: Sage.

Skinner, B. F. (1957). *Verbal behavior*. New York: Appleton-Century-Crofts.

Smalarz, Laura, Kyle Scherr, & Saul Kassin (2016). Miranda at 50: A Psychological Analysis. *Current Directions in Psychological Science, 25*, 455–460.

Snyder, Mark & William Swann, Jr. (1978). Behavioral confirmation in social interaction: From social perception to social reality. *Journal of Personality and Social Psychology, 36*, 1202–1212.

Stalnaker, Robert. (2002). Common ground. *Linguistics and Philosophy 25*, 701–721.

Sullivan, Thomas. (2019). Current report on recording custodial interrogations in the United States. *The Champion*, April, 54–55.

Sullivan, Thomas, Andrew Vail & Howard Anderson. (2008). The case for recording police interrogation. *Litigation, 34*, 1–8.

Trainum, James (2014). "'I did it' – Confession contamination and evaluation. *The Police Chief, 81*, June 2014: (Web-only).

Trainum, Jim. (2008). The case for videotaping interrogations: A suspect's false confession to a murder opened an officer's eyes. *The Los Angeles Times*, October 24, 2008.

Vrij, Aldert, Maria Hartwig & Pär Anders Granhag. (2019). Reading lies: Nonverbal communication and deception. *Annual Review of Psychology, 70*, 295–337.

Vrij, Aldert, Samantha Mann & Ronald Fisher. (2006). An empirical test of the Behaviour Analysis Interview. *Law and Human Behavior, 30*, 329–345.

Wallace, D. Brian & Saul Kassin. (2012). Harmless error analysis: How do judges respond to confession errors? *Law and Human Behavior, 36*, 151–157.

Weisselberg, Charles. (2008). Mourning Miranda. *California Law Review, 96*, 1521–1601.

White, Welsh. (2001). *Miranda's Waning Protections: Police Interrogation Practices after Dickerson*. Ann Arbor, MI: University of Michigan Press.

Wigmore, John Henry. (1985). *Wigmore on evidence* (4th ed.). Boston, MA: Little, Brown. (Original work published 1904.)

Izabela Skoczeń, Aleksander Smywiński-Pohl

The context of mistrust: Perjury ascriptions in the courtroom

Abstract: Classical theories of linguistic pragmatics focus on communication oriented solely to the exchange of information. In such communication, beliefs about what is said are naturally intertwined with beliefs about the world. However, in linguistic exchanges oriented at attaining concrete social goals, where speakers are less trustworthy, the interactions between the mentioned beliefs are more complex. We investigate these interactions with reference to a series of experiments in a courtroom setting providing empirical support for the strategic speech inferential framework. We argue that it is the strategic context, more than the role of the speaker, which governs inferential content. We also argue that in mistrust contexts, if a statement does not conform to the state of the world, participants judge it as perjurious; by contrast, if it is objectively true, it is judged as not being a lie irrespective of the knowledge and intention to deceive attributed to the speaker.

1 Introduction

The legacy of Paul Grice, both in philosophical and psycholinguistic work, resulted in a systematic inquiry of cooperative speech processing oriented at transmitting information (cf. for an overview Cummins & Katsos 2019). Moreover, recent scholarship has focused on violations of the Gricean cooperative speech framework resulting in lies (cf. for instance Chisholm & Feehan 1977; Adler 1997; Meibauer 2005; Saul 2012; Horn 2018; Wiegmann, Willemsen & Meibauer 2021). Since lying is a folk concept in the sense that we ascribe responsibility for lies to our interlocutors, the psycholinguistic inquiry into the mechanisms of lying ascriptions appears vital. This is especially pressing in the non-cooperative courtroom setting, as in this

Note: This research was funded by the Polish National Science Centre, Preludium Grant No. 2015/19/N/HS5/00029 we would like to thank Larry Horn, Krzysztof Kasparek, Anna Drożdżowicz, Katarzyna Kijania-Placek and Krzysztof Posłajko for comments on drafts. Appendix and data can be found in the repository: https://osf.io/j83ex/?view_only=42fbe334029d4caab212926ae905ffed

Izabela Skoczeń, Jagiellonian Centre for Law, Jagiellonian University,
e-mail: izaskoczen@gmail.com
Aleksander Smywiński-Pohl, AGH University of Science and Technology,
e-mail: apohllo@agh.edu.pl

https://doi.org/10.1515/9783110733730-013

special context of mistrust one can be ascribed criminal responsibility for willful lying. Willful lies in the courtroom, if they are 'material', i.e. relevant to the case, can be labeled perjury and can entail serious legal consequences. In the recent years, a debate has emerged over how consistent perjury ascriptions in the courtroom are (cf. for instance Solan 2018; Skoczeń 2021). The present paper aims at a systematic, experimental inquiry of the mechanisms of speech processing in mistrust contexts such as the courtroom context, as well as at a psycholinguistic inquiry into the patterns of lying ascriptions in such contexts. Let us now proceed to a discussion of our experimental hypotheses and study design.

1.1 Ex ante and ex post probability distributions (all experiments)

The courtroom context differs from communication contexts described by Gricean and neo-Gricean pragmatics in two main ways. First, there is not one 'accepted purpose or direction of the talk exchange' to quote the Gricean cooperative principle (Grice 1989). The goals of different interlocutors diverge. The goal of the speaker is not simply to transmit information. Second, the level of trust between interlocutors is lower than in every-day speech contexts.

For the above reasons, we hypothesize that since the goals of a lawyer and the goals of a lay person differ in the courtroom, these two groups might exhibit different pragmatic inference patterns.

In our experiments, we will employ the notion of scalar implicature in order to show the difference in pragmatic inference patterns. A scalar implicature can be illustrated with the following sentence:

(1) Some of the students passed the exam.

If you hear the above sentence, you (usually) infer that not all of the students have passed. This is a two-step procedure: by virtue of not using the stronger "all", the speaker implicates that for all she knows not all the students passed, which is a first stage implicature. Next, given that we assume the speaker's competence, the second-stage implicature is derived: the speaker knows/believes that not all the students passed. (cf. Horn 1984; Sauerland 2012; Geurts 2010).

If you know or believe that the speaker does not know how many students passed, then your inference might change. The 'not all' inference is an implicature because if you consider the sentence:

(2) If some of the students fail, then the teacher will be fired.

you infer that if all of the students fail, then the teacher will naturally also be fired. Thus, the meaning of 'some' is 'some and maybe all' instead of 'some but not all'. There is a heated debate on the exact nature of this sort of inference; however, since this debate is not directly relevant to the present considerations, we will assume a pragmatic account of scalar implicature.[1]

In court, the goal of the lawyer is to win the case, and thus she might employ a 'strategic inferential framework' and ascribe to her interlocutor the communicated content that will best fulfill the lawyer's goal and win the case instead of trying to grasp the speaker's genuine communicative intention (Marmor, 2014; Skoczeń, 2019; Struchiner et al., 2020). If this is the case, the legal expert group, as strategic hearers, might for instance infer the scalar implicature 'not all' from the use of the quantifier term 'some'. They might do so even while not being sure if not all of the objects at stake have the relevant property. They might do so in order to ascribe as much content to the interlocutor as possible since this makes it easier to attack the opponent's claims.

This means that since, quite often, lawyers in the courtroom have a contrary goal to their courtroom opponent, they will ascribe implicatures even if the opponent later denies having wanted to communicate them, so as to win the case. For example, if the lawyer argues that her client lent money to X, while X claims the sum was a donation, if X says that he received money from the lawyer's client, the lawyer might want to claim that X's utterance implicates this was a loan rather than a donation. This can be the case even though the opposing counsel could, and should, object to such a claim not supported by evidence.

Our experimental design concentrates only on lawyers (or lay people) assessing others' utterances. Note that the situation might be different if a lawyer assesses her own utterance – then she might have an incentive to admit as little content communicated as possible so as to be held responsible for as little as possible. In such a situation, a lawyer uttering 'some' would probably deny that her

---

1 In Polish, the word 'niektóre' is the counterpart of the quantificational uses of the English word 'some' (see for instance (Piasecki, 2002)). 'Niektóre' can mean either 'przynajmniej niektóre' (at least some) or 'tylko niektóre' (some but not all). Moreover, in an ordinary conversation 'niektóre' is understood as 'tylko niektóre' (some but not all) and this inference is contextually cancellable. However, just as the French counterparts of the quantifier term 'some' are 'certains (de)' or 'quelques' (see for instance (Pouscoulous et al., 2007)), the Polish 'niektóre' is plural in the sense that it is used to denote at least two objects, contrary to the word 'some,' which can be used to refer to a single object. Thus, while logicians are fully warranted to label the English 'some' as a linguistic counterpart of existential quantification leading to the inference that there exists at least one object that has a property, this inference is not warranted as far as the Polish word 'niektóre' is concerned.

utterance meant 'not all', but we will not test this hypothesis due to the structure of our experimental design.

By contrast, we hypothesize that a lay person's goal will be to refrain from implicature inference out of caution due to the low level of trust toward the interlocutor, who might be either an expert witness, the opposing party, or a lawyer. As a result of a low level of trust the hearer does not treat the utterance as information and thus is in an uncertainty context. This is a result predicted by the Rational Speech Act (RSA) model in an uncertainty situation (cf. Goodman & Stuhlmüller 2013), as well as a low-trust situation such as polite speech[2] (cf. Yoon et al. forthcoming). The RSA model predicts that human inferencing is based on a Bayesian mechanism – agents update their beliefs about the likelihood of an event based on the received information. A distrustful utterance is not treated as information – as a result of hearing it, the agent does not update his ex ante probability distribution on an event happening. If there are two possible world states (either not all of the objects have a property or all of the objects have the relevant property) then the agent infers from the utterance 'some of the objects have the property' that it is equally likely that not all and all of the objects have the property. In other words, the agent does not infer the scalar implicature 'not all' from the quantifier word 'some'.

To be more precise, according to RSA, a listener infers the state of the world given the heard words and considerations about how much information the speaker has about the state of the world and, in the case of the RSA model tailored to politeness (cf. Yoon et al., forthcoming), how willing is the speaker to share information. The listener models the speaker as choosing words depending on available information about the state of the world or willingness to share information given the hearer's prior probability distribution of the likelihood of a world state. Moreover, the speaker is modeled as choosing words or sentences given a utility function that assigns values to particular utterances in particular contexts. Thus, the utility function is designed to guarantee a congruence between communicated content and inferred state of the world (we test whether this is the case with the control question on communicated content described later).

We test whether the RSA predictions also hold in the case in a special kind of uncertainty context, namely a low trust or mistrust context. Next we test whether legal experts also conform to the RSA model predictions, or rather, since their goal is to win the case, whether they infer the scalar implicature 'not

---

2 If you are at a party and the food is awful, but you want to be polite, you usually thank your host for the wonderful meal. Thus, your host has reasons to disbelieve you – in this sense polite speech is speech in which we do not trust our interlocutors.

all'. We check whether this is due to a purposive neglect of the mistrust that enables one to hold the speaker responsible for as much content as possible.

Consequently, we suggest the following divergent hypotheses to be tested in a series of experiments:

> **Hypothesis for the non-legal-expert group in courtroom context:** The level of trust affects pragmatic inference. The scalar implicature of 'some', namely 'not all', is *not* inferred (maximize non-inference; minimize inference) in the courtroom context.

> **Hypotheses for the legal expert group in courtroom context:** The divergence of goals of interlocutors affects pragmatic inference. The scalar implicature of 'some', namely 'not all', is inferred (maximize inference; minimize non-inference) in the courtroom context.

In order to test the inference of the scalar implicature of 'some' in the courtroom we need in our experimental design a set of objects that could have a relevant property and a hearer that does not know how many objects have the relevant property. We need the latter so as to reflect the situation in a courtroom, which, we hypothesize, often involves a special kind of uncertainty, namely mistrust. The hearer mistrusts the speaker's utterance regardless of whether she attributes knowledge to the speaker. Consequently, as described above, we hypothesize that the patterns of inference based on mistrusted utterances might be different for lay people. We hypothesize that a hearer either will behave just as if there were no utterance at all or will retrieve her prior probability distribution on an event so as to infer from the utterance that it is equally likely that only some of the objects have the relevant property in question or that all do. By contrast, lawyers might exploit the mistrust and thus infer the implicature 'not all' so as to win the case. What differentiates uncertainty and mistrust inferences is, we hypothesize, that the mistrust inference patterns will hold regardless of whether the hearer ascribes knowledge to the speaker.

In order to control that it is the context that influences the patterns of inference we add a standard certainty (control) scenario where we hypothesize the scalar implicature 'not all' is inferred.

Furthermore, to allow participants to provide in the experiment an answer that 'at least some and maybe all' (no implicature) of the objects have the relevant property, we need a betting measure.[3] Thus, participants have a sum of 100

---

3 We use the betting measure with t-tests on compositional data following an established paradigm in this type of study (cf. Senn & Manley 1966; Hauser 1991; Desarbo, Ramaswamy & Chatterjee 1995; Frank et al. 2009; Frank & Goodman 2012; Bergen, Levy & Goodman 2016; Schuster & Degen 2019).

units, which they divide between objects just as in Goodman & Stuhlmüller (2013). In other words, they bet on zero, one, two, three, four or five objects having the property at stake. In order to distinguish an implicature from an a prior (ex ante) belief that it is unlikely for all objects to have the property, we first tell participants that the objects 'almost always' have the property, for example, example taken from Skoczeń (2021):

> Invoices for more than 1,000,000 USD found in the documentation of companies in bankruptcy proceedings are almost always unpaid. The court is considering five invoices for more than 1,000,000 USD in a bankruptcy proceeding.

Subsequently, we measure an ex ante probability distribution of the objects having the relevant property:

> How many of the invoices, do you think, are unpaid?

Only next we present participants with the utterance and another, ex post, betting measure question:

> The main accountant of the company says in reply to the judge's question:
> Some of the invoices are unpaid.
> Now, how many of the invoices, do you think, are unpaid?

Before proceeding, one caveat: in the RSA framework the implicature is a matter of degree. Thus, although we speak of inference or lack of inference of implicature, we mean that there might be more or less of an inference, rather than a binary yes or no judgment.

1.2 Control question on communicated content (experiment 4 and 5)

In order to make sure that the answer on the ex post betting measure question conforms to participants' beliefs about communicated content (and not what they overall find most likely to be true given their world knowledge), we ask a subsequent question:

'What did the main accountant want to communicate with the utterance 'Some of the invoices are unpaid'?
a) Only some but not all of the invoices are unpaid.
b) At least some and maybe all of the invoices are unpaid.

The hypotheses motivating posing this question are the following:

> **Tested hypothesis**: The answers to the question on ex post probability distribution do not conform to the answers on the communicated content. If the bets on 5 objects having the relevant property are low, then participants choose the answer 'at least some and maybe all' in the question on the communicated content. Alternatively, if the bets on 5 objects having the relevant property are high, then participants choose the answer 'some but not all' in the question on the communicated meaning.

If we find evidence for the tested hypothesis, then we will conclude that low trust contexts with divergent goals of interlocutors result in a divergence of beliefs about the state of the world and beliefs about communicated content (cf. Papafragou & Musolino 2003). This would mean that the group dependent patterns of inference hypothesized in the question on ex post probability distribution are in fact inferences about the state of the world rather than patterns of linguistic pragmatic inference.

1.3 The knowledge question (all experiments)

After the above-described scenario, we pose the following question:

> 'Do you think the main accountant knows exactly how many invoices are unpaid? (yes/no)'[4]

In experiments 1 and 2 we treat the knowledge question merely as a comprehension control question; however, for experiments 3–5 we analyze all answers to this question as we predict it provides interesting insights on the lying question, which was posed only in experiments 3–5 (see section 1.5 of the present paper).

We predict that the above-described patterns in answers on the ex post probability will persist independently of whether participants will ascribe knowledge about the state of the world to the speaker. This is because we predict that non-legal-experts will be cautious both with speakers to whom they ascribe knowledge (thinking that they might be under-informative) and speakers to whom they do not ascribe knowledge (since they might be bluffing). By contrast, legal experts might

---

4 To infer the implicature 'not all' from the utterance 'some of the invoices are unpaid' one does not need to know that the speaker knows exactly how many invoices are unpaid, one only needs to know that the speaker knows that there is at least one unpaid invoice. However, since we used the betting measure, in which each participant had to rate each possibility, we had to ask whether participants judged that the speaker knew *exactly* how many invoices were unpaid, as otherwise the betting measure would not be correctly understood.

want to ascribe as much content as possible and attack the opponents claims independently of whether the opponent has knowledge of the relevant facts.

Tested hypothesis: knowledge ascriptions to the speaker do not affect patterns of implicature inference in courtroom context.

1.4 The trust question (experiment 5)

In order to check whether the level of trust in the courtroom scenario is indeed low we add an additional question for instance: 'To what extent is respondent's utterance trustworthy?' Participants answer the question on a 7 point Likert scale ranging from 'not at all trustworthy' to 'completely trustworthy'. We hypothesize that it is due to the low level of trust that participants refrain from implicature inference:

Tested hypothesis: the level of trust affects implicature inference.

We believe that the present study might constitute a small step toward building a conceptual model of interpersonal trust, which is acknowledged as needed in the literature, as there is the need to make a coherent story from both empirical survey answers and neuroscience data (cf. Cramer et al., 2009; Krueger & Meyer-Lindenberg, 2019).

1.5 The question on lying (experiments 2,3,4 and 5)

Table 1: Conceptual relation between ascribed state of the world, communicated content and knowledge.

		Communicated content			
		All		**Not all**	
Knowledge		✓	✗	✓	✗
Ascribed State of the world	All	Not a lie	Guess/bluff – lie?	(Meibauer 2014): lie (Saul 2012): not a lie	Guess/bluff – lie?
	Not all	Is an under-informative statement a lie in courtroom?	Guess/bluff – lie?	not a lie	Guess/bluff – lie?

We hypothesize that the different possible answers to the above questions on probability distributions (ascribed state of the world), communicated content, and knowledge might in turn influence ascriptions of lying, which bears a direct practical influence on determining what is perjury in court. This is because, in common law systems, it is often the jury that decides whether a crime has been committed. Consequently, it is often the jury that decides whether an utterance was perjurious. The jury is instructed to do so following a folk concept of lying and it is precisely the folk concept of lying that this study purports to investigate. The parameters of the present study are involved in statutory definitions of perjury. For example, the 1911 U.K. Perjury statute and its 1975 revision states that:

'If any person, in giving any testimony (either orally or in writing) otherwise than on oath, where required to do so by an order under section 2 of the Evidence (Proceedings in Other Jurisdictions) Act 1975, makes a statement –
(a) which he knows to be false in a material particular, or
(b) which is false in a material particular and which he does not believe to be true, he shall be guilty of an offence . . . '.[5]

The Polish Criminal Code, in its article 233 § 1, states that perjury can be committed when a person either states something false or withholds the truth. Withholding the truth does not appear in the U.S. or U.K. statutory language; however, this difference is irrelevant for the present studies, which do not test cases of withholding truth.

Jennifer Saul (2012) argues that a false 'not all' implicature is not a lie even if the speaker knows that all of the objects have the property. She claims it is merely misleading behavior. A series of empirical studies provide support for this claim as far as the quantifier term 'some' is concerned (cf. Doran et al., 2012; Weissman & Terkourafi, 2019; for a claim that the inference is dependent on ethical views cf: Willemsen & Wiegmann, 2017).

We test whether this is the case in a (special) courtroom context through an additional lying question. We pose the following question:

'Do you think the main accountant's utterance is a lie? (1 = not a lie; 7 = completely a lie)'

Here are the hypotheses:

Null hypothesis: When the speaker knows the 'some' implicature is false she is not lying.   (cf. Saul, 2012)

---

5 The selection of legal rules has been taken from a lecture by Laurence Horn.

Tested hypothesis: When the speaker knows the 'some' implicature is false she is lying.

(cf. Meibauer, 2005, Meibauer, 2014)

We also predict that regardless of answers in the question on ex post probability and communication, if participants rate a speaker as not knowing the state of the world, they will not label her utterance a lie since it is only a guess or bluff. Here is the hypothesis:

Tested hypothesis: Guessing or bluffing in courtroom is not a lie.

Finally, we predict that participants that point to implicature in questions on ex post probability distribution (low bets on 5 objects having the relevant property) and choose the answer 'at least some and maybe all' in the question on communicated content as well as ascribe knowledge of the state of the world to the speaker do not rate the utterance as a lie since it is merely an under-informative statement.

Tested hypothesis: under-informative statements are not lies in courtroom.

2 Experiment 1a

2.1 Participants

62 participants (31 of whom were female) were recruited through the Polish crowd-sourcing service 'Research Online' and completed the experiment for a small payment. Each participant had 6 trials, so we collected 372 trials in total. We filtered out participants who were non-native speakers of the Polish language as well as those who had studied law, so as to receive a sample unconnected with any legal profession or education, leaving a sample of 40 participants. The mean age was 40.5 years.

2.2 Methods and materials

We employed a similar scenario structure as Goodman and Stuhlmüller (2013). Six scenarios were presented in a random order: unpaid invoices, forged signatures, defective products, unlicensed programs, illegally employed workers, and rent arrears (cf. Appendix). The scenarios were written in light of the Polish legal system in the Polish language. The study began with two control questions checking the general attention of the participants and familiarizing them with the betting measure (each participant had a fictive sum of 100 PLN –

Polish currency) which they divided through betting, when deciding between six options: whether zero, one, two, three, four, or five objects exhibit a property in question). Each scenario began with an increase in the prior probability that *all* the objects have the salient property, for example:

> *Invoices for more than 1M PLN found in the documentation of companies in bankruptcy proceedings, are almost always unpaid.* (cf. Skoczeń, 2021)

Next, we provided information stating that the total number of objects is five. We also posed a question regarding the prior probability distribution of the objects exhibiting the property in question, for example:

> *The court is examining five such invoices. How many of the invoices, do you think, are unpaid?*

This ex ante probability measurement was made to later 'distinguish an implicature from a prior belief that it is unlikely for all objects to have the property.' (Goodman and Stuhlmüller, 2013).

As mentioned, each participant had a sum of 100PLN (Polish currency), which could be divided (through betting) when deciding between six options: whether zero, one, two, three, four, and five objects exhibit the property in question. We displayed an utterance containing the information that 'some' of the objects have the property in question. In the first experiment, the speakers needed to be perceived as those with a high probability of possessing precise knowledge concerning the exact number of objects in question exhibiting the relevant property. Thus, the speakers were presented as claimants or witnesses in civil proceedings. Criminal case proceedings contexts, where the speaker is the accused, were avoided, as the accused might well lie in self-defense. The aim of the present study is not to investigate outright lies, but rather manipulative behavior of the kind where the speaker intends to convey as little information as possible so as to be held responsible for as little as possible, while at the same time not putting themselves at risk of being accused of perjury. Hence, the speaker will not utter propositions she believes to be false but will simply seek to utter less-than-fully informative statements. For example, one of the displayed utterances was:

> *The chief accountant of the company says in reply to the judge's question: 'Some of the invoices are unpaid.'*

This was then followed by a question about the posterior probability distribution of the objects featuring the relevant property:

> *Now, how many of the invoices, do you think, are unpaid?*

The last question posed was a control question:

Do you think the chief accountant knows exactly how many invoices are unpaid?

As a response to this last question, participants distributed the sum of 100 PLN between yes and no answers. We filtered out participants who bet larger sums on the response 'no' rather than on 'yes'. This was done to collect responses only from those participants who believed that the speaker indeed knew exactly how many objects were characterized by the property but wished to convey as little information as possible. We chose speakers in the scenarios who would typically possess all the relevant information.

2.3 Results

Since this is a per-trial experiment, we rejected any trial where a participant bet less than 70 PLN on the response 'yes' in the control question, as this meant they did not understand the scenario – in this experiment we were interested in answers of participants who ascribed knowledge to the protagonist. As a result of this filtering, 173 trials out of 372 remained. We also performed a within-subjects analysis to check whether the per trial measure gives robust results. Since each participant had 6 trials (as there were 6 scenarios), we rejected any participant who answered the control question incorrectly at least once, resulting in 13 participants remaining out of 62. However, no important differences were found between the per-trial and the within-subjects analyses. Thus, we present in detail only the per-trial analysis, as it is the commonly adopted paradigm for this type of experiment (cf. Goodman and Stuhlmüller, 2013). At the end of the results section, only within-subject t-tests for the first scenario are presented.

The mean bets in the control question regarding the response 'yes' was $M = 74.74$, $SD = 33.79$, while the mean bets on the response 'no' were $M = 25.26$, $SD = 33.79$. As there was no effect on implicature formation stemming from any particular scenario, we collapsed the results across this factor: in a one-way ANOVA with all priors and posteriors as within-subject factors and scenarios as between subject factors, there was an effect of scenario only on the prior bets that all objects had the property: $p = .019$; all the remaining p values were above .090.

The prior elicitation was as predicted (see Figure 1), namely, the mean bets on 5 objects having the property were higher than bets on 4 and so on: 5>4>3>2>1

The exact values can be found in Table 2 below.

We performed a paired-samples t-test; bets on 4 and 5 were similar: $t (172) = -.36$, $p = .72$, $d = .61$. Moreover, bets on 4 were greater than bets on 3: $t (172) = -3.76$, $p < .001$, $d = 5.92$.

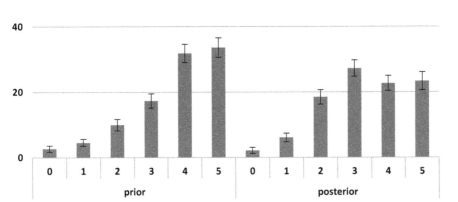

Figure 1: Mean bets in the responses to the prior and posterior probability distributions (for lay people); error bars denote the standard error of the mean.

Table 2: Mean responses to the Prior Probability Distribution Question for Lay participants; SE denotes the standard error of the mean.

Prior	0	1	2	3	4	5
Mean	2.66	4.59	9.95	17.32	31.87	33.61
SE	.95	1.07	1.74	2.14	2.74	2.98

The responses to the posterior probability distribution were also as predicted (See Table 3). The scalar implicature was not inferred since the bets on 3, 4 and 5 objects exhibiting the property were similar. We performed a paired samples t-test; bets on 3 and 4 were similar: t (172) = 1.191, p = .235, d = 1.88. Bets on 4 and 5 were also similar: t (172) = −.17, p = .865, d = .27.

Table 3: Mean responses to the Question regarding the Posterior Probability Distribution; SE denotes the standard error of the mean.

Posterior	0	1	2	3	4	5
Mean	2.14	6.12	18.47	27.24	22.67	23.36
SE	.92	1.31	2.21	2.53	2.32	2.81

The within-subjects (rather than per-trial) results showed the same trend in the means. We also performed the relevant t-test. In the responses to the prior probability distribution, bets on 3 and 4 differed: $t(12) = -2.51$, $p = .027$, $d = .95$, while the responses to 4 and 5 were similar: $t(12) = .29$, $p = .779$, $d = .14$.

In the responses to the posterior probability distribution, bets on 3 and 4 did not differ: $t(12) = .75$, $p = .470$, $d = .24$, the responses to 4 and 5 were also similar: $t(12) = -.32$, $p = .758$, $d = .12$.

2.4 Discussion

As predicted, in the strategic courtroom context, an utterance did not give rise to a scalar implicature generated by the quantifier term 'some', namely 'not all' (see also a similar study, though in this study the speakers were explicitly described as untrustworthy: Grodner & Sedivy, 2011).

What we discuss next is whether the same effect would occur with lawyers (rather than lay people) as participants, which we now proceed to investigate in experiment 1b.

3 Experiment 1b

3.1 Participants

52 participants (23 of whom were female), practicing lawyers (attorneys) and trainees, were recruited through their exclusive Facebook group. Each participant had 6 trials, thus we collected 312 trials in total. We included only native speakers of Polish with a Master's degree from law school and therefore rejected two participants. The mean age was 34.5 years.

3.2 Methods and materials

We used the same scenarios and methods as in experiment 1a.

3.3 Results

Since this was a per-trial experiment, we rejected each trial where a participant bet less than 70PLN on the response 'yes' in the control question, with 219 trials out of 312 remaining as a result of this filtering. We also performed a within-subjects analysis: since each participant had 6 trials (as there were 6 scenarios), we rejected any participant who answered the control question incorrectly at least once, leaving us with 16 participants out of 52. However, we did not find any important differences between the per-trial and the within-subjects analyses. Thus, we present only the per-trial analysis, as this is the typical design parameter for this type of experiment (cf. Goodman and Stuhlmüller, 2013).

The mean bets in the control question regarding the response 'yes' were M = 77.04, SD = 30.61, and the mean bets on the response 'no' were M = 22.96, SD = 30.61.

There was no effect of any particular scenario on implicature inference, so we collapsed results across this factor: in a one-way ANOVA, with all priors and posteriors as within-subject factors and scenarios as between-subject factors, there was an effect of scenario only on the prior bets that all objects had the property: $p = .003$. All the remaining p values were non-significant: all p values > .091.

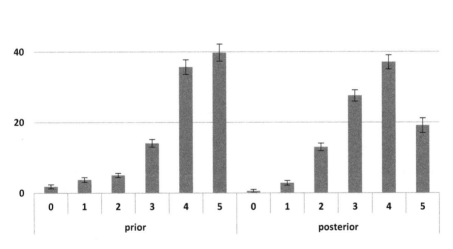

Figure 2: Mean Responses to prior and posterior probability distributions (lawyers); error bars denote the standard error of the mean.

The prior elicitation was as predicted (see Figure 2), namely, the mean bets on 5 objects having the property were higher than bets on 4 etc.: 5>4>3>2>1. The exact values can be found in Table 4 below:

Table 4: Responses to the question regarding the prior probability distribution, SE denotes the standard error of the mean.

Prior	0	1	2	3	4	5
Mean	1.81	3.70	4.99	14.06	35.67	39.75
SE	.59	.70	.59	1.09	2.07	2.43

We performed a paired-samples t-test: the bets on 5 were similar to the bets on 4: $t(218) = -.99$, $p = .322$, $d = .12$. The bets on 3 were significantly lower than the bets on 4: $t(218) = -9.13$, $p < .001$, $d = .88$.

Crucially, the responses to the posterior probability distribution differed from the results collected in experiment 1a, and were as predicted. This is because the bets on 5 were lower in number than the bets on 4. We performed a paired-samples t-test: $t(218) = -5.32$, $p < .001$, $d = .59$. Moreover, the bets on 3 were also lower than the bets on 4: $t(218) = -3.25$, $p = .001$, $d = .36$. Thus, this provided prima facie evidence that participants inferred the scalar implicature.

Table 5: Responses to the question regarding the posterior probability distribution; SE denotes the standard error of the mean.

Posterior	0	1	2	3	4	5
Mean	.53	2.81	12.95	27.54	37.07	19.1
SE	.46	0.66	1.1	1.59	2.0	2.1

The within-subjects results (rather than per-trial results) for the first scenario showed the same trend as in the means. We also performed the relevant t-test and found that for the responses to the prior probability distribution, bets on 3 and 4 differed: $t(15) = -5.21$, $p < .001$, $d = 1.92$, while the responses to 4 and 5 were similar: $t(15) = 1.21$, $p = .244$, $d = .60$.

In the responses to the posterior probability distribution, bets on 3 and 4 did not differ: $t(15) = -1.13$, $p = .278$, $d = .51$; by contrast the responses to 4 and 5 were different: $t(15) = 3.51$, $p = .003$, $d = 1.33$.

3.4 Discussion

Lawyers were more likely to infer the scalar implicature than lay participants. Future studies will be needed to investigate the reasons for the divergence. One potential reason is strategic: ascribing responsibility for more pragmatic content to your courtroom adversary is conducive to vigorously challenging their claims.

One could argue that lawyers understand 'some' as 'some but not all' because it is a practice of legislative drafting to write explicitly 'some or all' in the legal rule if the rule is obliged to include the 'all' case. Otherwise, the single 'some' is always understood as 'some but not all'. However, while this is a standard practice in common law, it is not established as robust practice in continental law. Since we carried out the experiment in the Polish language with the participation of Polish lawyers, this argument seems unfounded.[6]

---

6 In fact, in common law, statutory language and/or precedent often makes the alternative interpretation explicit, as in the case dubbed by L. Horn 'Grice v. California'. This is a case where the discussion concentrated on the decision as to whether Gricean maxims could be applied, while interpreting documents relevant to applying the law in California. (Horn, 2017a, p. 36):

'The courts recognize that employers do not have any duty to disclose information about their employees. However, if an employer chooses to provide a reference or recommendation, the reference giver must include factual negative information that may be material to the applicant's fitness for employment in addition to any positive information. Campus managers and supervisors who provide employment references on current or former employees must be aware that untrue, incomplete or misleading information may cause a different liability – negligent referral. The court in Randi M. v. Livingston Union School District, 1995 Cal. App. LEXIS 1230 (Dec. 15, 1995), found that, 'A statement that contains only favorable matters and omits all reference to unfavorable matters is as much a false representation as if all the facts stated were untrue.' (http://shr.ucsc.edu/procedures/reference_check/index.html)' (Horn, 2017b).

In the abovementioned case, if the employer does not explicitly write all of the negative facts he or she is aware of concerning the employee, this could be treated as a breach of the law. By contrast, if the law took only Gricean maxims under consideration, the problem would not arise. Paul Grice gives the example of a letter of recommendation written by a Professor recommending his former student for a Philosophy job, stating merely that 'Mr. X's command of English is excellent and he is never late for class'. For Grice, this clearly implicates, especially by virtue of the maxim of relevance, that X is not a good philosopher, and thus, not a good candidate for the job. In contrast, the law in California demands that this conclusion be explicitly formulated.

Another common law case where 'what is implicated' is incorporated into 'what is said' by a legal rule is the distinction between trying and attempting: trying merely implicates not

4 General discussion – experiments 1a and 1b

In the courtroom context, there is extremely little informative speech as the speaker wants to communicate as little as possible so as to be held responsible for as little as possible; the hearer ascribes to the speaker communicated content that the hearer believes to conform to the speaker's communicative intention.

In experiment 1a, we show that the strategic context can discourage lay hearers from engaging in implicature inference. This is in conformity with the pragmatic model of scalar implicatures, which emphasizes the role of context as the crucial factor that governs such inferences. This claim is strengthened by our results in experiment 1b, since in this study we hypothesize that lawyers' behavior arises from pragmatic considerations of courtroom-winning strategies.

We also suggest that lay participants (non-lawyers) treated speech as 'less-than-fully trustworthy speech' where the speaker wishes to maximize the amount of information withheld from the hearer and where the cautious hearer ascribes as little content to the speaker as possible. In other words, the hearer's interpretation is a literal interpretation because the hearer knows that the speaker might seek to deny the content she seemed to implicate.

In experiment 1b, we believe that lawyers treated speech as strategic speech, where the speaker wishes to maximize the amount of information uncommunicated to the hearer and the hearer wishes to ascribe to the speaker as much communicated content as possible, so as to be able to hold him responsible for as much communicated content as possible. This in turn enables the hearer to question the speaker's claims, and thus possibly win the case.

To sum up, in experiment 1a, in the strategic courtroom context, an utterance by a potentially untrustworthy speaker did not give rise to the scalar implicature of the quantifier term 'some', namely 'not all' (cf. Grodner & Sedivy, 2011). To gain additional support for this claim, and to investigate whether it is indeed the lack of trust or rather the strategic nature of the context that has the dominant effect, we designed experiment 2, in which we manipulated the speaker in the experimental design. In each scenario, we randomly displayed information regarding either a speaker who can be usually trusted in the concrete context, or a speaker that cannot be fully trusted (exactly as in the previous experiments described in this section). We predicted that lay people will derive the standard

---

succeeding (Grice, 1989); however, attempting, by virtue of legal rules, is necessarily failing (Horn, 2017a, p. 36).

To sum up, while in common law, making the alternative interpretation (or the upper lexical bound of the scale) explicit is a robust practice, this is less frequently the case in continental law.

scalar implicature based on the utterance of a trusted person, and refrain from implicature inference when an untrustworthy person speaks.

5 Misleading versus lying

At the outset of this article, we assumed that the example utterance: 'Some of the invoices were unpaid' is not a lie, but that it is merely under-informative; after all, it could be the case that indeed all of the invoices are unpaid. Yet if the utterance is merely misleading and not a lie, can it be perjury (a special, willful lie in court)? The concept of lying is a folk concept determined by the linguistic practices of a society cf. (A. Turri & Turri, 2015; Willemsen & Wiegmann, 2017; Weissman & Terkourafi, 2019; Wiegmann et al. 2021; Reins & Wiegmann 2021). The standard definition of a full-fledged lie requires 4 conditions to be fulfilled:

(1) Criteria for lying:
    (C1) S says/asserts that p
    (C2) S believes that p is false
    (C3) p is false
    (C4) S intends to deceive H[7] (Horn 2017a)

It is disputable whether all of the above elements are necessary for lying. However, what is strictly relevant to our present considerations is C1: for perjury to occur, the speaker needs to commit herself to her statement; in other words, he or she must perform the speech act of assertion (Solan & Tiersma 2005, Horn 2017a,b). What constitutes an assertion (for example what is a norm of assertion) is also an empirical issue – much discussion ensues as to whether this norm is the requirement of knowledge, or rather a 'mere' justified belief (cf. Turri 2015; Pagin 2016; Kneer 2018).

While the norm of assertion is disputed, the majority of scholars agree that asserted content is expressed rather than implicated. Thus, since an implicature does not enter the truth conditions of the expressed proposition, it is controversial as to whether a false implicature (inferred from a true statement) can be labeled a lie, or should merely be termed misleading behavior. Moreover, this is crucial for defining perjury, since it is the speaker who is to be blamed for a lie,

---

7 The requirement that lying involves more than just the objective falsity of the statement dates back at least to Augustine (cf. Chisholm & Feehan 1977; Adler 1997; Saul 2012; Horn 2017a). Turri & Turri (2015) find that folk ascriptions of lying require objective falsity; however their methodology is criticized by Wiegmann et al. 2016 and Horn 2017b). What we argue in the present paper that the strong influence of objective falsity on lying judgments is a bias (cf. Skoczeń 2021).

while the blame for misleading statements can be shared by the speaker and the hearer, seeing as the hearer could have been more cautious in his inferences (Laurence R. Horn 2017a).

In the philosophical literature, there are two opposing stances on the issue. Very roughly, the first stance (represented for instance by Jörg Meibauer) states that false implicatures are lies because they are encompassed by the speaker's intention to lie (Meibauer, 2005, 2014). The second stance, roughly, claims that false implicatures are not lies but merely mislead for the reason that implicatures do not enter the truth conditions of the utterance (Chisholm & Feehan 1977; Adler 1997; Saul 2012). In fact, according to this view, they express a separate proposition, which is separately truth evaluable. In other words, the truth or falsity of the uttered proposition is independent of the truth or falsity of the implicature. At least in the case of particularized conversational implicatures, one can usually only hold a speaker responsible for what he or she said, rather than what was implicated (Saul 2012). In the case of generalized conversational implicatures and, a fortiori, conventional implicatures, such a responsibility might arise, although the stakes are lower. This view is also shared by others: (Dynel, 2011; Fallis, 2009; Michaelson, 2016; Horn, 2017a; Solan, 2018).

Horn argues, contra Meibauer, that conventional implicatures, which are part of encoded meaning but not part of what is said, are not relevant to whether the speaker lied, but they are clearly part of what the speaker is responsible for. However, this is a different degree of responsibility, and the nature of the violation is different as well. Consequently, this might not license perjury (Meibauer, 2014; Horn, 2017b, p.165).

Experimental studies point to the fact that people's judgments on whether a false implicature is a lie or not are dependent on their moral views (see Willemsen & Wiegmann 2017, Weissman & Terkourafi 2019). Assessing whether a false implicature is a lie can also depend on whether the listener respects the speaker's 'face', which includes for instance the speaker's political views (Sanderson 1995; Bonnefon, Feeney & Villejoubert 2009; Weissman, 2019). In addition to all this, experimental studies confirm that the speaker's presumed competence or impairment also influences utterance interpretation in a systematic way (Grodner & Sedivy 2011). Thus, perceiving an utterance as a lie may be dependent on a wide variety of specific characteristics of the speaker.

In the present experimental design in a courtroom, the differences in the assessments as to whether there was an occurrence of lying cannot be explained exclusively in terms of 'face', since the same people may take on various roles. For example, an attorney can represent, in different cases, parties with completely contrary goals and aims. Thus, courtroom speech is more a game of goals than a game of particular individuals. We now turn to an experimental design which attempts to capture differences in implicature derivation depending on whether the speaker is to be trusted or not, along with differences in judgment as to whether the implicated content constitutes lying, depending on whether the speaker is to be trusted.

Before proceeding, let us note that we will collect and compare cross-linguistic data concerning lying ascriptions in the courtroom both in the Polish and English languages. We do so because the semantics of the English verb 'to lie' are similar to the semantics of the Polish verb 'kłamać'. Equally, the English verb 'to mislead' corresponds to the Polish verb 'zwodzić'. The definitions of perjury are slightly different in the common law versus Polish legal systems, however we will test cases in which this difference does not matter (see section 1.5). (The distinction between lying and misleading is a focus of several other papers in this volume.)

6 Experiment 2

6.1 Participants

59 participants (37 of whom were female), unconnected with any legal profession or education, were recruited through the Polish crowd-sourcing service 'Research Online' and completed the experiment for a small payment. The mean age was 39.92 years. Each participant had 6 trials so we collected 356 trials in total. We filtered out participants who were non-native speakers of the Polish language as well as those who had studied law; we also measured participants' time of reaction and filtered outliers. This resulted in 54 participants remaining.

6.2 Methods and materials

The scenarios and methods were identical to those in experiments 1a and 1b, with one exception. Namely, in experiment 2, we varied the choice of speaker in the following way: a trustworthy or untrustworthy speaker was assigned to each of the six scenarios (cf. Appendix) in random order. The trustworthy speakers were either government representatives or expert witnesses. Furthermore, we investigated the folk concept of lying by adding a statement detailing the true state of affairs at the end of each scenario. We stated: 'It turns out that all of the invoices were unpaid'. This was followed up by a question: 'Do you think that the utterance was a lie?'

6.3 Results

Since this is a per-trial experiment, we rejected each trial in which a participant bet less than 70PLN on the response 'yes' to the control question. 249 trials out of 356 remained. We also performed a within-subjects analysis: since each participant had 6 trials (as there were 6 scenarios), we rejected each participant who answered the control question incorrectly at least once. This meant that 30 participants out of 59 then remained. However, we did not find important differences between the per-trial and the within-subject analyses. Thus, we present only the per-trial analysis, as is typical for this kind of experiment (cf. Goodman and Stuhlmüller, 2013).

The mean bets for the control question regarding the response 'yes' was $M = 80.79$, $SD = 28.16$, while the mean bets for the response 'no' were $M = 19.21$, $SD = 28.16$. There was no effect produced by a change in scenario, so we collapsed results across this factor: in a one-way ANOVA, with all priors and posteriors as within-subjects factors and scenarios as between-subject factors, only bets on the priors 0 and 5 were significantly influenced by scenarios: for bets on the prior 0 the p value was .042; for bets on the prior 5 the p value was .011. All the remaining values for both priors and posteriors were nonsignificant: all p values above .058 (.058 is for bets on the prior 3).

The prior probability distribution elicitation was as predicted: the highest bets were on all of the objects possessing the property (see Figure 3). The exact mean bets on each option in the untrustworthy speaker condition (zero, one, two, three, four or five objects exhibiting the property) are in Table 6.

The mean bets on each option for the trusted speaker condition are in Table 7.

The responses to the posterior probability distribution were only partially as predicted. The prediction that in a special, strategic context the implicature is not inferred was borne out since the highest bets were on the option that 'five' objects have the property (i.e. all of them). However, the responses did not conform to the prediction that the implicature would arise if a trustworthy speaker produces the utterance at stake in the strategic context. This is because the bets on 5 were highest for both the trusted speaker and untrustworthy speaker conditions.

In the untrustworthy speaker condition, the mean bets on each option (zero, one, two, three, four or five objects having the property) are shown in Table 8.

The bets on 4 were not greater than the bets on 5; we performed a paired-samples t-test: $t(127) = -1.83$, $p = .069$, $d = .27$. Thus, the implicature was not inferred.

In the trusted speaker condition, the mean bets on each option (zero, one, two, three, four, five objects with the property) are shown in Table 9.

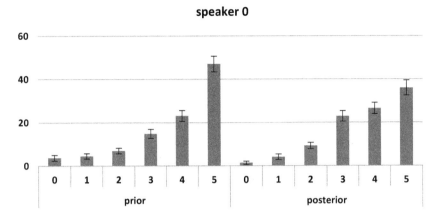

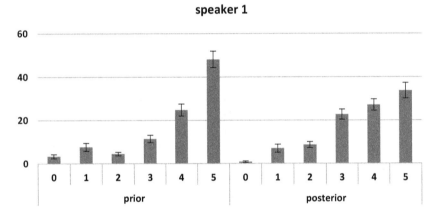

Figure 3: Responses to prior and posterior probability distributions for trusted (speaker 1) and untrustworthy (speaker 0) speakers; error bars denote the standard error of the mean.

Table 6: Mean bets on each option in the question regarding prior probability distribution in the untrustworthy speaker condition; SE denotes the standard error of the mean.

Untrusted speaker prior	0	1	2	3	4	5
M	3.63	4.47	6.94	14.81	23.09	46.98
SE	1.35	1.23	1.28	2.05	2.49	3.63

Table 7: Mean bets on each option in the question regarding prior probability distribution in the trusted speaker condition; SE denotes the standard error of the mean.

Trusted speaker prior	0	1	2	3	4	5
M	3.40	7.65	4.56	11.48	24.75	48.15
SE	.9	1.81	0.83	1.69	2.78	3.75

Table 8: Mean bets on each option in the question regarding posterior probability distribution in the untrustworthy speaker condition; SE demotes the standard error of the mean.

Untrusted speaker posterior	0	1	2	3	4	5
M	1.34	4.01	9.22	22.92	26.56	35.95
SE	.75	1.29	1.43	2.5	2.59	3.46

Table 9: Mean bets on each option in the question regarding posterior probability distribution in the trusted speaker condition; SE denotes the standard error of the mean.

Trusted speaker posterior	0	1	2	3	4	5
M	.84	7.06	8.68	22.70	27.06	33.66
SE	.38	1.81	1.42	2.35	2.61	3.63

The bets on 4 were not greater than the bets on 5; we performed a paired samples t-test: $t(120) = -1.22$, $p = 0.225$, $d = .19$. Thus, the implicature was not inferred, even though the speaker was trusted.

However, it is notable that there was a significant difference between the mean bets on 5 for the question regarding prior probability distribution across conditions and the mean bets on 5 for the question regarding posterior probability distribution across conditions; we performed a paired samples t-test: $t(248) = 5.96$, $p < .001$, $d = .32$. Thus, participants did note a difference in degree between the formulations 'almost always' and 'some'.

When participants answered the control question positively, in other words, they believed that the speaker was aware of the actual number of objects exhibiting the property, while stating 'some' (even though all the objects possessed the property), responses to the question about whether the utterance was a lie were positive in both conditions. The mean responses to the 1–7 Likert scale are presented in Table 10.

Table 10: Mean responses to the question whether the tested utterance was a lie; SD denotes the standard deviation from the mean.

Condition	Untrusted speaker	Trusted speaker
M	5.55	5.51
SD	2.07	2.00

There was no difference in the assessment of lying between the trusted and untrusted speakers conditions – perhaps our manipulation was unsuccessful. We performed an independent samples t-test: t (247) = −.17, p = .869, d = .03. The within-subjects (rather than per-trial) for the first scenario results depicted the same trend in the means. We also performed the relevant t-test across the two conditions (trusted vs. untrusted speaker). In the responses to the prior probability distribution, bets on 3 and 4 differed: t (29) =−2.95, p = .006, d= .72, while the responses to 4 and 5 were similar: t (29) =−1.20, p = .241, d=.40. In the responses to the posterior probability distribution, bets on 3 and 4 were similar: t (29) =−.63, p = .534, d= .19, just as the responses to 4 and 5 were similar: t (29) =.02, p = .987, d= .01.

6.4 Discussion

The participants did not infer the implicature 'not all' regardless of whether the utterance was produced by a trusted or non-trusted speaker. Thus, the psychological effect of a strategic context is strong enough to prompt the subjects to refrain from inferring implicatures even if the utterance is produced by a speaker who, prima facie, has no reason to mislead or lie.

After the participants received the information that all of the objects instantiated the property and were asked the question whether the speaker's utterance was a lie, contra other studies, (Saul, 2012, Dynel, 2011;Fallis, 2009; Michaelson, 2016; Horn, 2017a; Solan, 2018), we found that subjects take false scalar implicatures of the word 'some', namely 'not all' to be lies. Perhaps this is because, when discovering the state of the world, participants retrospectively blame the protagonist for trying to deceive them (see Skoczeń, 2021).

If the speaker's manipulation has no influence in the courtroom on the extent of the implicature inference from 'some' and the assessment of lying, we next consider whether it is the hearer's manipulation that has an impact, or whether it is in fact the special context that is decisive rather than the speaker's or hearer's manipulation.

7 Experiment 3

7.1 Participants

63 participants, 30 of whom were female, unconnected with any legal profession or education, were recruited through the Polish crowd-sourcing service 'Research Online' and completed the experiment for a small payment. Their mean age was 41.25. Each participant had 6 trials, so we collected a total of 384 trials. We filtered out participants who were non-native speakers of the Polish language as well as those who had studied law. We also measured the participants' reaction time and removed outliers; 52 participants remained.

7.2 Methods and materials

We used identical scenarios to experiment 2 with one modification – the following information was added at the beginning of each scenario: 'Imagine you are the judge/claimant's representative/respondent's representative in the present case'. The purpose of the addition was to observe whether role- and goal-switching has any effect on utterance interpretation on the part of the hearer.

7.3 Results

As in experiments 1 and 2, since this is a per-trial experiment, we rejected any trial where a participant bet less than 70 PLN on the response 'yes' for the control question, which left 294 trials out of 384. We also performed a within-subjects analysis: since each participant had 6 trials (there were 6 scenarios), we rejected any participant who answered the control question incorrectly at least once. This left 33 participants out of 63. Again, as in previous experiments, we did not find important differences between the per-trial and the within-subjects analyses. Thus, we present only the per-trial analysis, as is usually the case in this type of experiment (cf. Goodman and Stuhlmüller, 2013).

The mean bets in the control question regarding the response 'yes' were $M = 84.41$, $SD = 24.6$, and the mean bets on the response 'no' were $M = 15.59$, $SD = 24.6$. There was no effect in terms of the scenario, so we collapsed across this factor: in a one-way ANOVA with all priors and posteriors taken as within-subject factors, and the scenarios taken as between-subject factors, across all conditions, all the p values were above .123.

The prior elicitation was as predicted, the highest bets were on 5 objects having the property (cf. Figure 4). We performed a paired samples t-test comparing the prior bets on 4 and 5, and found there was a significant difference: t (293) =7.80, p<.001, d=.46.

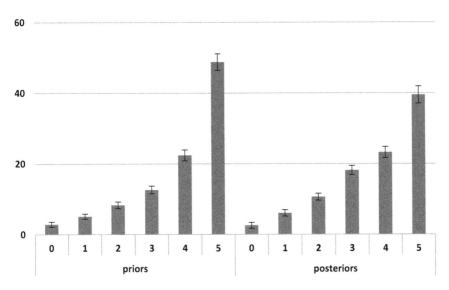

Figure 4: Mean bets in the question regarding the prior and posterior probability distributions across all conditions; error bars denote the standard error of the mean.

We performed a t-test comparing the posterior bets on 4 and 5 and found a significant difference: t (293)= 4.78, p<.001, d=.28.

The implicature was not inferred in all conditions, independent of the identity of the hearer, as in all conditions the bets on 5 were highest (cf. Figure 5).

In all four conditions, participants responded to the question about whether the speaker's utterance was a lie in the positive (see Figure 6). We performed a one sample t-test: there was a significant difference in the average responses and the mid-point (4): t (293) =13.26, p<.001.

We also checked the results of participants that answered the control question in the negative – namely, they believed the speaker did not know how many of the objects exhibited the property. Interestingly, those participants also considered the speaker's utterance a lie; see Figure 7. We performed a one sample t-test: there was a significant difference between the average response and the mid-point (4): t (15) = 2.68, p = .017.

For the first scenario, the within-subjects (rather than per-trial) results depicted the same trend in the means. We also performed the relevant t-test across

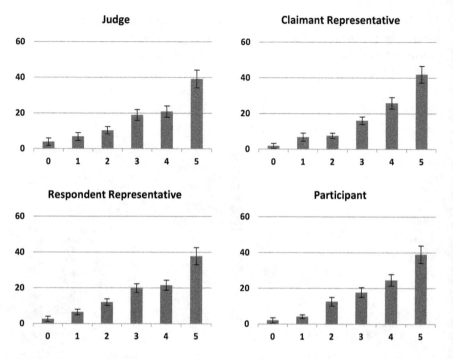

Figure 5: Mean posterior bets in all four conditions for four different hearer roles; error bars denote the standard error of the mean.

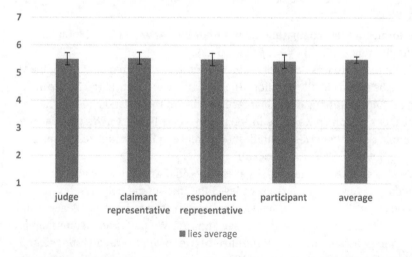

Figure 6: Mean responses to the question about whether the utterance tested was a lie; error bars denote the standard error of the mean.

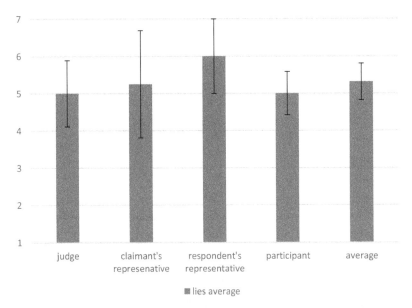

Figure 7: Mean responses to the question as to whether the utterance tested was a lie according to participants who judged that the speaker did not know the actual state of the world; error bars denote the standard error of the mean.

the four conditions (all hearers). In the responses to the prior probability distribution, bets on 3 and 4 were similar: t (32) = −1.74, p = .091, d= .44; the responses to 4 and 5 were different (but as predicted higher bets on 5): t (32) = −2.55, p = .016, d= .79. In the responses to the posterior probability distribution, bets on 3 and 4 were similar: t (32) = −.64, p = .529, d= .15, just as the responses to 4 and 5 were borderline similar: t (32) = −2.10, p = .044, d= .65.

7.4 Discussion

Contrary to our predictions, the role of the hearer did not in fact influence utterance interpretation – the implicature was not inferred. In courtroom contexts, levels of trust are so low that venturing to make pragmatic inferences is seen as perilous – and thus, resorting to the literal meaning seems to be considered the safer bet in every instance.

Moreover, participants assessed the utterance as a lie in all four conditions. This is inconsistent: participants first understand the utterance as 'some and perhaps all of the invoices are unpaid' and next, when it turns out that 'all of the invoices are unpaid', they retrospectively judge the utterance as a lie. As extensively argued in (Skoczeń 2021) this effect replicates and may be considered a cognitive bias.

Even though the courtroom context is one where the probability of attempts at deception is extremely high, participants still assessed false implicatures as lies, contrary to claims arising from other theoretical studies in this area (Saul, 2012; Dynel; 2011; Michaelson, 2016; Horn, 2017a; Solan, 2018).

8 Experiment 4

In this experiment we test whether the results replicate in the English language with non-legal experts. We also test whether participants' judgments of communicated content conform to their answers on ex post probability distributions. If these are divergent, then it might be that participants' beliefs about what is said and the state of the world also diverge. We test again the interactions between state of the world, knowledge and lies. This time we ask the question on lying without stating that 'It turns out that all the invoices were unpaid' so as not to prime participants with the state of the world in their assessment of lying. This is because, we predict that the affirmative results on the lying question in the previous experiments are due to a bias: people do not look at the intention to deceive but at whether the claim is objectively true or false.

8.1 Participants

497 participants were recruited online via Amazon Mechanical Turk and performed the survey on the Qualtrics platform. The IP address location was restricted to the USA. Participants who failed the attention check were excluded, leaving a sample of 381 participants (36% of participants were female; the mean age was 40.6 years). The study was pre-registered.[8]

8.2 Methods and materials

After being presented with an attention check, participants were presented with one out of five scenarios (see Appendix for all scenarios), here is an example:

---

[8] Link to preregistration: https://aspredicted.org/blind.php?x=an5du3; link to repository with all data: https://osf.io/j83ex/.

Invoices for more than $1M USD found in the documentation of companies in bankruptcy proceedings are almost always unpaid. The court is considering five such invoices.

This formulation was intended to boost the ex ante probability of all of the invoices being unpaid. Next, participants answered the following question on the prior (ex ante) probability distribution:

How many of the invoices, do you think, are unpaid?

Thereafter, participants were presented with the following information:

The main accountant of the company says in reply to the judge's question: 'Some of the invoices are unpaid.'

Subsequently, they were asked a question on the posterior (ex post) probability distribution:

Now, how many of the invoices, do you think, are unpaid?

Next followed questions on knowledge, meaning and lies:

Do you think the main accountant knows exactly how many invoices are unpaid? (yes/no)

What does the main accountant's utterance mean?
(a)   Only some but not all of the invoices are unpaid.
(b)   At least some and maybe all of the invoices are unpaid.

Do you think the main accountant's utterance is a lie?
(1 = not a lie; 7 = completely a lie)

This time we asked the question on lying without stating that 'it turns out that all the invoices were unpaid' so as not to prime participants with the state of the world in their assessment of lying.

8.3 Results

The prior elicitation was as predicted, the highest bets were on 5 objects having the property (see Figure 8). We performed a paired samples t-test comparing the prior bets on 4 and 5, and found there was a significant difference: $t(380) = -5.22$, $p < .001$, $d = .44$.

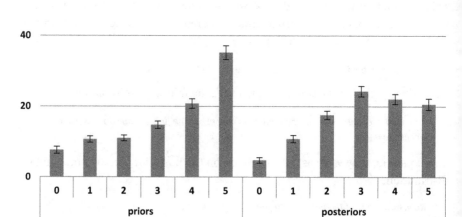

Figure 8: Mean bets on ex ante and ex post probability distributions averaged over all 6 scenarios; error bars represent the standard error of the mean.

We performed a t-test comparing the posterior bets on 4 and 5 and found no significant difference: t (380)=.62, p=.539, d=.05. Thus on average the bets on 5 were comparable to bets on 4 (see Figure 8).

There was an effect of scenario: we performed a repeated measures ANOVA with posterior bets on 4 and 5 objects having the property as within subject factor and scenario as between subjects factor: F(5) = 2.689, p=.021. 232 participants answered that the speaker knew how many objects had the property, while 149 answered this question in the negative.

Interestingly, answers on the question on meaning were different than the bets on the ex post probability distribution – most participants judged that 'some' in the utterance meant 'some but not all' just as if they had inferred a scalar implicature (see Figure 9).

The participants responded to the question about whether the speaker's utterance was a lie in the negative independently of whether they judged the speaker as knowing how many of the objects had the property in question (see Figure 10). All answers were below the midpoint 4 We performed a one sample t-test: there was a significant difference in the average responses and the midpoint (4): t (380) =–9.01, p < .001.

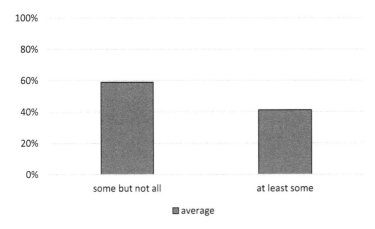

Figure 9: Percentage of participants who chose each of the two answers ('some but not all' versus 'at least some') in the question on the meaning of the utterance.

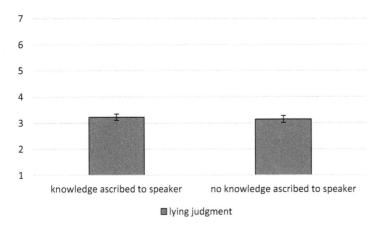

Figure 10: Mean answers in the lying question depending whether participants judged the protagonist as knowing how many of the objects have the relevant property; error bars depict the standard error of the mean.

8.4 Discussion

We found that the ex post probability distribution was flat – answers on 4 and 5 objects having the same property did not differ significantly. However, in the question on meaning, a majority of participants judged 'some' as meaning 'some but not all'. This hints toward a potential divergence between beliefs about the world (some and maybe all of the objects have the relevant property) and beliefs about what is said (some but not all of the objects have the property). There remains the question of whether these answers are indeed due to a divergence of beliefs or rather to the fact that a betting measure is difficult to understand, as well as the fact that we used technical legal language. For this reason, in the last experiment, we use an additional attention check to familiarize participants with the betting measure and we simplify the language of the scenarios to explore whether our results will be robust.

Interestingly, contrary to previous findings (experiments 2 and 3), this time participants judged the protagonist as not lying, independently of whether they judged him as knowledgeable. The difference between the present experiment and experiments 2 and 3 was that in the present experiment we did not add before the question on lying the information 'It turns out that all of the objects had the relevant property'. Consequently, we hypothesize that lying judgments in courtroom are biased by information about the objective state of the world (cf. Guthrie et al., 2001; Kneer & Machery, 2019; Turri & Turri, 2015; Wiegmann, Samland, Waldmann, 2016). In other words, participants do not look at a potential belief of the speaker that the asserted proposition is false in their assessments of a statement as a lie, but rather they look at what the objective state of the world is (see Figure 11). This is striking and goes against the theoretical tradition, which claims that the speaker's belief that the asserted proposition is false is required to lie, while the state of the world is irrelevant (Coleman & Kay 1981; Carson 2012). We suspect that participants are biased by the state of the world, which makes participant ascribe blame to the speaker (for data supporting the hypothesis, see Skoczeń 2021).

The bias view:

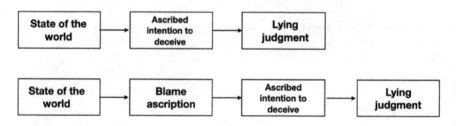

Figure 11: The bias view with regard to lying ascriptions.

It remains a project for a future study to investigate whether the objective state of the world drives blame ascriptions, which in turn drive deceptive intentions ascriptions, which finally drive lying judgments (Skoczeń 2021). We test in the next experiment whether the findings concerning lies from experiment 4 are robust.

9 Experiment 5

In the last experiment we test whether the results from experiment 4 are robust. We employ the design of experiment 4 with a couple of alterations. We add an additional attention check just as in Goodman & Stuhlmüller, 2013 in order to familiarize participants with the betting measure. We also employ less legal and technical language – we reformulate scenarios in everyday speech (for exact wording see Appendix section 1.2). Moreover, we switch the formulation of the question on meaning so as to be sure that this is a question on communicated content rather than on dictionary meaning. We also add an additional question on the level of trust to check whether our scenarios generated low trust ascriptions. Finally, we add a non-courtroom control scenario to investigate whether the levels of trust and patterns of implicature ascription will differ according to predictions.

9.1 Participants

694 participants were recruited online via Amazon Mechanical Turk and performed the pre-registered survey[9] on the Qualtrics platform. The IP address location was restricted to the USA. Participants who failed the attention check were excluded, leaving a sample of 568 participants (female: 53%; mean age: 40.7 years).

9.2 Methods and materials

The design was analogous to experiment 4 (for exact scenarios see Appendix section 1.2) with content modifications. We added a non-courtroom control scenario (labels in bold omitted):

> 'The apples collected on John's farm are almost always red. John just received and opened a basket containing five apples from his farm.
> **Ex ante**: How many of the apples, do you think, are red?
> John checks the five apples and says: 'Some of the apples are red.'
> **Ex post**: Now, how many of the apples, do you think, are red?'

---

9 Link to preregistration: https://aspredicted.org/blind.php?x=7dm4ia.

In all scenarios we added an altered question on meaning, so as to make sure that partic- ipants asses the communicated content rather than dictionary meaning:

Communication: 'When uttering 'some of the apples are red' John wants to communicate that:

a)   Only some but not all of the apples are red.

b)   At least some and maybe all of the apples are red.'

We also added an additional question on trust to check whether the trust ma- nipulation was successful. All questions listed below were presented in ran- domized order in *each* scenario:

'**trust**: To what extent is John's utterance trustworthy? (1 = not at all trustworthy; 7 = completely trustworthy)

Knowledge: Do you think John knows exactly how many apples are red? (yes/no)

Lying: Do you think John's utterance was a lie?'

All the remaining scenarios were courtroom scenarios (cf. Appendix section 1.2).

9.3 Results

The prior elicitation was as predicted, the highest bets were on 5 objects having the property (for courtroom scenarios see Figure 12, while for control scenario see Figure 13). We performed a paired samples t-test (control scenario excluded) comparing the prior bets on 4 and 5, and found there was a significant differ- ence: t (488) =−5.84, p<.001, d=.40.

We performed a t-test comparing the posterior bets on 4 and 5 (control sce- nario excluded) and found a significant difference: t (488)= −4.34, p<.001, d=.32.

On average the bets on 5 were lower than on 4 (see Figure 13). There was an effect of scenario: we performed a repeated measures ANOVA with posterior bets on 4 and 5 objects having the property as within subject factor and scenario as between subjects factor F(5)=7.558, p<.001. The bets on 4 and 5 objects having the property were similar in courtroom scenarios 2 and 7 (p values below .065 in paired samples t-test comparing posterior bets on 4 and 5). Scenario 6 was incon- clusive (borderline significance, p = .056 in paired samples t-test comparing pos- terior bets on 4 and 5). The difference was significant in scenarios 3, 4 and 5 (all *p* values below .015 in paired samples t-test comparing posterior bets on 4 and 5).

The trust manipulation was unsuccessful: the mean bets on the trust ques- tion (control scenario excluded) were 4.57 (SD = 1.443). All answers were above the midpoint 4. We performed a one sample t-test: there was a significant differ- ence in the average responses and the mid-point (4): t (488) =8.75, p<.001. Thus, contrary to predictions, participants trusted the speaker's utterance.

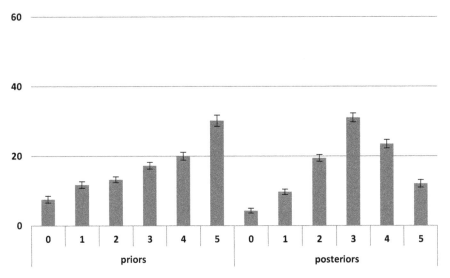

Figure 12: Mean bets on ex ante and ex post probability distributions averaged over all courtroom scenarios; error bars denote standard error of the mean.

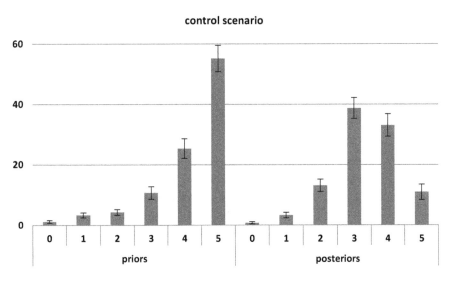

Figure 13: Mean bets on ex ante and ex post probability distributions in control scenario; error bars denote the standard error of the mean.

In courtroom scenarios (control scenario excluded) 377 participants answered that the speaker knew how many objects had the property, while 191 answered this question in the negative. In a paired samples t-test there was a significant difference between ex post bets on 4 and 5 objects having the property (all *p* values below .001).

Interestingly, this time the answers on the communication question were the same as the bets on the ex post probability distribution – most participants judged that some meant 'some but not all' just as if there was a scalar implicature (see Figure 14).

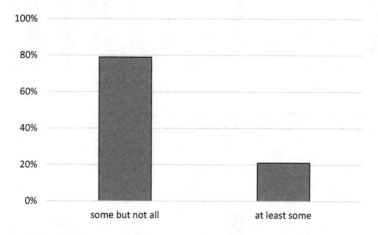

Figure 14: Proportions of participants that chose each of the two answers in the question on communication (control scenario excluded).

Again, just as in experiment 4, the participants responded to the question about whether the speaker's utterance was a lie in the negative, independently of whether they judged the speaker as knowing how many of the objects had the property in question (see Figure 15). All answers were below the midpoint 4. We performed a one sample t-test: there was a significant difference in the average responses and the mid-point (4): t (488) =−17.91, p<.001.

9.4 Discussion

Since in the present experiment our manipulation of trust turned out not as predicted (participants ascribed high instead of low trust levels), we cannot state conclusively whether the level of trust affects pragmatic inference. The results indicate that with a high level of trust the implicature was inferred (in all scenarios,

Figure 15: Mean answers in the lying question depending whether participants judged the protagonist as knowing how many of the objects have the relevant property; error bars depict the standard error of the mean.

including the control scenario). It remains to be investigated whether low levels of trust would make participants refrain from implicature inference (Skoczeń 2021). The answers on the question on the posterior probability distribution conformed to the answers on the question on communicated content. There was no divergence between beliefs about the world and beliefs about what is said. We hypothesize that this was again due to the high level of trust. Knowledge ascriptions did not affect the patterns of bets on ex post probability distributions. Trust was revealed to affect the relation between ex post bets on 4 and 5 objects having the property in question.

Interestingly, participants judged the protagonist as not lying, independently of whether they judged the protagonist as knowing how many objects had the property. Thus, guessing or bluffing in courtroom context is not a lie. Since participants judged that not all of the objects had the property in question and that the protagonists' communicated content conformed to this, they naturally judged her as not lying. For this reason, we could not further investigate the bias hypothesis in the present experiment (for further bias investigation cf. Skoczeń, 2021).

It might have been that the trust manipulation was unsuccessful this time since we used less technical language than in experiment 4. It remains a project for future studies, funding permitted, to investigate whether the technical language in experiment 4 indeed generated low trust levels. It also remains a future project to devise a non-legal language scenario with low trust levels and to observe the patterns of implicature inference and lying judgments.

10 General discussion

We find that the scalar implicature of 'some', namely 'not all', is weakened in uncertainty contexts just as the RSA predicts (maximize non-inference; minimize inference). Moreover, the divergence of interlocutors' goals affects pragmatic inference. The scalar implicature of 'some', namely 'not all', is stronger (maximize inference; minimize non-inference) in the courtroom context if the hearer has legal training. Thus, we provide preliminary empirical support for the strategic speech inferential framework.

In low trust contexts, the answers to the question on ex post probability distribution do not conform to the answers on the communicated content. If the ex post bets on 5 objects having the relevant property are high, then participants choose the answer 'some but not all' in the question on the communicated meaning. Thus, in low trust contexts the beliefs about what is said and the beliefs about the world diverge and lawyers exploit this divergence to achieve their discursive goals. We also find that ascribing knowledge to the speaker does not affect patterns of implicature inference in courtroom context. Rather, these patterns are affected by trust ascriptions.

Finally, guessing or bluffing in courtroom is not considered a lie: in experiment 4, even though participants' beliefs about what is said and the state of the world diverged, they did not treat the speaker as lying, even if they did not ascribe knowledge to the speaker. Interestingly, they treated a knowledgeable speaker as also not lying – the level of responsibility for the scalar implicature of 'some' in the courtroom, just as Jennifer Saul predicts, is low. At this point, we refrain from generalizing this conclusion to other types of implicatures since, as numerous previous studies have shown (cf. Doran et al, 2012, van Tiel et al., 2014), people rate different types of implicatures very differently in terms of whether they are lies or not. Van Tiel et al. (2014) investigated the potential reasons for this, finding that it is the distinctness of the scale mates which accounts for this variability. If the beliefs about what is communicated and the beliefs about the world do not diverge (cf. experiment 5) the speaker is also not treated as lying, independently of knowledge ascriptions.

Interestingly, the lying ascription patterns differ when participants are told that 'It turns out that all of the objects have the property in question', in other words, they are told that the implicature is objectively false (cf. experiment 2 and 3). In this case, they judge the speaker as lying independently of whether they ascribe knowledge to the speaker. In these cases, a bluffing speaker (who is not ascribed knowledge of how many objects have the property in question) is treated as lying to the same extent as a speaker who is ascribed knowledge (and therefore probably intending to deceive). We take it as a bias: it is the state of the world, rather than knowledge ascriptions or ascriptions of the intention to deceive, that drive lying judgments. We treat this as a cognitive bias because we think that it is the intention to deceive rather than the objective state of the world that should determine perjury (or willful lying in court) ascriptions. The courtroom mistrust and uncertainty context is a context where guessing or bluffing often takes place and we should be cautious in holding speakers extensively responsible for content, which is not conveyed explicitly. Moreover, when participants are not primed with information about the state of the world, their answers diverge. For further evidence of the bias hypothesis, see Skoczeń 2021.

It remains to be investigated in further studies whether it is blame that mediates the relation between the objective state of the world and the lying judgment (cf. Horn, 2017b; Kneer & Machery, 2019; Skoczeń, 2021).

11 Conclusion

Classical theories of linguistic pragmatics describe communication oriented solely toward the exchange of information. In such communication, beliefs about what is said are naturally intertwined with beliefs about the world. However, in linguistic exchanges oriented at attaining concrete social goals such as winning the case, where speakers are less trustworthy, the interactions between the mentioned beliefs are more complex. We investigated these interactions with reference to a series of experiments in a courtroom setting. We argued that it is the strategic context, more than the role of the speaker, which governs inferential content. We also argued that in mistrust contexts, if a statement does not conform to the state of the world, participants judge it as perjurious, by contrast, if it is objectively true, it is judged as not being a lie irrespective of the knowledge and intention to deceive attributed to the speaker. We thus find that perjury ascriptions are inconsistent or biased since they are driven more by the objective state of the world rather than by the mental states ascribed to the speaker. We hypothesize that this is due to retrospective judgments of blame toward the protagonist driven by the discovered state of the word. These blame judgments could in turn fuel unfounded ascriptions of intent to deceive to the speaker (see Skoczeń, 2021).

Finally, one caveat applies to our findings: it is not clear to what extent the subjects were aware that they were assessing not just whether someone lied but whether they committed perjury, which requires a different (and higher) standard of proof. It could be the case that participants would be more lenient in assessing someone as responsible for perjury rather than for uttering lies. This question remains an avenue for future studies.

References

Adler, Jonathan. 1997. Lying, deceiving, or falsely implicating. *Journal of Philosophy 94*. 435–452.

Bergen, Leon, Roger Levy & Noah Goodman. 2016. Pragmatic reasoning through semantic inference. *Semantics and Pragmatics 9*. 10.3765/sp.9.20. http://semprag.org/article/view/sp.9.20 (29 January, 2018).

Bonnefon, Jean-François, Aidan Feeney & Gaëlle Villejoubert. 2009. When some is actually all: Scalar inferences in face-threatening contexts. *Cognition* 112(2). 249–258. https://doi.org/10.1016/j.cognition.2009.05.005.

Carson, Thomas L. 2012. *Lying and deception: theory and practice*. Oxford: Oxford University Press.

Chisholm, Roderick M. & Thomas D. Feehan. 1977. The Intent to Deceive. *The Journal of Philosophy* 74(3). 143. https://doi.org/10.2307/2025605.

Coleman, Linda & Paul Kay. 1981. Prototype Semantics: The English Word Lie. *Language* 57(1). 26–44.

Cramer, Robert J., Stanley L. Brodsky & Jamie DeCoster. 2009. Expert witness confidence and juror personality: their impact on credibility and persuasion in the courtroom. *The Journal of the American Academy of Psychiatry and the Law* 37(1). 63–74.

Cummins, Chris & Napoleon Katsos (eds.). 2019. *The Oxford handbook of experimental semantics and pragmatics* (Oxford Handbooks in Linguistics). Oxford: Oxford University Press.

Desarbo, Wayne S., Venkatram Ramaswamy & Rabikar Chatterjee. 1995. Analyzing Constant-Sum Multiple Criterion Data: A Segment-Level Approach. *Journal of Marketing Research* 32(2). 222. https://doi.org/10.2307/3152050.

Doran, Ryan, Gregory Ward, Yaron McNabb & Rachel Baker. 2012. A novel paradigm for distinguishing between what is said and what is implicated. *Language* 88. 124–154.

Dynel, Marta. 2011. A Web of Deceit: A Neo-Gricean View on Types of Verbal Deception. *International Review of Pragmatics* 3(2). 139–167. https://doi.org/10.1163/187731011X597497.

Fallis, Don. 2009. What Is Lying?: *Journal of Philosophy* 106(1). 29–56. https://doi.org/10.5840/jphil200910612.

Frank, Michael C. & Noel D. Goodman. 2012. Predicting Pragmatic Reasoning in Language Games. *Science* 336(6084). 998–998. https://doi.org/10.1126/science.1218633.

Frank, Michael C., Noah D. Goodman, Peter Lai & Joshua Tenenbaum. 2009. Informative Communication in Word Production and Word Learning. http://langcog.stanford.edu/papers/FGLT-cogsci2009.pdf.

Geurts, Bart. 2010. *Quantity implicatures*. Cambridge; New York: Cambridge University Press.

Goodman, Noah D. & Andreas Stuhlmüller. 2013. Knowledge and Implicature: Modeling Language Understanding as Social Cognition. *Topics in Cognitive Science* 5(1). 173–184. https://doi.org/10.1111/tops.12007.

Grice, H. P. 1989. *Studies in the way of words*. Cambridge, Mass.: Harvard University Press.

Grodner, Daniel & Julie C. Sedivy. 2011. The Effect of Speaker-Specific Information on Pragmatic Inferences. In Edward A. Gibson & Neal J. Pearlmutter (eds.), *The Processing and Acquisition of Reference*, 239–272. Cambridge, MA: MIT Press.

Guthrie, Chris, Jeffrey J. Rachlinsky & Andrew J. Wistrich. 2001. Inside the Judicial Mind. *Cornell Law Faculty Publications* (814). http://scholarship.law.cornell.edu/facpub/814.

Hauser, John. 1991. Comparison of importance measurement methodologies and their relationship to consumer satisfaction. *MIT Marketing Center Working Papers*.

Horn, Laurence R. 1984. Toward a new taxonomy for pragmatic inference: Q-based and R-based implicature. In Deborah Schiffrin, Round Table on Languages and Linguistics, & Georgetown University (eds.), *Meaning, form, and use in context: linguistic applications*, 11–42. Washington, DC: Georgetown University Press.

Horn, Laurence R. 2017a. Telling it slant: Toward a taxonomy of deception. In Janet Giltrow & Dieter Stein (eds.), *The Pragmatic Turn in Law*, 23–55. Berlin: De Gruyter. https://doi.org/10.1515/9781501504723-002.

Horn, Laurence R. 2017b. What lies beyond: untangling the web: In Rachel Giora & Michael Haugh (eds.), *Doing Pragmatics Interculturally*, 151–174. Berlin: De Gruyter.

Horn, Laurence R. 2018. Perjury, puffery, and the presidency: Two ways to nor (quite) lie. Presented at the Pragmatics of Legal Discourse Workshop, AMPRA 4, Albany.

Kneer, Markus. 2018. The norm of assertion: Empirical data. *Cognition* 177. 165–171. https://doi.org/10.1016/j.cognition.2018.03.020.

Kneer, Markus & Edouard Machery. 2019. No luck for moral luck. *Cognition* 182. 331–348. https://doi.org/10.1016/j.cognition.2018.09.003.

Krueger, Frank & Andreas Meyer-Lindenberg. 2019. Toward a Model of Interpersonal Trust Drawn from Neuroscience, Psychology, and Economics. *Trends in Neurosciences* 42(2). 92–101. https://doi.org/10.1016/j.tins.2018.10.004.

Maciej Piasecki. 2002. *Język Modelowania Znaczenia Polskiej Frazy Nominalnej* [The Language of Modeling the Meaning of the Polish Nominal Phrase]. Doctoral thesis.

Marmor, Andrei. 2014. *The language of law*. First edition. Oxford: Oxford University Press.

Meibauer, Jörg. 2005. Lying and falsely implicating. *Journal of Pragmatics* 37(9). 1373–1399. https://doi.org/10.1016/j.pragma.2004.12.007.

Meibauer, Jörg. 2014. *Lying at the Semantics-Pragmatics Interface*. Berlin: De Gruyter.

Michaelson, Eliot. 2016. The Lying Test. *Mind & Language* 31(4). 470–499. https://doi.org/10.1111/mila.12115.

Nausicaa Pouscoulous, Noveck I, Politzer G, Bastide A. 2007. A Developmental Investigation of Processing Costs in Implicature Production. *Language Acquisition* 14(4). 347–375. doi:10.1080/10489220701600457.

Pagin, Peter. 2016. Assertion. In Ed Zalta (ed.), *Stanford Encyclopedia of Philosophy*. https://plato.stanford.edu/archives/win2016/entries/assertion/.

Papafragou, Anna & Julien Musolino. 2003. Scalar implicatures: experiments at the semantics–pragmatics interface. *Cognition* 86(3). 253–282. https://doi.org/10.1016/S0010-0277(02)00179-8.

Reins, Louisa M. & Alex Wiegmann. 2021. Is Lying Bound to Commitment? Empirically Investigating Deceptive Presuppositions, Implicatures, and Actions. *Cognitive Science* 45(2). https://doi.org/10.1111/cogs.12936.

Sanderson, Linda. 1995. Linguistic Contradiction: Power and Politeness in Courtroom Discourse. *Canadian Journal for Studies in Discourse and Writing/Rédactologie* 12(2). 24-Jan. https://doi.org/10.31468/cjsdwr.397.

Sauerland, Uli. 2012. The Computation of Scalar Implicatures: Pragmatic, Lexical or Grammatical?: Computation of Scalar Implicatures. *Language and Linguistics Compass* 6(1). 36–49. https://doi.org/10.1002/lnc3.321.

Saul, Jennifer Mather. 2012. *Lying, Misleading, and What is Said: An Exploration in Philosophy of Language and in Ethics*. Oxford University Press.

Schuster, Sebastian & Judith Degen. 2019. *I know what you're probably going to say: Listener adaptation to variable use of uncertainty expressions*. Preprint. PsyArXiv. https://doi.org/10.31234/osf.io/8w6xc. https://osf.io/8w6xc (31 May, 2021).

Senn, David J. & Myron B. Manley. 1966. Comparison of Scaling Methods: Paired Comparisons versus Constant-Sum. *Perceptual and Motor Skills* 22(3). 911–918. https://doi.org/10.2466/pms.1966.22.3.911.

Skoczeń, Izabela. 2019. *Implicatures within legal language* (Law and Philosophy Library). Berlin: Springer.

Skoczeń, Izabela. 2021. Modelling Perjury: Between Trust and Blame. *International Journal for the Semiotics of Law – Revue internationale de Sémiotique juridique*. https://doi.org/10.1007/s11196-021-09818-w.

Solan, Lawrence M. 2018. Lies, Deceit, and Bullshit in Law. *Duquesne Law Review* 56.

Solan, Lawrence & Peter Tiersma. 2005. *Speaking of Crime: The Language of Criminal Justice*. Chicago: University of Chicago Press.

Struchiner, Noel, Ivar Hannikainen & Guilherme da F.C.F. Almeida. 2020. An experimental guide to vehicles in the park. *Judgment and Decision Making* 15. 312–329.

Tiel, Bob van, Emile van Miltenburg, Natalia Zevakhina & Bart Geurts. 2014. Scalar Diversity. *Journal of Semantics* (33). 137–175.

Turri, Angelo & John Turri. 2015. The truth about lying. *Cognition* (138). 161–168.

Turri, John. 2015. Knowledge and the norm of assertion: a simple test. *Synthese* (192). 385–392.

Weissman, Benjamin. 2019. False Implicatures and Presidents: How the Relationship Between Speaker and Judger Influences Lie Judgments. Unpublished paper.

Weissman, Benjamin & Marina Terkourafi. 2019. Are false implicatures lies? An empirical investigation. *Mind & Language*. https://doi.org/10.1111/mila.12212.

Wiegmann, Alex, Jana Samland & Michael R. Waldmann. 2016. Lying despite telling the truth. *Cognition* 150. 37–42. https://doi.org/10.1016/j.cognition.2016.01.017.

Wiegmann, Alex, Pascale Willemsen & Jörg Meibauer. 2021. *Lying, Deceptive Implicatures, and Commitment*. Preprint. PsyArXiv. https://doi.org/10.31234/osf.io/n96eb.

Willemsen, Pascale & Alexander Wiegmann. 2017. How the truth can make a great lie: An empirical investigation of the folk concept of lying by falsely implicating. *39th Annual Meeting of the Cognitive Science Society*. https://mindmodeling.org/cogsci2017/papers/0663/paper0663.pdf.

Yoon, Erica, Michael Henry Tessler, Noah D. Goodman & Michael C. Frank. forthcoming. Polite speech emerges from competing social goals.

Benjamin Weissman

What counts as a lie in and out of the courtroom? The effect of discourse genre on lie judgments

Abstract: A statement can be technically true but carry a false implicature which, if it were said explicitly, would meet all the criteria of a lie. Much of the literature has discussed whether a statement like this can be a lie: either it cannot, since the false content isn't literally said, or is just as much of a lie as its literally-said counterpart. This chapter explores whether situational context can have any effect on the perception of such statements, specifically investigating the effects of discourse genre. The experiment probes whether judgments of false generalized and particularized conversational implicatures as lies change based on the setting and standards of discourse of the interaction. Several different stories were manipulated to fit three different contexts – a witness testifying under oath in a courtroom, a politician giving a public address, or a casual conversation between two friends. Results indicate that participants judged false GCIs to be less of a lie when coming from a witness testifying under oath in a courtroom. On average, GCIs in all contexts were rated precisely at the midpoint between a lie and not a lie, while PCIs were treated as not lies. These results are discussed further with respect to the standards of discourse in a courtroom scenario and how those standards encourage people to adopt more literal utterance interpretations than they would in a different context. The present study also connects to theories of linguistic meaning and commitment in order to account for the nuances in the evaluation of false implicatures as lies.

1 Introduction

This study investigates the effect of genre on lie judgments. Genre here refers to the societally-shared discourse expectations and standards that accompany the setting of an interaction. In the experiment presented here, the critical items are false implicatures in three different settings, situated within the courtroom, political, and casual genres. By manipulating the genre in which a conversation occurs, this experiment aims to determine whether the standards for what

Benjamin Weissman, Rensselaer Polytechnic Institute, e-mail: weissb2@rpi.edu

https://doi.org/10.1515/9783110733730-014

counts as a lie differ between conversational settings.[1] By investigating the effects of context on lie judgments, this contribution will help to make sense of the thus-inconclusive findings regarding whether false implicatures are lies. In addition, the results will be connected to a theory of linguistic meaning, namely the Privileged Interactional Interpretation (Ariel 2002a, 2002b) that can account for the nuance and variation present in studies of false implicature.

1.1 Lying and linguistic meaning

There is a quickly becoming classic debate within the literature regarding the nature of the relationship between linguistic meaning and lying. Everyday definitions of lying often refer to *saying* something false, but the colloquial version of *saying* is rather imprecise when compared to the philosophy of language version. For a more rigorous definition of lying, then, the meaning of *saying* is a focal point. García-Carpintero (2021) provides an account of lying that hinges upon exactly this distinction – it is only conventionally semantic content that can be a lie. Within the debate, however, there are some who hold that an enriched version of *saying* is relevant for lying, allowing that false content delivered through false implicature could count as a lie (e.g., Vincent and Castelfranchi 1981; Meibauer 2011, 2014, 2018; Dynel 2020). Others, and there are more in this camp, maintain that false content must be strictly *said* in order to be able to count as a lie, rejecting the premise that false implicatures are lies and instead treating those as instances of misleading or some kind of non-lie deception (e.g., Adler 1997; Saul 2002; Fallis 2009; Stokke 2013; Horn 2017; Marsili 2017).

Along the way, some have acknowledged that a blanket statement covering all implicatures is unlikely to work (Dynel 2015; Meibauer 2016; Weissman 2019; Marsili 2020; Thalmann, Chen, Müller, Paluch & Antomo 2021). Given the documented diversity of implicatures as a class in general (Doran, Ward, Larson, McNabb & Baker 2012; Sternau, Ariel, Giora & Fein 2015; van Tiel, van Miltenburg, Zevakhina & Geurts 2016; Terkourafi, Weissman & Roy 2020), more nuance is likely required to get the most accurate picture. Different perspectives on how to define linguistic meaning itself yield different conclusions here. Within a Neo-Gricean view, the case has been made that perhaps *generalized* conversational implicatures (GCIs) can be lies but *particularized* conversational implicatures (PCIs)[2]

---

1 The work presented here is adapted from author's dissertation (Weissman 2019).

2 The general premise behind such a distinction is that particularized conversational implicatures are derived due to the specific context of the conversation whereas generalized conversational implicatures are derived based on the language itself and are relatively context-

cannot (Thalmann, Chen, Müller, Paluch & Antomo 2021). From a Relevance Theoretic perspective, Kisielewska-Krysiuk (2016) has argued that explicatures can be lies but implicatures cannot. Other theories of linguistic meaning, such as Privileged Interactional Interpretation theory (Ariel 2002a, 2002b) or Default Semantics (Jaszczolt 2005) allow for variation in what counts as being *said* in different contexts; theories like these have been used to claim that whether a false implicature can count as a lie may vary based on context (Weissman 2019).

This study will utilize Neo-Gricean GCI/PCI terminology for consistency with other experimental work on false implicatures as lies, though Ariel's Privileged Interactional Interpretation theory will be invoked to make sense of the results. Ariel (2002a) proposes that there are actually multiple levels of so-called "minimal meaning" that may all be available: Literal$_1$, the bare (linguistic) meaning, Literal$_2$, the most salient meaning (psycholinguistically), and Literal$_3$, the "basic content" meaning taken as binding the speaker. The third level, which she deems the "privileged interactional interpretation" (henceforth PII), is necessarily grounded in context and, as such, may vary across scenarios; Ariel is clear that the PII works as a *description*, not a definition. She posits that any of bare linguistic meaning, explicature, and conversational implicature can, in certain contexts, be taken as the PII. Even in the same context, two interlocutors may well have different PIIs of the same utterance (Ariel 2002b). This theory provides the flexibility necessary to account for the many subtleties involved with the discussion of false implicatures as lies.

1.2 Commitment

More recently, the notion of commitment has been invoked as a way to account for the nuance in this discussion. This comes in parallel with a broader push for a commitment- rather than intention-based approach to speech acts, implicature, and pragmatics as a whole (see, for example, work by Moeschler 2013; Geurts 2019; Krifka 2019). Under this view, the extent to which a speaker is committed to content constitutes the extent to which that content, if the speaker believes it to be false, could be considered a lie.[3] This could also allow for more gradation in assessing false implicatures as lies. Speakers are not equally committed to implicated content;

---

independent (Levinson 2000). It is worth noting that some theories, such as Relevance Theory (e.g., Sperber & Wilson 1986/1995; Carston 2002), dispute a theoretical distinction between GCIs and PCIs and have different terms and concepts to describe different sorts of meaning relations.

3 See García-Carpintero (2021) for arguments against a commitment-based account of lying.

therefore it will not be the case that all false implicatures are or are not lies. False implicatures that speakers are committed to could be lies, and false implicatures that speakers are not committed to could be non-lies. Indeed, early experimental evidence (Reins and Wiegmann 2021; Wiegmann, Willemsen & Meibauer forthcoming) and theoretical work (Marsili 2020) seem to support this conclusion. This idea, then, is consistent with the idea that from any utterance, the PII is what is judged as a lie or not a lie, as that is the meaning that is taken as binding (i.e., committing) the speaker. This commitment-based approach also upholds a distinction between lying (involving commitment) and non-lying deception, a distinction which many (e.g., Krifka 2019) wish to uphold and which an intention-based account would run the risk of erasing.

Determining speaker commitment, however, is not itself an easy task, and there does not yet exist a consensus around how to assess perception of speaker commitment. Experimental investigations into commitment have asked participants a range of questions, including the extents to which the speaker should be punished, the speaker can be trusted (Mazzarella, Reinecke, Noveck & Mercier 2018), the speaker owes an apology, the hearer would rely on the speaker, the speaker has violated a promise (Bonalumi, Scott-Phillips, Tacha & Heintz 2020), or the speaker has deniability (Boogaart, Jansen & van Leeuwen 2021), as well as the extents to which the implicated content is cancelable, calculable, non-reinforceable, and added to common ground (Reins and Wiegmann 2021). It seems apparent that more work is required to determine exactly what can be asked to determine commitment.

This commitment mechanism could also be used to explain the effects of context on lie judgments. There is certainly a linguistic component to commitment (see Moeschler 2013 for such an explanation), but there could be a contextual component as well. Elements of surrounding context could result in a speaker being more or less committed to the same implicated content in different scenarios – again consistent with the PII theory. Such a premise would provide an explanation for differing lie judgments for the same utterance in different contexts (Weissman 2019).

1.3 Experimental work on lying

Once scarce, empirical investigations testing lie judgments are now plentiful, including several specifically probing false implicatures; results from these have been largely inconclusive thus far. Some have found that false implicatures are judged as lies (Willemsen and Wiegmann 2017; Antomo, Müller, Paul, Paluch & Thalmann 2018 (both GCIs and PCIs); Reins and Wiegmann 2021; Skoczeń 2021; Thalmann, Chen, Müller, Paluch & Antomo 2021 (GCIs only); Wiegmann, Willemsen & Meibauer forthcoming), while others have found that participants tended to not count false implicatures as lies (Weissman and Terkourafi 2019; Thalmann, Chen, Müller, Paluch & Antomo 2021 (PCIs only); Viebahn, Wiegmann, Engelmann & Willemsen forthcoming). Despite the inconclusive results, further investigation of the stimuli used to draw these conclusions can provide some enlightenment.

Willemsen and Wiegmann (2017) tested "half-truths," corresponding to false PCIs based on the Gricean Maxims of Quantity, Relation, and Manner. They found that all scenarios based on all three Maxims are rated as lies by participants in a binary forced-choice (yes/no) task and take this as evidence that false implicatures can indeed be lies. The following is an example used in this study, based on Coleman and Kay's (1981: 31) Story VI:

> Peter and Jane have been a couple for a year now. They are very happy and just moved in together. Peter trusts Jane, but he knows about her ex-fiancé Steven who still tries to win Jane back. Thus, Peter is very jealous and does not like Jane meeting Steven. Jane is sometimes thinking about getting back together with Steven. As they work in the same company, they have coffee from time to time to talk about their joint projects. Today, Jane and Steven have coffee after lunch to finalize a cost calculation they are supposed to send to their client the next morning. After a few minutes, Steven asks Jane if they could talk about each other and getting back together. Jane tells Steven that they don't have much time and need to focus on the project. Steven has been sick the whole week, but he has nevertheless been at work.
>
> In the evening, Peter and Jane have dinner. Peter asks Jane:
> "You told me about this project with your ex-fiancé. Did you see him today?"
> To avoid confirming that she saw Steven during lunch, Jane says:
> "Steven has been sick the whole week."
> Just as Jane intended, Peter does not believe Jane met with Steven.
> (Willemsen and Wiegmann 2017: 3518)

A binary forced-choice task was given to participants here, in which 81% of people judged Jane's utterance to be a lie.

A similar story was used in Weissman and Terkourafi (2019):

> John and Mary have recently started dating. Gabe is Mary's ex-boyfriend. Gabe has been sick with the flu for the past two weeks, but went on a date with Mary last night.
>
> John: "Have you seen Gabe this week?"
>
> Mary: "Gabe has been sick with the flu for two weeks."
> (Weissman and Terkourafi 2019: supplemental materials)

This study used a different task, asking participants to rate whether Mary has lied on a 1–7 scale (in which 1 = not a lie and 7 = lie). Despite the different task, the disparity in results is obvious: in this study, the mean rating for this story was a

2.8 out of 7, and only 10% of participants judged it to be a lie. The utterance tested was essentially the same, so there must have been some other task-related or stimulus-related difference driving the disparity in results. The striking divergence between the two is the amount of information conveyed in the story. Weissman and Terkourafi (2019) furnished a minimal context and the dialogue, whereas Willemsen and Wiegmann (2017) provided a much more robust context, and, crucially, explicitly mentioned the speaker's intention and the interlocutor's uptake of the utterance. In this story, the one judging knew that Jane's utterance was intended to "avoid confirming that she saw Steven during lunch" and also that "just as Jane intended, Peter does not believe Jane met with Steven." It is possible that the inclusion of this information in Willemsen and Wiegmann (2017) led to the utterance being judged as a lie; correspondingly, the exclusion of this information in Weissman and Terkourafi (2019) led to the utterance being judged as not a lie; indeed, two experiments directly testing whether including a highlighted intent to deceive (Weissman 2019) or an explicit motive to lie (Skoczeń 2021) both found that such an inclusion yielded an increase in lie ratings. Such observations can help to make sense of the mixed bag of results observed in these studies thus far and understand how elements of context can affect judgments in these tasks.

In addition to testing motive, Skoczeń's (2021) experiment included a comparison between courtroom and casual contexts. This experiment tested lie judgments (as well as several other judgments, such as trustworthiness and blameworthiness) to a classic some-but-not-all scenario: if someone says "some of the invoices are unpaid" when it turns out that all of the invoices are unpaid, has that someone lied? Results indicated no significant difference in the likelihood that participants would draw the "some but not all" inference in a courtroom context versus a casual one. Within the group of participants that did draw that scalar inference, there was no significant difference in lie ratings in the courtroom context versus the casual. Skoczeń concludes that "participants [do] not make a sharp distinction between casual and courtroom contexts" and that "there is a robust folk intuition that a false implicature in the courtroom is a lie."[4] (See also Skoczeń and Smywiński-Pohl's chapter in this volume.)

---

[4] Two notes of caution may be added to the second of these conclusions, that false implicatures in the courtroom are robustly considered lies. The conclusion is based on the finding that lie ratings were significantly higher than the midpoint among participants who drew the "some but not all" inference; there was still a non-trivial portion of participants (31%) who did not draw this inference, and lie ratings in this group (3.67) were not significantly higher than the midpoint. Second, the utterance tested contained a scalar implicature, which would be a GCI; this finding does not necessarily extend to PCIs or other GCIs.

The present study will continue to investigate genre (including the casual vs. courtroom comparison) as one of many elements of context that could affect perceptions of whether false implicatures are lies (see Weissman 2019 for more details).

1.4 Genre

1.4.1 The linguistic concept of genre

The notion of genre in linguistic analysis refers to the interrelationships between utterance, setting, and society, "the specifically discoursal aspect of ways of acting and interacting in the course of social events" (Fairclough 2003: 65). The sorts of genre that will be manipulated here are what Fairclough refers to as "situated genres . . . which are specific to particular networks of practices" (69). Fairclough describes the key components of genre: activity, which refers to the purpose(s) of discourse within that context; generic structure, which refers to ritualized structural staging within the discourse; the social relation of the interactants, which includes power dynamics and discursive roles; and communication technologies, which refers to the medium through which the interaction plays out. These components together describe a genre and can be discussed independently or together.

Bakhtin (1986) also writes about the idea of speech genres, which he summarizes as "relatively stable types of . . . utterances" that are each developed within different "spheres of language" (Bakhtin 1986: 60). He stresses that speech genres are societally developed, and genre-related language behavior exists at a macro level, shared by members of that culture:

> Speech genres organize our speech in almost the same way as grammatical (syntactical) forms do. We learn to cast our speech in generic forms and, when hearing others' speech, we guess its genre from the very first words; we predict . . . a certain compositional structure . . . If speech genres did not exist and we had not mastered them, if we had to originate them during the speech process and construct each utterance at will for the first time, speech communication would be almost impossible.   (Bakhtin 1986: 78–79)

The genre governing a conversation or an utterance not only guides the speaker but the listener, too; this extension suggests that knowledge of genres influences the interpretation of utterances in real time. Genre research in linguistics tends to focus on formal aspects of language as opposed to any sort of experimental approach to utterance interpretation, but the notion of context systematically constraining discourse (and its interpretation) is afforded by genre in such a way that it may be applicable in an experimental study of lying and linguistic meaning.[5]

With respect to lying, Williams (2002) wonders whether there are certain "relations . . . situations or domains" that may come with an increased level of distrust ("as it is often said that no sensible person expects to hear the truth when buying a used car from a dealer") (Williams 2002: 110). There may be certain genres that engender different standards regarding the concepts of *lying* and *saying* from those in casual everyday speech.

The genres examined in this experiment – courtroom, political, and casual – are more likely what Giltrow and Stein (2009: 10) refer to as "hyper-genres", a superordinate level encompassing multiple related genres. For example, the political hyper-genre would include the genres of political speech, election posters, debates, press conferences, etc. (Cap and Okulska 2013). Others (e.g., van Eemeren 2010) use different terminology to refer to the same notion, like "genre" for the supercategory and "activity type" (per Levinson 1979) for the subcategory.

1.4.2 Lying in the courtroom

Horn (2017: 39) directly addresses the notion of false implicatures in the courtroom. Discussing cases like the notorious Bronston example and Bill Clinton's impeachment proceedings, Horn notes that the Literal Truth Defense – that what is on record is *literally* true, despite any false inferences it may give rise to – has proved "exceptionally useful" for a variety of lawyers and defendants in court. Indeed, with the Bronston case as the Supreme Court precedent, the American legal system (and the Australian and English as well) explicitly differentiates between deceptions which are literally said (perjury) and those which are not (non-perjury) (Green 2018; see also Chapters 1, 8, 14). When Clinton "offered an evasive, non-responsive, and factually true reply to the question posed . . . he did not actually lie" (Green 2018: 3).

Based on several examples and the official legal precedent, Horn sees the legal context as one that clearly and strictly differentiates between lying and intentionally misleading. Horn and Green both predict that examples of false implicatures in a courtroom context are not lies, on the basis that the utterances are literally true. If this distinction is as pertinent to courtroom discourse and this understanding as societally shared as Horn and Green expect, it is hypothesized that most people will

---

5 It is important to note that all discussions of genre here refer to the versions within the greater United States society; a courtroom encounter refers to a typical courtroom encounter in the U.S. legal system while recognizing that judicial encounters in different societies and cultures are entirely different genres.

judge these examples to not be lies as well. Regarding the present experiment, lie ratings for witness utterances are hypothesized to be lower than lie ratings for utterances in the casual conversation genre, the latter being an unmarked case in which the distinction between lying and non-lying verbal deception is neither as relevant nor as attended to as it is in the courtroom.

There are several subgenres within the hyper-genre of courtroom discourse (Danet 1980), including lawyer arguments, lawyer-client interactions, the reading of official documents, witness testimony, etc. As witnesses are sworn in under oath to tell "the truth, the whole truth, and nothing but the truth," notions of truth and lying are of the utmost relevance for witness testimony. This leads Green and Horn to discuss false implicatures in the courtroom primarily in terms of witnesses; as such, the witness testimony genre will be utilized in this experiment.

1.4.3 Lying in politics

Both scholarly and popular conversations within political discourse have focused on the role of truth in political speech; see, for example, "Many Politicians Lie. But Trump Has Elevated the Art of Fabrication," in the *New York Times* (Stolberg 2017), or "How to Address the Epidemic of Lies in Politics" from *Scientific American* (Tsipursky 2017). Ralph Keyes (and many others) have referred to the "post-truth era" (Keyes 2004) in reference to modern times of a "lie tolerance" that permeates society, most notably and visibly in politics. Discussions of the "post-truth" political landscape and the associated "malaise" (Lewandowsky, Ecker & Cook 2017) stress the idea that truth is less relevant than ever in the political speech genre. "Dishonesty has come to feel less like the exception and more like the norm" (Keyes 2004: 12); certainly, the sentiment has increased manifold since Keyes noted this in 2004. But the notion of politicians being deceptive is seemingly as old as politics itself.[6] As Hannah Arendt writes, "truthfulness has never been counted among the political virtues" (Arendt 1971: 4).

Bakir, Robinson, Herring and colleagues (Robinson, Miller, Herring & Bakir 2018; Bakir, Herring & Robinson 2019) frame the discussion of lying in politics within their Organized Persuasive Communication (OPC) framework, in which manipulative (propaganda) and non-manipulative (public relations) strategic maneuvers are considered under the same umbrella. OPC refers to "organized

---

6 These discussions can be traced through much of the history of Western philosophy, from Aristotle to Kant. For a more recent example, see Eric Alterman's book *When Presidents Lie* (Alterman 2005) for perspective on US presidential lies throughout the 20th century.

activities aimed at influencing beliefs, attitudes, and behavior" (Robinson, Miller, Herring & Bakir 2018: 7) and includes deceptive and non-deceptive acts. Their work does not explicitly predict whether a deceptive utterance (including, but not limited to, false implicatures) will be seen as more of a lie if it comes from a politician, but they do note that OPC, including deceptive OPC, has "become part of the political environment and central to the exercise of power" (Robinson, Miller, Herring & Bakir 2018: 11). Within the category of deceptive OPC exist the subcategories of lying and several types of deception without lying; the authors note that the latter can be just as serious as overt lying and that their subtlety can make them more effective.

While existing theories and frameworks, including OPC, do not make explicit predictions about lie ratings of political utterances, this experiment can be framed with respect to related work. Recent studies on fact-checking have investigated whether finding out that a politician's statement is false influences peoples' attitudes towards that politician; the results from this research have supported different conclusions. Aird and colleagues (Aird, Ecker, Swire, Berinsky & Lewandowsky 2018) found that encountering multiple fact-checked false statements decreased participants' support for a political figure, suggesting that a politician's veracity does matter to people. This does not necessarily mean that people will expect a politician not to lie, but they at least hope that the politician speaks truthfully and care when they do not. Two other studies (Swire, Berinsky & Ecker 2017; Nyhan, Porter, Reifler & Wood 2019), on the other hand, found that fact-checked false statements had no effect on participants' attitudes towards the political figure in their experiment.[7]

From both scholarly work and discussions in popular media, it is apparent that people expect politicians to be deceptive. Whether this deception is justified is the subject of rigorous debate, but the expectations that come with the role are essentially built into our society. Regarding the present experiment, this expectation could manifest in several directions. One possibility is that false implicature lie ratings will be lower in the political genre than the casual conversation genre. If societal expectations are that politicians either are expected to speak deceptively or are justified in doing so, then false implicatures could be seen as "part of the job" and not as much of a lie. Another potential outcome is that people fall under the camp of Kantian idealists and view a politician's deception as worse than a non-politician's. In this case, lie ratings would be higher in the political

---

7 It is worth noting that these two studies used Donald Trump as the speaker in their stories and the null effect of falsehoods on support was found among Trump supporters. It may be the case that this case and sample are not wholly representative.

genre. A third possibility is that laypeople, like scholars, are split on this matter, with some viewing the politician's deception as justified and others viewing it as egregious. In this case, ratings in the political genre would be more polarized than those in the casual or courtroom genres.

As with the courtroom genre, there are several sub-genres within the broader classification of political discourse (Cap and Okulska 2013), including political speeches, campaign advertisements, debates, late night TV interviews, press conferences, etc. As the press conference is an occasion in which politicians, speaking directly to both the media and the public, are especially accountable for what they say (Ekström 2009). The notions of truth and lying are relevant in such a context, and the press conference genre will be used in this experiment.

1.4.4 Casual speech

The last genre investigated in this experiment will be "casual conversation" (Eggins and Slade 1997). This sort of genre is seen as being less "purpose-driven" (Fairclough 2003: 71) than, for example, courtroom or public political discourse scenarios; there are, of course, purposes in interactions with friends or peers, but they are not as "clearly tied to broadly recognized social purposes" as others. It is worth noting another dimension that differs between these interactions and those in the other two genres: these conversations are private whereas the other two are public. The purpose of exploring this genre experimentally is to provide a relatively non-descript "baseline" condition to which the other two can be compared, in this case ordinary daily conversations between two people who are given relationship roles (e.g., friends, peers, spouses) but not societal roles. This was done to limit the impact of any societal role-based genre influence and ensure there was no overt difference in power between the two interlocutors in these scenarios.

1.5 Research questions

This study addresses the following specific research questions:

Does the genre in which an interaction occurs influence lie ratings of false implicature utterances?

Are there differences between generalized and particularized conversational implicatures, regarding both overall lie status and any potential effects of genre?

Is there internal consistency within the categories of GCI and PCI with respect to lie status?

It is hypothesized that witnesses' false implicature utterances in the courtroom genre will be rated as less of a lie than those utterances spoken in a casual conversation due to the important legal distinction between lies and non-lie deception in the courtroom. There are several potential hypotheses regarding whether politicians' false implicature utterances in the press conference genre are treated as more of a lie than in casual conversation: they may be rated as more of a lie, due to people holding politicians to a higher standard of truth-telling, they may be rated as less of a lie, due to peoples' expectations that politicians will lie, or both may be true and ratings will be more polarized in this condition. There are no specific hypotheses regarding the casual conversation genre; this condition will be treated as a baseline to which the other two conditions can be compared. In line with previous work from various domains (e.g., Moeschler 2013; Sternau, Ariel, Giora & Fein 2015; Thalmann, Chen, Müller, Paluch & Antomo 2021), it is expected that GCIs will garner higher lie ratings than PCIs (though Antomo, Müller, Paul, Paluch & Thalmann 2018 found no significant difference between the two types). Lastly, a lack of internal consistency within the categories of GCI and PCI would further establish the need for nuance in discussions of false implicatures as lies. So, too, would any significant effects of genre.

2 Methods

2.1 Procedure

This experiment utilized a Lie Rating Task in which participants saw a story that ends with the critical line of dialogue for them to judge. They did so on a lie rating sliding scale in response to the question "Is this statement a lie?" This scale ran from −3 to 3, with lower ratings indicating not a lie and higher ratings indicating a lie.

While most lying experiments have utilized an integer-based Likert scale (i.e., 7 discrete choices on the scale), the present experiment used a more continuous sliding scale. This setup allowed participants to slide freely along a continuum that abstractly represents the two poles: absolutely not a lie and absolutely a lie. The endpoints and midpoint of the scale were labeled to provide anchors, but the actual value of the selection was not revealed to the participant. The underlying value was still recorded (down to two decimal places, in this case), providing a more continuous measure of the same type of rating. This sliding scale provides results essentially identical to an ordinal Likert scale (Weissman 2019; Kemper, Campbell, Earleywine & Newheiser 2020) and allows for less controversial use of parametric data analysis (Allen and Seaman 2007; Sullivan and Artino 2013).

In this experiment, participants saw a total of three items. The first item was the critical item – the one which includes a false implicature.[8] Following the critical item were two control items – a straightforward truth and a straightforward lie – presented in random order. Unlike other lie judgment surveys, participants in this experiment saw and rated only one target item. This sidesteps a potential issue of surveys in which participants see and rate multiple critical items, which could result in participants (implicitly or explicitly) comparing stories to each other and providing relative ratings as opposed to a single, more naïve rating.

2.2 Materials

In this experiment, three versions were crafted for each target expression: one in a courtroom setting, one in a political speech setting, and one in a casual setting. The political genre featured political figures making public appearances at press conferences in front of the media; by featuring, for example, a governor giving a press conference, it was ensured that the governor was acting publicly as governor and the discourse is within the political genre. Speakers in the courtroom genre were always witnesses in a trial and the story clarified that their utterance was delivered under oath. Speakers in the casual genre were proper-named characters; example scenarios include ordinary conversations with friends and acquaintances (in line with the examples of casual conversation analyzed in Eggins and Slade 1997).

The implicatures used here were based on examples from various sources in the literature, including Coleman and Kay (1981), McQuarrie and Phillips (2005), Meibauer (2005), Sternau (2013), and Sternau, Ariel, Giora & Fein (2017).[9]

2.3 Participants

The dataset comprised responses from 642 native English speaking participants (average age = 31.82; 332 male, 299 female, 4 non-binary, 7 did not report gender) recruited from Amazon MTurk.

---

8 The critical item was always first so as to avoid participants comparing the critical item rating directly to their straightforward lie or truth item ratings.
9 The full set of stimuli can be found at: https://osf.io/t7md9/?view_only=9010e52a1d6d4 d7194ae9f9cd7868dd3.

3 Results

This portion of the study will first investigate overall effects, followed by a more detailed look within each category of linguistic meaning within each category of linguistic meaning. Results sorted by discourse genre and implicature type are presented in Figure 1 left and right, respectively.

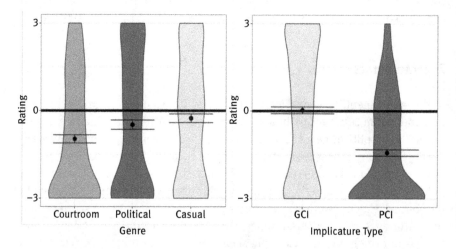

Figure 1: Violin plot[10] of lie ratings to all critical items, split by Genre on left and Implicature Type on right. Black circles with error bars in each column show the condition mean with standard error. Horizontal bar across at 0 displays scale midpoint.

A linear model was run on the data with Genre, Implicature Type, and their interaction as fixed effects. Results of the model are presented in Table 1.

Table 1: Linear model on entire dataset.

Parameter	SumSq	Df	F value	Pr(>F)
(Intercept)	16.08	1	3.836	0.051
Genre	80.49	2	9.604	< 0.001 ***
Implicature Type	130.84	1	31.226	< 0.001 ***
Genre:Implicature Type	33.59	2	4.008	0.019 *

---

10 A violin plot is preferable here to a traditional box plot because it includes a representation of probability density, represented by horizontal width; this provides a more robust depiction of the distribution of the data.

Across all ten critical items, there were significant main effects of Genre, Implicature Type, and their interaction. Notably, courtroom ratings were lower than the other two genres and PCI ratings were lower than GCI ratings. In order to understand the nature of the interaction, the rest of this section will examine each level of Implicature Type separately, reviewing both the effects of Genre and the within-category consistency of each.

3.1 GCIs

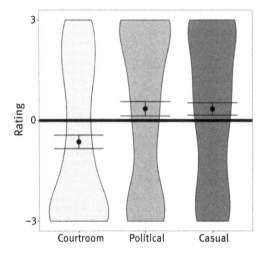

Figure 2: Violin plot of lie ratings to GCI items, split by Genre. Black circles with error bars in each column show the mean with standard error. Horizontal bar across at 0 displays scale midpoint.

To investigate effects specifically within the category of GCIs, a linear mixed effects model was run on only the six GCI items, with Genre as a fixed effect and random intercepts for Expression. Casual was the reference level in this model. Results are presented in Table 2.

Ratings to GCIs in the courtroom genre were significantly lower than ratings to GCIs in the other two genres (Figure 2), as confirmed by an estimated marginal means post-hoc test (Tukey-corrected p-values of 0.04 (Courtroom – Political) and < 0.001 (Courtroom – Casual)). There was no significant difference in ratings between GCIs in the political and the casual genres (p = 0.51).

To investigate within-category consistency for GCIs, a new model was run with Expression as a fixed effect instead of a random effect; the Genre:Expression

Table 2: Linear mixed effects model on GCI items.

Genre	Estimate	Std. Error	t value	Pr(>|t|)
(Intercept)	0.380	0.418	0.909	0.396
Courtroom	−1.000	0.254	−3.941	< 0.001
Political	−0.083	0.257	−0.323	0.747

interaction term was included as well. Results of the model are presented below in Table 3.

Table 3: Linear model on GCI items, including Expression as fixed effect.

Parameter	SumSq	Df	F value	Pr(>F)
(Intercept)	14.92	1	3.824	0.051
Genre	36.97	2	4.740	0.009 **
Expression	169.00	5	8.667	< 0.001 ***
Genre:Expression	157.29	10	4.033	< 0.001 ***

There were significant effects of Genre, Expression, and their interaction; expression accounted for more variance than Genre. The interaction was driven by ratings for the default enrichment utterance,[11] in which the difference between the courtroom utterance and the other two was larger than for other GCIs. The results, by Genre, for each Expression, are shown in Figure 3.

For the sake of clarity, examples of the target utterance and corresponding implicature from each story are reproduced below in Table 4. Full stimuli (with all three contexts) can be found at the link provided in the supplementary materials.

---

11 It should be noted here that the utterance in the courtroom context was different from the utterances in the political and casual contexts due to naturalness of constraints in crafting the stimuli. The item exhibits the same default enrichment GCI type, but the exact lexical items used differ; it is not unreasonable to wonder whether two different types of default enrichment GCI may indeed behave differently based on e.g., the salience of that default, the relevance in the context. The interaction based on this one item should not be over-interpreted but rather accepted with this caveat.

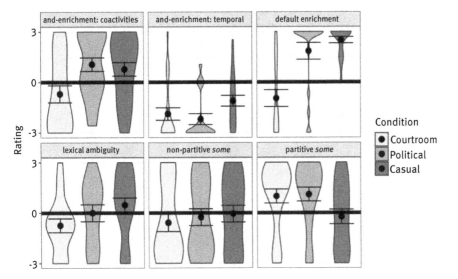

Figure 3: Violin plot of lie ratings to GCI items, split by Genre and sorted by Expression. Black circles with error bars in each column show the mean with standard error. Horizontal bar across at 0 displays scale midpoint.

Table 4: Utterances and corresponding GCIs.

Type	Utterance	Implicature
and-enrichment: coactivities	"Jerry and I played golf last Saturday"	Jerry and I played golf together last Saturday
and-enrichment: temporal	"Last year, I got married and had a baby"	Last year, I got married and had a baby in that order
default enrichment	"I've been elected governor"	I've been elected governor in today's election
lexical ambiguity	"My wife is a doctor"	My wife is a medical doctor
non-partitive *some*	"Some committee members disagreed with my proposal"	Some but not all committee members disagreed with my proposal
partitive *some*	"Some of the squirrels they tested had rabies"	Some but not all of the squirrels they tested had rabies

3.2 PCIs

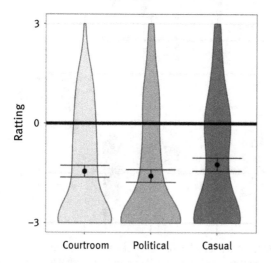

Figure 4: Violin plot of lie ratings to PCIs, split by Genre and sorted by Expression. Black circles with error bars in each column show the mean with standard error. Horizontal bar across at 0 displays scale midpoint.

To investigate effects specifically within the category of PCIs, a linear mixed effects model was run on only the implicature items, with Genre as a fixed effect and random intercepts for Expression. Casual was the reference level in this model. Results are presented in Table 5.

Table 5: Linear mixed effects model on PCI items.

Genre	Estimate	Std. Error	t value	Pr(>|t|)
(Intercept)	−1.259	0.268	−4.704	0.002 **
Courtroom	−0.188	0.262	−0.716	0.475
Political	−0.342	0.261	−1.310	0.191

There were no significant effects of genre on ratings to PCIs (Figure 4).

To investigate within-category consistency for PCIs, a new model was run with Expression as a fixed effect instead of a random effect; the Genre:Expression interaction term was included as well. Results of the model are presented below in Table 6.

Table 6: Linear model on PCI items, including Expression as fixed effect.

Parameter	SumSq	Df	F value	Pr(>F)
(Intercept)	27.30	1	9.508	0.002
Genre	4.60	2	0.801	0.450
Expression	1.99	3	0.231	0.875
Genre:Expression	33.50	6	1.944	0.07 ^

There was a small, marginally-significant interaction between Expression and Genre. The interaction was influenced by the "all-natural ingredients" implicature, in which ratings to the casual genre item were significantly higher than the other genres, and the "let's just say all the food was eaten" implicature, in which ratings to the political genre item were significantly lower than the other genres (Figure 5).

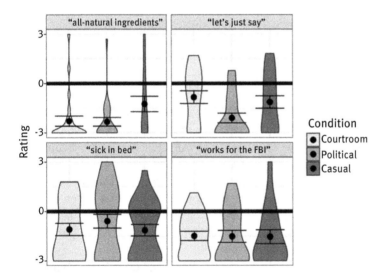

Figure 5: Violin plot of lie ratings to PCI items, split by Genre and sorted by Expression. Black circles with error bars in each column show the mean with standard error. Horizontal bar across at 0 displays scale midpoint.

Table 7 shows the target utterances and contextually-derived PCIs from each story.

Table 7: Utterances and corresponding PCIs.

Utterance	Implicature
"Product X contained 100% all-natural ingredients"	Product X was environmentally-friendly
"Let's just say all the food was eaten"	The event was well-attended
"The CEO has been sick in bed with pneumonia for the past two weeks"	I did not see the CEO last week
"My father works for the FBI"	My father is an FBI agent

4 Discussion

There was a significant effect of genre on lie ratings, specifically pertaining to the courtroom genre. Lie ratings were lower for utterances spoken by a witness under oath during a trial than for the same types of utterances spoken by a politician in a political speech or someone speaking casually to a friend. This finding is consistent with the predictions made by Horn and Green as well as the standards of the U.S. legal system, whose operationalization of perjury differentiates between that which is literally said and that which is indirectly conveyed. Because the difference between lying and non-lying deception is so pertinent within the courtroom context, participants are especially cognizant of this distinction when making lie judgments, resulting in lower ratings in this context. It may be the case that the concept of perjury plays a role in people's awareness of the distinction. In a casual, everyday encounter, liars are not at risk of being convicted of perjury, whereas those sworn in under oath in a trial are. The risk of perjury – with the legal precedent that it applies only to that which is literally true (established in the Bronston case) – may cause judgers to be hyper-aware of the distinction between that which is literally uttered and that which is otherwise conveyed.

Within the PII theoretical framework (Ariel 2002a, 2002b), this result can be attributed to the selection of a stricter level of linguistic meaning in a courtroom context than in the others. The *and*-enrichment: coactivities utterance will be examined here as a case study to exemplify the phenomenon.

"Jerry and I played golf on Saturday"

In all three versions of the pre-utterance context, it is revealed that the speaker and Jerry each played golf on Saturday but did not play together – they went and played separately, at different courses. The bare linguistic meaning of the utterance is that (a) Jerry played golf on Saturday and (b) the speaker played golf on Saturday; the inference being (c) that they played golf together. In the courtroom context, the mean lie rating (on a −3 to 3 scale) for this utterance was −0.696, compared to 1.058 in the political context and 0.772 in the casual context. Not only was the courtroom utterance significantly lower, it fell on the opposite side of the lie/not lie dividing line as well. In fact, converting the scale-based ratings to a binary lie/not-lie reveals that 35% of people rated it a lie on the scale, compared to 77% for the political utterance and 64% for the casual utterance.

In the current framework, there are two potential meanings of the utterance – the bare meaning that Jerry played golf and the speaker played golf, and the enriched meaning that they played golf together. The selection of one of these meanings as the PII is dependent on the surrounding context and the individual interpreter. In this courtroom context, stricter (more bare) interpretations were favored, compared to the more enriched interpretations preferred in the other contexts; the stricter interpretations here resulted in lower overall ratings. This context-based difference surfaced at an overall level with a large enough sample, though it is not the case that every individual rated the courtroom utterance as a lie and the others as not a lie. In addition to contextual genre-related differences, there are individual-level differences – the PII framework can account for this variation as well via the same selection mechanic.

The present findings can be related to recent work in the subjects of lying and linguistic meaning regarding commitment. The idea that lie judgments are tied to levels of speaker commitment to the implicated content is not incompatible with the results presented here. As the idea of commitment receives more attention in the literature, it may become clear that different contexts constitute different standards of speaker commitment. In a casual context, people may be more likely to perceive the speaker as committed to implicated content, but they may be less likely to do so if the utterance takes place in a courtroom context, for example. Similarly, the process of making a lie judgment may in and of itself be a context that changes how people attribute commitment to speakers. Lastly, participants are not necessarily uniform in how they attribute commitment to speakers; accepting this allows for better understanding of the range of responses that emerge in experiments like these. Future study could investigate what drives these individual differences themselves.

While this context-based difference was significant for GCIs, such was not the case for PCIs; there was no significant effect of genre for PCIs. Apparent in Figure 4, few ratings were above the midline; only 25% of people considered any false PCI to be a lie of any strength. With 75% of ratings in the not-lie category, there was a high level of agreement that such cases are not lies. PCIs exhibited less variance (3.05) and a smaller interquartile range (3.03) than did GCIs (5.18, 4.45). Analyses conducted on means within PCI items revealed no significant effect of genre, but it is worth noting that the median rating in the political genre was lower than in the other two. This can be attributed to positive skew of the data in this condition: there were higher proportions of ratings in both the upper half of the scale and the extremely low range of the scale for political genre utterances, resulting in a lower median than mean.

There was no significant difference in lie ratings between the political genre and the casual genre, but there was some evidence, related to the positive skew of the data, to suggest ratings were more polarized in the political genre than in the other two. The interquartile range of ratings to political genre utterances was 4.59, compared to 3.90 and 4.06 to the courtroom and casual genre utterances, respectively. The political genre ratings also exhibited greater variance (5.32 compared to 4.16 and 4.69). These metrics indicate that results were more widespread, exhibiting greater variance, for the political utterances. It was neither the case that idealist hope for honest politicians caused lie ratings to be higher nor the case that realist expectations that politicians will lie caused lie ratings to be lower. These sentiments may still have been present, and the coexistence of both among the population may have driven the increased spread in ratings; another task would be necessary to assess their presence.[12]

The PII theoretical framework adds nuance to the discussion of false implicatures as lies in multiple ways, specifically by accounting for both context and inter-individual differences. A less flexible account of linguistic meaning will struggle to explain how in certain contexts a statement with a false implicature could count as a lie while in other contexts it will not. Any account that claims it is uniformly the level of bare linguistic meaning, or GCI, or PCI, or explicature that the speaker is committed to and therefore is judged as a lie will fail. Instead, the relevant meaning that is judged as a lie is the PII, which itself varies between different contexts.

Especially apparent in the violin plots from the results section is a range of responses from participants; not all implicatures tested reflected a high degree of agreement as a lie or not a lie. Take, for example, the non-partitive *some* GCI ("some committee members disagreed with my proposal" when it was the case that all committee members disagreed with the proposal). Across the three

---

12 The politicians in the present study were made-up characters, but in situations involving real politicians, it has been demonstrated that lie ratings are significantly higher when the rater disagrees with the politician in question than when the rater supports them (Weissman 2019).

versions of this utterance, 55% of participants rated this below the midline (i.e., not a lie) and 45% of participants rated it above the midline (i.e., a lie). It is evident from cases like this that participants did not necessarily select the same level of linguistic meaning for judgment. A flexible theory like PII allows for the idea that given the same utterance in the same context, people may still select different levels of meaning as binding the speaker. It may also be the case that different respondents were operating with different thresholds for what counts as a lie; perhaps participants assigned different weights to the ethical and moral considerations versus those related to truth and belief.

In addition to exhibiting the tendency to be affected by discourse genre, GCIs also demonstrated less within-category consistency than did PCIs. The average lie ratings for the six GCI utterances (averaged across genres) were −1.66 (*and*-enrichment: temporal relation), −0.290 (non-partitive *some* non-partitive), −0.05 (lexical disambiguation), 0.42 (*and*-enrichment: coactivities), 0.63 (partitive *some*), and 1.11 (default enrichment). While the mean for all GCI ratings is 0.02 (at the midpoint), a further look reveals that this was made up of four utterances that closely reflect this midpoint-centralized tendency, one significantly above the others, and one significantly below the others. This is consistent with the suggestion from Doran, Ward, Larson, McNabb & Baker (2012) and Sternau, Ariel, Giora & Fein (2017) that there is diversity within the category of GCIs (or explicatures).

PCIs, on the other hand, exhibited more consistency. The average lie ratings for the four implicature utterances (averaged across genres; scaled −3 to 3) were −0.93 ("sick in bed"), −1.36 ("let's just say"), −1.51 ("worked for the FBI"), and −1.97 ("all-natural ingredients"); there were no significant differences between these. This suggests that the group theoretically defined as PCIs was more consistent than the group theoretically defined as GCIs with respect to lie ratings, specifically due to the fact that they were rarely given ratings above the midpoint and thus treated as lies.

5 Conclusions

Overall, this experiment observed a significant effect of genre on lie ratings, specifically for GCIs. Ratings on false implicature utterances in the political genre were more polarized than those in the casual conversation baseline condition, hinting at the possibility that the population is split regarding its standards and expectations for politicians and the truth of their public claims. In addition, this experiment supports Horn's hypothesis that false implicatures in the courtroom are regarded differently from those in a casual, unmarked genre. Consistent with Horn's hypothesis and the precedent in the U.S. legal system, false GCIs were less readily judged as lies if they appeared in a courtroom genre than if they appeared in a political or casual genre. This constitutes evidence that the genre in which a conversation is situated influences people's perceptions of what counts as a lie; within the present framework, the courtroom genre drives the selection of a more minimal PII. PCIs, however, were rather uniformly considered to not be lies and thus were not significantly affected by genre overall.

The fact that the difference between lying and non-lie verbal deception (i.e., by false implicature) is pertinent in the courtroom inspires Horn to defend the distinction at a theoretical level. Horn writes (2017: 51):

> Meibauer's approach [treating false implicatures as lies] risks the blurring of important conceptual (and legal) distinctions; it strikes me as sounder to describe the properties that related categories have in common than to efface a distinction between categories that must be recovered in some – arguably most – contexts. To subsume knowingly false assertions and knowingly false implicatures under a single banner would be akin to branding all types of intentional killing as first degree murder and then having to reconstitute the category of manslaughter when we need it.

The results of this experiment, in line with Horn's above point, argue against a theory of lying that does not distinguish between straightforward lie and false implicature. In fact, that false PCIs were uniformly rated as not lies runs counter to Meibauer's claim that false PCIs are lies. The central tendency of GCIs was precisely at the midpoint between lie and not-lie, suggesting that neither is it the case that only bare linguistic meaning falsehoods are considered lies. These results support theories of lying that promote a focus on speaker commitment and theories of linguistic meaning that allow for flexibility and variability across both contexts and individuals.

References

Adler, Jonathan E. 1997. Lying, deceiving, or falsely implicating. *The Journal of Philosophy* 94(9). 435–452.

Aird, Michael J, Ullrich K H Ecker, Briony Swire, Adam J Berinsky & Stephan Lewandowsky. 2018. Does truth matter to voters ? The effects of correcting political misinformation in an Australian sample. *Royal Society open science* 5(12). doi: 10.1098/rsos.180593

Allen, I Elaine & Christopher A Seaman. 2007. Likert Scales and Data Analyses. *Quality Progress* 40. http://rube.asq.org/quality-progress/2007/07/statistics/likert-scales-and-data-analyses.html.

Alterman, Eric. 2005. *When presidents lie: A history of official deception and its consequences.* New York: Penguin.

Antomo, Mailin, Susanne Müller, Katharina Paul, Markus Paluch & Maik Thalmann. 2018. When children aren't more logical than adults : An empirical investigation of lying by falsely implicating. *Journal of Pragmatics* 138. 135–148. https://doi.org/10.1016/j.pragma.2018.09.010.

Arendt, Hannah. 1971. Lying in Politics. In *Crises of the Republic*, 3–47. New York: Harcourt Brace Jovanovich.

Ariel, Mira. 2002a. The demise of a unique concept of literal meaning. *Journal of Pragmatics* 34(4). 361–402. https://doi.org/10.1016/S0378-2166(01)00043-1.

Ariel, Mira. 2002b. Privileged interactional interpretations. *Journal of Pragmatics* 34(8). 1003–1044. https://doi.org/10.1016/S0378-2166(01)00061-3.

Bakhtin, Mikhail M. 1986. The Bildungsroman and its Significance in the History of Realism. *Speech genres and other late essays.* Austin: University of Texas Press.

Bakir, Vian, Eric Herring & Piers Robinson. 2019. Organized Persuasive Communication: A new conceptual framework for research on public relations, propaganda and promotional culture. *Critical Sociology* 45(3). 311–328. https://doi.org/10.1177/0896920518764586.

Bonalumi, Francesca, Thom Scott-Phillips, Julius Tacha & Christophe Heintz. 2020. Commitment and communication: Are we committed to what we mean, or what we say? *Language and Cognition* 12(2). 360–384. https://doi.org/10.1017/langcog.2020.2.

Boogaart, Ronny, Henrike Jansen & Maarten van Leeuwen. 2021. "Those are your words, not mine!" Defence strategies for denying speaker commitment. *Argumentation* 35(2). 209–235. https://doi.org/10.1007/s10503-020-09521-3.

Carston, Robyn. 2002. Linguistic meaning, communicated meaning and cognitive pragmatics. *Mind & Language2* 17(1–2). 127–148.

Cap, Piotr & Urszula Okulska. 2013. Analyzing genres in political communication: An introduction. In Piotr Cap & Urszula Okulska (eds.), *Analyzing genres in political communication.* Amsterdam: John Benjamins Publishing.

Coleman, Linda & Paul Kay. 1981. Prototype semantics: The English word lie. *Language* 57(1). 26–44. https://doi.org/10.1353/lan.1981.0002.

Danet, Brenda. 1980. Language in the legal process. *Law & Society Review* 14(3): 445–564.

Doran, Ryan, Gregory Ward, Meredith Larson, Yaron McNabb & Rachel E. Baker. 2012. A novel experimental paradigm for distinguishing between what is said and what is implicated. *Language* 88(1). 124–154. https://doi.org/10.1353/lan.2012.0008.

Dynel, Marta. 2015. Intention to deceive, bald-faced lies, and deceptive implicature: Insights into Lying at the semantics-pragmatics interface. *Intercultural Pragmatics* 12(3). 309–332. https://doi.org/10.1515/ip-2015-0016.

Dynel, Marta. 2020. To say the least: Where deceptively withholding information ends and lying begins. *Topics in Cognitive Science* 12(2). 555–582. https://doi.org/10.1111/tops.12379.

Eemeren, Frans H van. 2010. *Strategic maneuvering in argumentative discourse: Extending the pragma-dialectical theory of argumentation.* Amsterdam: John Benjamins Publishing.

Eggins, Suzanne & Diana Slade. 1997. *Analysing Casual Conversation.* Washington: Cassell.

Ekström, Mats. 2009. Power and affiliation in presidential press conferences: A study on interruptions, jokes and laughter. *Journal of Language and Politics.* John Benjamins 8(3). 386–415.

Fairclough, Norman. 2003. *Analysing Discourse: Textual Analysis for Social Research.* London: Routledge.

Fallis, Don. 2009. What Is Lying? *The Journal of Philosophy* 106(1). 29–56.

García-Carpintero, Manuel. 2021. Lying vs. misleading: The adverbial account. *Intercultural Pragmatics* 18(3). 391–413.

Geurts, Bart. 2019. Communication as commitment sharing: Speech acts, implicatures, common ground. *Theoretical Linguistics* 45(1–2). 1–30.

Giltrow, Janet & Dieter Stein (eds.). 2009. *Genres in the Internet: Issues in the Theory of Genre*. Amsterdam: John Benjamins.

Green, Stuart P. 2018. Lying and the Law. In Jörg Meibauer (ed.), *Oxford Handbook of Lying*, 483–494. Oxford: Oxford University Press.

Grice, H. Paul. 1989. *Studies in the Way of Words*. Cambridge, MA: Harvard University Press.

Horn, Laurence R. 2017. Telling it slant: Toward a taxonomy of deception. In Janet Giltrow & Dieter Stein (eds.), *The pragmatic turn in law: inference and interpretation in legal discourse*, vol. 18, 23–55. Berlin: De Gruyter Mouton. https://doi.org/10.1515/9781501504723-002.

Jaszczolt, Kasia M. 2005. *Default semantics: Foundations of a compositional theory of acts of communication*. Oxford: Oxford University Press.

Kemper, Nathan S., Dylan S. Campbell, Mitch Earleywine & Anna-Kaisa Newheiser. 2020. Likert, slider, or text? Reassurances about response format effects. *Addiction Research and Theory* 28(5). 406–414. https://doi.org/10.1080/16066359.2019.1676892.

Keyes, Ralph. 2004. *The post-truth era: dishonesty and deception in contemporary life*. New York: St. Martin's Press.

Kisielewska-Krysiuk, Marta. 2016. Lying and the relevance-theoretic explicit/implicit distinction. *ANGLICA-An International Journal of English Studies* 25(2). 73–86.

Krifka, Manfred. 2019. Indicative and subjunctive conditionals in commitment spaces. In *Proceedings of the 22nd Amsterdam Colloquium*, 248–258.

Levinson, Stephen C. 1979. Activity types and language. *Linguistics* 17. 365–399.

Levinson, Stephen C. 2000. *Presumptive Meanings*. Cambridge, MA: MIT Press.

Lewandowsky, Stephan, Ullrich K. H. Ecker & John Cook. 2017. Beyond misinformation: understanding and coping with the "post-truth" era. *Journal of Applied Research in Memory and Cognition* 6(4). 353–369. https://doi.org/10.1016/j.jarmac.2017.07.008.

Marsili, Neri. 2017. *You don't say! Lying, asserting and insincerity*. PhD dissertation, University of Sheffield.

Marsili, Neri. 2020. Lying, speech acts, and commitment. *Synthese*. https://doi.org/10.1007/s11229-020-02933-4.

Mazzarella, Diana, Robert Reinecke, Ira Noveck & Hugo Mercier. 2018. Saying, presupposing and implicating: How pragmatics modulates commitment. *Journal of Pragmatics* 133 (June). 15–27. https://doi.org/10.1016/j.pragma.2018.05.009.

McQuarrie, Edward F & Barbara J Phillips. 2005. Indirect persuasion in advertising: How consumers process metaphors presented in pictures and words. *Journal of Advertising* 34(2). 7–20. https://doi.org/10.1080/00913367.2005.10639188.

Meibauer, Jörg. 2005. Lying and falsely implicating. *Journal of Pragmatics* 37(9). 1373–1399. https://doi.org/10.1016/j.pragma.2004.12.007.

Meibauer, Jörg. 2011. On lying: Intentionality, implicature, and imprecision. *Intercultural Pragmatics* 8(2). 277–292. https://doi.org/10.1515/IPRG.2011.013.

Meibauer, Jörg. 2014. *Lying at the Semantics-Pragmatics Interface*. Berlin: De Gruyter Mouton.

Meibauer, Jörg. 2016. Topics in the linguistics of lying: A reply to Marta Dynel. *Intercultural Pragmatics* 13(1). 107–123. https://doi.org/10.1515/ip-2016-0004.

Meibauer, Jörg. 2018. The linguistics of lying. *Annual Review of Linguistics* 4. 357–377. https://doi.org/10.1146/annurev-linguistics-011817-045634.

Moeschler, Jacques. 2013. Is a speaker-based pragmatics possible ? Or how can a hearer infer a speaker ' s commitment ? *Journal of Pragmatics* 48(1). 84–97. https://doi.org/10.1016/j.pragma.2012.11.019.

Nyhan, Brendan, Ethan Porter, Jason Reifler & Thomas J Wood. 2019. Taking fact-checks literally but not seriously? The effects of journalistic fact-checking on factual beliefs and candidate favorability. *Political Behavior* 42(3). 939–960. https://doi.org/10.1007/s11109-019-09528-x.

Reins, Louisa M. & Alex Wiegmann. 2021. Is lying bound to commitment? Empirically investigating deceptive presuppositions, implicatures, and actions. *Cognitive Science* 45(2). https://doi.org/10.1111/cogs.12936.

Robinson, Piers, David Miller, Eric Herring & Vian Bakir. 2018. Lying and deception in politics. Online paper, https://doi.org/10.1093/oxfordhb/9780198736578.013.42. Revised version published in Jörg Meibauer (ed.), *Oxford Handbook of Lying*, 529–540. Oxford: Oxford University Press.

Saul, Jennifer M. 2002. What is said and psychological reality; Grice's project and relevance theorists' criticisms. *Linguistics and Philosophy* 25(3). 347–372.

Skoczeń, Izabela. 2021. Modelling perjury: Between trust and blame. *International Journal for the Semiotics of Law*. https://doi.org/10.1007/s11196-021-09818-w.

Sperber, Dan & Deirdre Wilson. *Relevance: Communication and Cognition*. 2nd edn. Oxford: Blackwell.

Sternau, Marit. 2013. *Levels of Interpretation: Linguistic Meaning and Inferences*. PhD dissertation, Tel Aviv University.

Sternau, Marit, Mira Ariel, Rachel Giora & Ofer Fein. 2015. Levels of interpretation: New tools for characterizing intended meanings. *Journal of Pragmatics* 84. 86–101. https://doi.org/10.1016/j.pragma.2015.05.002.

Sternau, Marit, Mira Ariel, Rachel Giora & Ofer Fein. 2017. Deniability and explicatures: cognitive, philosophical, and sociopragmatic perspectives. In Rachel Giora & Michael Haugh (eds.), *Doing Pragmatics Interculturally: Cognitive, Philosophical, and Sociopragmatic Perspectives*, 97–120. Berlin: De Gruyter. https://doi.org/10.1515/9783110546095-006.

Stokke, Andreas. 2013. Lying, deceiving, and misleading. *Philosophy Compass* 8(4). 348–359. https://doi.org/10.1111/phc3.12022.

Stolberg, Sheryl G. 2017. Many Politicians Lie. But Trump Has Elevated the Art of Fabrication. *The New York Times*, August 7, 2017. https://www.nytimes.com/2017/08/07/us/politics/lies-trump-obama-mislead.html

Sullivan, Gail M. & Anthony R. Artino. 2013. Analyzing and interpreting data from Likert-type scales. *Journal of Graduate Medical Education* 5(4). 541–542. https://doi.org/10.4300/JGME-5-4-18.

Swire, Briony, Adam J Berinsky & Ullrich K H Ecker. 2017. Processing political misinformation: comprehending the Trump phenomenon. *Royal Society Open Science* 4(3). https://doi.org/10.1098/rsos.160802.

Terkourafi, Marina, Benjamin Weissman & Joseph Roy. 2020. Different scalar terms are affected by face differently. *International Review of Pragmatics* 12(1). 1–43. https://doi.org/10.1163/18773109-01201103.

Thalmann, Maik, Yuqiu Chen, Susanne Müller, Markus Paluch & Mailin Antomo. 2021. Not all implicatures can be used to lie: Evidence from deceptive language in German and Chinese. *Linguistische Berichte* 267. https://www.researchgate.net/publication/348187499.

Tiel, Bob van, Emiel van Miltenburg, Natalia Zevakhina & Bart Geurts. 2016. Scalar diversity. *Journal of Semantics* 33(1). 137–175. https://doi.org/10.1093/jos/ffu017.

Tsipursky, Gleb. 2017. How to address the epidemic of lies in politics. *Scientific American*. https://blogs.scientificamerican.com/observations/how-to-address-the-epidemic-of-lies-in-politics/.

Viebahn, Emanuel, Alex Wiegmann, Neele Engelmann & Pascale Willemsen. Forthcoming. Can a question be a lie? An empirical investigation. To appear in *Ergo*. https://doi.org/10.31219/osf.io/jfyn8.

Vincent, Jocelyne M & Cristiano Castelfranchi. 1981. On the art of deception: how to lie while saying the truth. In Herman Parret, Marina Sbisà, and Jef Verschueren (eds.), *Possibilities and limitations of pragmatics*, 749–777. Amsterdam: John Benjamins.

Weissman, Benjamin. 2019. *The role of linguistic meaning in the concept of lying*. PhD dissertation, University of Illinois at Urbana-Champaign.

Weissman, Benjamin & Marina Terkourafi. 2019. Are false implicatures lies? An experimental investigation. *Mind & Language* 34(2). 221–246.

Wiegmann, Alex, Pascale Willemsen & Jörg Meibauer. Forthcoming. Lying, Deceptive Implicatures, and Commitment. To appear in *Ergo*. https://psyarxiv.com/n96eb.

Willemsen, Pascale & Alex Wiegmann. 2017. How the truth can make a great lie: An empirical investigation of the folk concept of lying by falsely implicating. *39th Annual Meeting of the Cognitive Science Society*. https://doi.org/10.5167/uzh-175885.

Williams, Bernard. 2002. *Truth and Truthfulness: An Essay in Genealogy*. Princeton: Princeton University Press.

7 Supplemental materials

All versions of the stories used in this experiment, as well as the dataset used for analysis, can be found at: https://osf.io/t7md9/?view_only=9010e52a1d6d4d7194ae9f9cd7868dd3

Lawrence M. Solan
Lies, deception, and bullshit in law

Abstract: In 1973, the Supreme Court established a "literal truth" defense to perjury prosecutions. Samuel Bronston, a film producer, had filed for bankruptcy. When asked under oath if he had ever held bank accounts in Switzerland, he testified, "The company had an account there for about six months, in Zurich." That much was true. But it was also true that Bronston himself previously had bank accounts in Switzerland and wished to conceal these assets from his creditors. The Supreme Court unanimously held that Bronston did not commit perjury because he didn't lie. Rather, he merely deceived, and the perjury statute does not cover deception committed by telling half-truths. It was up to the prosecutor to discover the deception. This narrative raises both legal and moral issues. This chapter will explore both.

As a legal matter, the Supreme Court acknowledged that determining whether a statement is "literally true" requires the decision maker to determine how the witness understood the question that was asked. What should happen (as n *U.S. v. DeZarn* (1998)) when the questioner misspeaks, but the witness understands what the questioner meant and answers falsely not to the question that was actually asked but to the question that the lawyer intended? As a moral question, the Supreme Court's perjury jurisprudence appears very weak. Philosophers are not in accord as to whether lying is invariably worse than deceiving through intentionally misleading statements. Add to this mix Harry Frankfurt's judgment that bullshit is worse than a lie because while the liar must respect the truth in order to violate it, the bullshitter will say whatever is advantageous to him without knowing or even caring if the statement is true. But lying is certainly antithetical to the

Note: An earlier draft of this essay was written while I was a Visiting Fellow in 2017 at the Käte Hamburger Center for Advanced Study in the Humanities, an institute of the University of Bonn. I am grateful to the Directors, Fellows, and other members of the community there for all their support and for the wonderful intellectual atmosphere that helped me enormously in preparing this piece. The current essay appeared in a slightly different form as Solan 2018 (= 56 Duq. L. Rev. 73 (2018)), and I am indebted to the participants and organizers of the Duquesne symposium on "Resurrecting Truth in American Law and Public Discourse" which inspired this work and especially to the Duquesne Law Review for generously granting permission to reprint it here in revised form. I am also grateful to Laurence Horn and Jason Stanley for valuable insights into these issues.

Lawrence M. Solan, Brooklyn College of Law, e-mail: larry.solan@brooklaw.edu

https://doi.org/10.1515/9783110733730-015

truth-seeking goals of the law, and drawing a clear line between lying and other offenses, including deception, bullshit, and fraud, can serve a valid purpose within the judicial system.

1 Introduction

Gerald Shargel, a prominent New York criminal attorney, has written, "A trial may be a search for the truth, but I – as a defense attorney – am not part of the search party" (Shargel 2007: 1267). This essay asks who is a member of the search party, and by what tactics parties and lawyers impede a successful search for the truth, both in the courtroom and in the interactions among people that set the stage for judicial intervention. In this effort, the essay distinguishes among three kinds of dishonesty: lies, deceit and bullshit.

The federal perjury statute criminalizes an assertion of a material fact that the speaker believes to be false, but which is asserted as true (18 U.S.C. §1621). Once one has taken an oath to tell the truth, it is a crime to willfully violate that oath by testifying to a material fact that one believes to be false. The law purports to disapprove of lying. By and large it does, but not always. For example, the legal system gives law enforcement officers license to lie both during the interrogation of witnesses and during sting operations, and subsequently permits prosecutors to take advantage of these lies. The prosecutors themselves may not lie, however. For that matter, sometimes laypeople acting as testers, especially in housing discrimination investigations, are permitted to misrepresent their identities. Moreover, there is well-studied tolerance by judges of police officers lying about the circumstances under which they seized evidence or interrogated a suspect (for discussion, see Green 2019 and Chapter 11).

Apart from such selective tolerance, conceptual questions about lies arise from time to time. May a witness who intended to lie be saved from a perjury conviction if the testimony turns out to be true by some kind of fluke? For example, what if the witness was mistaken about the facts and what he intended as a lie was really true? Another issue asks whether the witness must intend that the false statement be believed. In the film *Casablanca*, Humphrey Bogart's character, Rick, is asked his nationality and answers, "I'm a drunkard."[1] Whether that was a lie or not depends upon whether an intention to deceive is part of the definition of lying. In civil litigation, a plaintiff who claims to have been damaged by having relied on a false statement must demonstrate that the reliance was reasonable. Whether

---

1 https://www.youtube.com/watch?v=ZkM6HegRk3A.

perjury requires that the speaker could reasonably expect to be believed is not well-established in the case law, suggesting that there are few if any prosecutions that raise the issue.

While lying is about both false testimony and the state of mind of the speaker, deceit is also about the state of mind of the recipient of the information. A speaker has deceived another when the speaker has led the hearer to come to believe something to be true that the speaker believes to be false. It makes no difference whether the speaker did this by means of making false assertions of fact, or by uttering half-truths or by other means of persuasion. Speech act theorists refer to a hearer-oriented element of an act of speech as the perlocutionary effect of the utterance – the effect it has on the state of mind of the hearer, rather than the communicative intent of the speaker (Austin 1962: 101). Verbs vary as to their focus in this regard. "Persuade," for example, holds when the perlocutionary effect of an assertion is to convince the hearer of a proposition.

As for *bullshit*, I intend that word to be understood as described by Harry Frankfurt, in his book *On Bullshit*. Frankfurt (2005) paints the bullshitter as an amoral person, not concerned about whether what he says is true or false. Thus, the bullshitter is not a liar, because the liar must say something he believes is false, and the bullshitter does not bother himself with such concerns. Whether the bullshitter engages in deceit is a different matter. The bullshitter may be concerned with the perlocutionary effect of his or her statements, but not with whether the statement is intended to convince the hearer of something true or false. As Frankfurt puts it (2005: 55) (but see also Chapters 1, 8, and 9):

> The fact about himself that the bullshitter hides . . . is that the truth-values of his statements are of no central interest to him; what we are not to understand is that his intention is neither to report the truth nor to conceal it . . . .It is impossible for someone to lie unless he thinks he knows the truth. Producing bullshit requires no such conviction.

In a number of circumstances, the law declares bullshit as unacceptable, recognizing that it would not be covered by the ordinary definitions of deceit or lying. Illustrations include Rule 11 of the Federal Rules of Civil Procedure, which requires that an attorney (or party) make adequate investigation of the facts underlying a submission to a federal court, or be subject to monetary or other sanctions. Of course, lawyers do sometimes include false allegations in a legal pleading. More often, however, a lawyer may simply intend to fill in the gaps in a narrative in which a number of the assertions required for the lawyer to succeed can be proven, but not all such assertions. When a lawyer takes liberties with these remaining facts, the lawyer is engaged in bullshitting. The same holds true for fraud, under a number of common law and statutory definitions. Asserting something as true without finding out whether it is true or not is considered fraudulent behavior in many circumstances.

The remainder of this essay explores the themes raised in this introduction, with examples from legal proceedings, business transactions (real and hypothetical) that may become the subject of such proceedings, and political discourse.

2 Lying

Lying is outlawed in one context after another. Lying under oath is perjury (18 U.S.C. §1621 (2016)). Lying to a government official is a federal crime (18 U.S.C. §1001 (2016)). Lying in a business transaction is a species of fraud if the party that was lied to reasonably relies on the lie to his or her detriment.[2] Lawyers may not lie in the course of representing a client.[3] Nor may they arrange to have a non-lawyer employee lie as their agent.[4]

2.1 What is a lie?

Horn (2017: 24–25) sets forth four criteria that have been proposed in defining what constitutes a lie (see also Chapter 1 for some complicating factors):

(C1) S says/asserts that p
(C2) S believes that p is false
(C3) p is false
(C4) S intends to deceive H

There is general agreement that a lie must be an assertion of some kind. An opinion, a question, a promise and other such speech acts do not have truth value, and therefore cannot be false.[5] Philosopher Don Fallis elaborates (2007: 33): "I

---

2 A classic example is Rule 10b-5 of the Securities and Exchange Commission (17 C.F.R. §240.10b-5 (2018)): It is fraudulent conduct "[t]o make any untrue statement of a material fact" on a securities transaction. While lying is sufficient to constitute fraud, it is not a necessary condition, in that the rule also outlaws other types of deceptive practice.
3 See, e.g., N.Y. Rules of Professional Conduct, rule 4.1: "In the course of representing a client, a lawyer shall not knowingly make a false statement of fact or law to a third party."
4 See e.g. N.Y. Rules of Professional Conduct, rule 5.3(b)(1) (2017).
5 Well, almost. One can lie about what one's opinion is, although as an opinion, its substance lacks truth value.

think that you assert something when (a) you say something and (b) you believe that you are in a situation where you should not say things that you believe to be false." Fallis, in turn, takes this condition on assertions to follow from Paul Grice's maxim of quality that we expect of our partners in conversation: "Do not say what you believe to be false" (Grice 1975: 46).[6]

There is also wide agreement that one does not lie if one says what one believes to be true, but is wrong. Such cases are matters of mistake. It is the last two criteria that create disagreement and some confusion. Does a statement have to be false for it to constitute a lie? Most say no, following the writings of St. Augustine in late antiquity (Augustine 420). If one intends to make a false statement, he is not rescued by the truth if he happens to have spoken truthfully because he mistook the facts. If I attempt to protect my friend by saying he was in Cleveland when a crime was committed although I am quite certain that he was in Pittsburgh committing the crime, I have lied even if it turns out that I was wrong and he really was in Cleveland. As we shall see, the law of perjury follows this tradition.

Finally, there is a question of whether a lie must be part of an effort to deceive. Those who argue that this is not required (although it is characteristic of most lies) cite examples such as the student who falsely claims to the school authorities that he did not commit plagiarism, knowing that the dean will punish him only if he goes on record to confess his guilt. The plagiarist knows that nobody, including the dean, will believe the denial but he stands his ground to escape expulsion. No doubt the student lied. Likewise, a witness afraid of repercussions may testify falsely to protect himself, knowing full well that he will fool no one, having already told authorities the true story before the trial began. Roy Sorensen (2007) refers to such assertions as bald-faced lies. In keeping with positions taken by Jennifer Saul (2012), Don Fallis (2009), and other philosophers, this essay proceeds on the claim that an attempt to deceive is a feature of the prototypical lie, but not necessarily a necessary condition for an assertion to be deemed a lie.

2.2 Perjury

As for perjury, the leading case is a 1973 unanimous Supreme Court decision, *Bronston v. United States* (409 U.S. 352, 354 (1973); see also Chapters 1, 2, 8, 9,

---

6 Grice also includes a maxim to the effect that one should avoid bullshit in conversation: "Do not say that for which you lack adequate evidence" (Quality-2).

13 for other reflections on *Bronston*). Bronston was a film producer who had filed for bankruptcy. Required to answer questions under oath from the creditors from whom he sought relief, the following colloquy took place:

> Q: Do you have any bank accounts in Swiss banks, Mr. Bronston?
>
> A: No sir.
>
> Q: Have you ever?
>
> A: The company had an account there for about six months, in Zurich.

It turned out that not only did the company have an account in Zurich in the past, but so did Bronston himself. As a result, he was prosecuted for perjury, and convicted. The perjury statute (18 U.S.C. §1621 (2016)) states in relevant part:

> Whoever – having taken an oath before a competent tribunal, officer, or person, in any case in which a law of the United States authorizes an oath to be administered, that he will testify, declare, depose, or certify truly, or that any written testimony, declaration, deposition, or certificate by him subscribed, is true, willfully and contrary to such oath states or subscribes any material matter which he does not believe to be true; . . . is guilty of perjury.

But the Supreme Court reversed the conviction, relying on a distinction between a false statement on the one hand and a true statement leading to a false inference on the other (409 U.S. 357–8 (1973)):

> The words of the statute confine the offense to the witness who "willfully . . . states . . . any material matter which he does not believe to be true." Beyond question, petitioner's answer to the crucial question was not responsive if we assume, as we do, that the first question was directed at personal bank accounts. There is, indeed, an implication in the answer to the second question that there was never a personal bank account; in casual conversation this interpretation might reasonably be drawn. But we are not dealing with casual conversation and the statute does not make it a criminal act for a witness to willfully state any material matter that *implies* any material matter that he does not believe to be true.

This has come to be known as the "literal truth defense" to perjury (see e.g. Posner 1999: 49). The Court noted that Bronston's answer was unresponsive, not false, and that an alert lawyer would be on sufficient notice to ask a follow-up question, such as, "Mr. Bronston, I didn't ask about your company, I asked about you." As Peter Tiersma and I have noted (2005), this holding, at least if taken at face value, sets a very low moral floor for witnesses who swear to tell the truth in an enterprise whose goal is to seek out and discover the truth. (Solan and Tiersma 2005; see also Solan 2012)

Yet the questions and answers in a courtroom or a deposition are not ordinary conversational exchanges. The philosopher Paul Grice famously wrote that conversation is a cooperative enterprise. When speaking with others, we typically abide by his cooperative principle, "Make your contribution such as is required, at the stage at which it occurs, by the accepted purpose or direction of the talk exchange in which you are engaged" (Grice 1975: 45). In litigation contexts, however, witnesses are instructed by their lawyers to answer the questions asked, and not to volunteer more information for the sake of being helpful. This instruction does not entirely flout the cooperative principle because witnesses must give answers that are both relevant and truthful. Grice lists four maxims as components of cooperation in conversation. Two are the maxim of relation (be relevant), and the maxim of quality (be truthful). Others, following Sperber and Wilson 1995, have elevated relevance to the principal component of conversational responsibility.

As for Bronston, the Court held in essence that by giving an answer that was literally both truthful and irrelevant, he had flouted the maxim of relation, but not the maxim of quality. But that, of course, is not all that Bronston did. In normal discourse, if a person says that his company had a Swiss bank account in response to a question about whether the witness himself had one, the normal inference is that the witness intends to convey, "No. I never had one, but . . ." Bronston thus succeeded in misleading the questioner into concluding that Bronston himself did not have one. If the questioner thought otherwise, he would indeed have asked the follow-up question necessary to button down the facts about what Bronston himself owned.

Without question, Bronston engaged in dishonest conduct. Some commentators, e.g. Gaines (2015: 235), believe that the case was wrongly decided for that reason. But if perjury is about lying, and the Court decided to articulate a bright line rule, then at first glance it seemed to have accomplished its goal. But the Court itself took a second glance, recognizing that whether an answer to a question is truthful requires not only analysis of the answer, but also analysis of the question. Because Bronston's response was so blatantly unresponsive, the Court reasoned, it was the questioner who should be held responsible for the truth not coming out. The Court thus distinguished Bronston's conduct from the conduct in a hypothetical case that the trial court had presented. It concerns another of Grice's maxims: the maxim of quantity (say whatever is necessary to make one's point, but not more; cf. Grice 1975: 45–46). The district court, which the Supreme Court quoted, had noted (409 U.S. 352, 355 n.3):

> [I]f it is material to ascertain how many times a person has entered a store on a given day and that person responds to such a question by saying five times when in fact he knows that he entered the store 50 times that day, that person may be guilty of perjury even though it is technically true that he entered the store five times.

The Supreme Court argued that the hypothetical situation was unlike that in *Bronston* because "the answer 'five times' is responsive to the hypothetical question, and contains nothing to alert the questioner that he may be side-tracked." The Court continued:

> Whether an answer is true must be determined with reference to the question it purports to answer, not in isolation. An unresponsive answer is unique in this respect, because its unresponsiveness, by definition, prevents its truthfulness from being tested in the context of the question – unless there is to be speculation as to what the unresponsive answer "implies."

The Court was correct in declaring that unresponsive answers may generate false inferences, but are not false answers to the questions in their own right. But the situation is a bit more complex. Ambiguous questions pose a similar problem. If a question is subject to more than one interpretation, a witness's answer may be truthful if the question is understood one way, false if it is understood another way. Generally, as the Court assumes, if we ask someone how many times he or she has been to a particular place, we mean to ask for the sum total of times. But this is not always true of quantitative inquiries. Consider this hypothetical. Two friends are taking a long walk, and one sees a beverage machine at a gas station that they pass. It requires inserting a one dollar bill and some coins. He has the coins, but not the dollar bill. He asks his friend, "how much cash do you have?" The friend, understanding the situation, responds, "I have a dollar." In fact, he has 32 dollars. Did he lie? No. He was merely trying to advance the conversation by giving a relevant response. What he meant was that he had at least the dollar required for the beverage, and he would be understood that way. By the same token, if a store has a special promotion for patrons who had been there at least five times in the past month, a person who had been there 50 times could enter the store and say forthrightly that he had been there five times when asked how many times he had been there.

If what I have said thus far is right, it presents a problem for the Court's analysis. The Court was correct in its assertion that construing an unresponsive answer as misleading requires it to speculate as to the inferences that a reasonable hearer would draw. However, it is also true that determining whether a seemingly responsive answer is true or false requires a court to speculate as to the inferences that the witness drew in understanding the question, at least in the examples that the Supreme Court used. Of course, some questions are sufficiently clear that this is not a problem. But many are not, and we routinely resolve ambiguity as we attempt to understand the discourse.

If the truth of an answer can be judged only with respect to the question that was asked, can a witness be saved from a perjury conviction if the questioner misstated the question, but both questioner and witness understood the question to mean what the questioner intended to ask? This inquiry may sound bizarre, but it is exactly what happened in *United States v. DeZarn* (157 F.3d 1042 (6th Cir. 1998)) and the answer the Sixth Circuit Court of Appeals gave was no: if you are under oath, and you answer a question in a manner that you believe to be false, then you have committed the crime of perjury even if your answer is literally true.

In 1990, Robert DeZarn, a retired officer in the Kentucky National Guard, attended and participated in a fundraising party for a political candidate running for governor of Kentucky. The party was held at the home of an officer in the Kentucky National Guard, General Wellman. The party was referred to as a "Preakness party," because it was held on the day of the annual Preakness horse race. It is illegal for officers to solicit such funds from military personnel. In 1991, that same officer held another, smaller party, this time on the day of the Kentucky Derby race. DeZarn attended that party as well. No fundraising took place at the 1991 event.

Because of the illegality of the fundraising by military personnel, an investigation ensued, once authorities heard about the incident. DeZarn was questioned by an officer, in relevant part as follows:

Q: Okay, sir. My question is going to deal with General Wellman, though. Was it traditional for General Wellman to hold parties at his home and invite Guardsman to attend?

A [by DeZarn]: Well, I suppose you could say that for a number of years that going back to the late 50s he has done this on occasion.

Q: Okay. In 1991, and I recognize this is in the period that you were retired, he [i.e., General Wellman, the host] held the Preakness Party at his home. Were you aware of that?

A: Yes.

Q: Did you attend?

A: Yes.

[. . .]

Q: Okay. Sir, was that a political fundraising activity?

A: Absolutely not.

Q: Okay. Did then Lieutenant Governor Jones, was he in attendance at the party?

A: I knew he was invited. I don't remember if he made an appearance or not.

Q: All right, sir. You said it was not a political fundraising activity. Were there any contributions to Governor Jones' campaign made at that activity?

A: I don't know.

Q: Okay. You did not see any, though?

A: No.

Q: And you were not aware of any?

A: No.

DeZarn was convicted of perjury for having given these answers. He appealed on the ground that the questioner placed the party in 1991, and in that year, there was no political fundraising. The jury, though, believed that DeZarn and the questioner both understood at the time that they were talking about the 1990 fundraising Preakness party that had occurred the year before, and that he had therefore testified falsely.

Had the questioner asked DeZarn about a 1991 Kentucky Derby party, there would have been little justification for the conviction. The testimony as it did occur, however, presents a thorny doctrinal question. Why is it that Bronston's answer *is not* perjurious because it requires the hearer to draw an inference that Bronston himself did not have a Swiss bank account, but DeZarn's testimony *is* perjurious, even though his answer requires the hearer to draw an inference that the questioner had mistakenly placed the Preakness party in the wrong year? Interestingly, in *Bronston*, the Court put the blame on the lawyer for not following up after receiving an unresponsive answer (Bronston, 409 U.S. 358):

> It is the responsibility of the lawyer to probe; testimonial interrogation, and cross examination in particular, is a probing, prying, pressing form of inquiry. If a witness evades, it is the lawyer's responsibility to recognize the evasion and to bring the witness back to the mark, to flush out the whole truth with the tools of adversary examination.

In *DeZarn*, in contrast, the questioner, rather than failing to follow up with the witness, simply asked the wrong question in the first place, leaving a degree of ambiguity that the witness attempted to leverage to create a misleading record. The court reasoned (157 F.3d 1046):

At trial, DeZarn testified that Colonel Tripp, by mistakenly setting the questions in his interview about the Preakness Party in 1991, rather than 1990, led him to answer the questions with reference to the 1991 dinner party, which was not a fundraiser and at which he did not collect any contributions.

Evidence was presented at trial, however, to establish that DeZarn was not misled by the 1991 date but had answered the investigators' questions as he had with intent to deceive them. Specifically, all of the individuals questioned by the investigators described the same party, even though some were questioned about a "Preakness Party", some were questioned about a "1990 Preakness Party", and some, like DeZarn, were questioned about a "1991 Preakness Party".

Let us assume that the court was accurate in its description of DeZarn's motives. The question then becomes what difference should DeZarn's motives make if he arguably did not answer falsely in light of the questioner's mistake in wording the question? After all, Bronston had bad motives too.

The perjury statute, read literally, does not have a literal truth defense (18 U.S.C. § 1621 (2016)). Bronston did not violate the law if we read the law as written. He did not say something that he did not believe to be true. What about DeZarn? If DeZarn believed that the question was asking about the 1990 Preakness party, then he did violate the law. But what if he was just being cagy? What if DeZarn saw an opening in the question that permitted him to answer as he did without actually lying? If so, he did this not because he was trying to be helpful and forthright, but rather because he wanted to take advantage of the lawyer's mistake and avoid having to say what really took place without perjuring himself. If that is what happened it is difficult to distinguish the two cases on their relevant facts.

The Supreme Court was certainly correct in concluding that one cannot assess the truthfulness of an answer without knowing what the question was. Yet it is not a simple matter to reconcile *Bronston* and *DeZarn*. There was only one Preakness party, and it was an illegal fundraising event. If DeZarn understood the question as referring to that party, then he committed perjury. By the same token, there was only one relevant party in 1991, and it was not a fundraising party. If DeZarn understood the question as referring to that party, then he did not commit perjury. The more difficult question is what should have happened if DeZarn recognized the error, and for the sake of obfuscating the facts, chose the 1991 date over the name of the horse race to accomplish this goal. Perhaps it was right to leave that decision to the jury. The rule of lenity tells us that ambiguities in law are to be resolved in favor of the defendant. But this, at least arguably, is not an ambiguity of law. Rather, it is a murkiness in the facts, regarding the defendant's state of mind.

Regardless, taken together the cases describe a rather simple story: If a person makes statement under oath that she believes to be false at the time she makes it, then that person has committed perjury. *Bronston* tells us that a false statement must be literally false – not a true statement that leads the hearer to infer something false. *DeZarn* tells us that the "literal truth" defense is a misnomer. More important than literal truth is the speaker's belief in the falsity of her statement, which is exactly how the perjury statute is worded.

Experimental work in the psychology of language suggests that native speakers' intuitions about what constitutes a lie match the holding of the *De-Zarn* court. Most notably, linguists Linda Coleman and Paul Kay (1981) set out to determine how people understand the concept of lying. Participants in a study were presented with vignettes that ended with a person making some kind of statement. The participants were then asked to rate the statement on a 1–7 scale, where 1 indicated "very sure" it is not a lie, 2 and 3, were "fairly sure" and "not too sure" it is not a lie, 4 was "can't say," and 5–7 went from "not too sure" it is a lie to "very sure" it is a lie.

The statements in the vignettes were varied systematically along three axes. First, the statement was either true or false. Second, the speaker either believed the statement to be true or believed it to be false. Third, the speaker either intended to deceive the hearer or not. These axes are the very features that Horn (2017: 25) attributes to the various definitions of lying, in addition to the requirement that a lie be an assertion.

Coleman and Kay hypothesized that these three factors each contributed to the meaning of the verb to lie, but that none is a necessary condition, some combination may be sufficient. They further hypothesized that participants would rate the statements with either all three or no elements to be the strongest, i.e., prototypical, examples of lying, with various combinations of features being less clear. And that is just what happened.

First, consider the all-or-nothing vignettes. Vignette (1) has all of the features of a prototypical lie, vignette (2) none of them:

(1)   Moe has eaten the cake Juliette was intending to serve to company. Juliette asks Moe, 'Did you eat the cake?' Moe says, 'No.' Did Moe lie?

(2)   Dick, John, and H.R. are playing golf. H.R. steps on Dick's ball. When Dick arrives and sees his ball mashed into the turf, he says, 'John, did you step on my ball?' John replies, 'No, H.R. did it.' Did John lie?

Both answers are self-serving, but only one is true and intended to convey the truth. Sure enough, Coleman and Kay's subjects almost universally reported that (1) contains a lie (6.96 average), and that (2) does not contain a lie (1.06 average).

The more interesting cases are ones in which some but not all of the three elements of lying are present. What do people think when a person makes a truthful statement, knowing it to be true, but with the intention of attempting to get the hearer to draw a false inference? This is the typical scenario of fraud without lying, discussed earlier. Below is a scenario that contains these conditions (Coleman and Kay 1981: 31, Scenario VI ; see also Chapter 4 on this scenario):

> John and Mary have recently started going together. Valentino is Mary's ex-boyfriend. One evening John asks Mary, 'Have you seen Valentino this week?' Mary answers, 'Valentino's been sick with mononucleosis for the past two weeks'. Valentino has in fact been sick with mononucleosis for the past two weeks, but it is also the case that Mary had a date with Valentino the night before. Did Mary lie?

The mean score on this question was 3.48, close to the midpoint of 4.00. This suggests that on the average, people did not consider this to be a lie, but it approaches being a lie. I have presented this scenario to law students who, when probed, typically agree with the statement: "I don't think Mary lied, but what she did was dishonest, and I'm uncomfortable saying it's not a lie, because that answer doesn't reflect my disapproval of her behavior."

In essence, this scenario is *Bronston*. Mary evaded answering the question directly so that she would not have to tell the whole story or take responsibility for having lied, neither of which was palatable under the circumstances. It also resembles President Bill Clinton's efforts to evade the truth without lying.[7] Clinton had been sued by Paula Jones, an employee of the state of Arkansas, for sexual harassment while Clinton was Governor of that state. Later, Kenneth Starr, a special prosecutor appointed to investigate whether the President or those close to him had committed any crimes in connection with a real estate investment called Whitewater, convened a grand jury to determine whether Clinton had perjured himself or obstructed justice when he testified in a deposition in the *Jones* litigation. Much of the questioning in the deposition was about his sexual relationship with Monica Lewinsky, which apparently caught him off guard. Before the grand jury, he testified in true Bronstonian fashion (Kuntz 1998: 361):

> Q: Was it your responsibility to answer those questions truthfully, Mr. President?
>
> A: It was. But it was not my responsibility, in the face of their repeated illegal leaking, it was not my responsibility to volunteer a lot of information.

---

7 For a fair account of the relevant facts, see Posner 1999. Both Peter Tiersma and I wrote about linguistic issues concerning the Clinton impeachment and the events leading up to it. See Solan 2002; Tiersma 2004; Solan and Tiersma 2005: 221–231. See also Carson's discussion in Chapter 1.

The House of Representatives voted to bring Articles of Impeachment against Clinton for lying to the grand jury, but not for lying in the Jones deposition. Both before the grand jury and at his deposition, Clinton refused to characterize his conduct with Lewinsky as "having sexual relations," because in his dialect of English the term is only applicable if the relationship includes sexual intercourse. In fact, it was not until his grand jury appearance that he admitted having a physical relationship with Lewinsky at all. Testifying about an affidavit that Lewinsky had sworn, Clinton said to the grand jury:

> I believe at the time that [Lewinsky] filled out this affidavit, if she believed that the definition of sexual relationship was two people having intercourse, then this is accurate. And I believe that is the definition that most ordinary Americans would give it.
>
> (Grand Jury Transcript at 473–75)

To Clinton, intercourse is a necessary element of the concept, *sexual relations*. Along these same lines, Clinton had famously told the press: "I did not have sexual relations with that woman, Ms. Lewinsky."[8] Whether he would agree that his own conduct may be within that term but not its prototype for some people is something we cannot know.

If the *Mary/Valentino* vignette and Clinton's statements resemble *Bronston's* approach to the truth, what do they say about *DeZarn*? Coleman and Kay's Scenario VIII portrays a person who thought he was lying, but later found out that he had spoken truthfully:

> Superfan has got tickets for the championship game and is very proud of them. He shows them to his boss, who says 'Listen, Superfan, any day you don't come to work, you better have a better excuse than that.' Superfan says, 'I will.' On the day of the game, Superfan calls in and says, 'I can't come to work today, Boss, because I'm sick.' Ironically, Superfan doesn't get to go to the game because the slight stomach ache he felt on arising turns out to be ptomaine poisoning. So Superfan was really sick when he said he was. Did Superfan lie?

Most people said he did. The mean score was 4.61, again fairly close to the midpoint of 4, but this time on the "lying" side of the line.

Other findings were interesting as well. When a person makes a false statement as a result of having mistaken the fact of the matter, participants did not call it a lie. But they did call it a lie when the speaker made a true statement as a result of having mistaken the facts in an effort to tell a lie. They also considered a polite statement from a guest to a host after a dismal party to be a lie. These results

---

8 Statement made January 26, 2017, https://www.youtube.com/watch?v=VBe_guezGGc.

reinforce the intuitive appeal of the perjury law, which focuses on the belief of the speaker, rather than on the speaker's factual accuracy. It also gives some credence to both *Bronston* and *DeZarn* as consistent with people's judgments about what constitutes a lie and what does not.

Coleman and Kay's results indeed suggest that we are more comfortable calling some statements lies than others, and that falsity is not the determining factor, at least not by itself. Rather, in keeping with the earlier work of Eleanor Rosch (e.g. Rosch 1975), we are more comfortable categorizing prototypical cases as members of a category than we are categorizing fringe cases as members of a category. Work by British psychologist James Hampton and his colleagues confirms that consensus about category membership dissipates as we stray from the prototype (cf. Hampton, Estes, and Simmons 2007). This explains why the scores get closer to the midpoint when some but not all of the features of a prototypical lie are present. Steven Winter (2001) develops the case for this approach impressively in his book, *A Clearing in the Forest*. However, it should be kept in mind that the means reported by Coleman and Kay are only partly informative. If half the participants are certain that a statement is a lie and the other half certain that it is not, the mean on a one-to-seven scale would be exactly 4 – the midpoint – even though there is no uncertainty about category membership – only sharp disagreement.

Also to be kept in mind are findings of Lila Gleitman and her colleagues. A study by Armstrong, Gleitman, and Gleitman (1983) found that while words indeed have prototypes, people use them more to sort out good and bad examples of a concept than they do in deciding category membership in the first place. For example, people agree that a robin is a better example of a bird than is a penguin. However, when asked, they also say that a penguin is no less a bird than is a robin. Regardless, with the law's concern about "ordinary meaning" in legal interpretation, it seems clear that prototype analysis has a place in legal argumentation.

2.3 Section 1001: Lying to a government official

Perjury is not the only crime that requires proof of a lie. It is also a crime to lie to a government official in the context of an official interaction even when not under oath. The relevant part of Section 1001 of the U.S. Criminal Code (18 U.S.C. §1001(a)) reads:

(a) Except as otherwise provided in this section, whoever, in any matter within the jurisdiction of the executive, legislative, or judicial branch of the Government of the United States, knowingly and willfully–

(1) falsifies, conceals, or covers up by any trick, scheme, or device a material fact;

(2) makes any materially false, fictitious, or fraudulent statement or representation; or

(3) makes or uses any false writing or document knowing the same to contain any materially false, fictitious, or fraudulent statement or entry;

shall be fined under this title, imprisoned not more than 5 years . . .

The law does not apply to false statements made by parties or their lawyers in judicial proceedings (18 USC § 1001(b)). Those are covered by the perjury and obstruction of justice laws, and by procedural rules that sanction parties who act dishonestly.

Section 1001 came into play famously during the Trump administration. Two members of President Trump's inner circle pleaded guilty to having violated this statute. In December 2017, former National Security Advisor Michael Flynn pleaded guilty to having lied to the FBI about contacts had had with former Russian ambassador to the United States, Sergey Kislyak, in violation of Section 1001.[9] The charges to which Flynn pleaded guilty alleged that he had falsely told the FBI that he did not ask the Russian Ambassador "to refrain from escalating the situation in response to sanctions that the United States had imposed against Russia . . . ;" that he did not remember being told that Russia agreed to moderate its response as a result of Flynn's request; that Flynn did not ask the Russian Ambassador to act with respect to a then pending UN Security Council resolution; and that the Russian Ambassador never conveyed to Flynn Russia's response to this request. The Flynn plea agreement required Flynn's cooperating with Special Council Robert Mueller in the investigation into Russian meddling in the 2016 presidential election.

About six weeks earlier, George Papadopoulos, who served as a foreign policy advisor to Donald Trump during his campaign, also pleaded guilty to having violated Section 1001 by lying of to the FBI about his interactions with individuals connected to the Russian government. He told the FBI that his contacts with these individuals was superficial and occurred before he joined the campaign, whereas in fact, the contacts were serious efforts to work with the Russian individuals, and

---

9 The relevant documents are available at https://www.lawfareblog.com/michael-flynn-plea-agreement-documents (last visited October 15, 2021).

occurred during his tenure with the Trump campaign.[10] After losing his bid for a second term in November 2020, Trump pardoned both of these former associates.

Section 1001 has been the source of another interesting interpretive issue: The "exculpatory no" defense. As noted, Section 1001 does not apply to statements made in judicial proceedings. This, of course, includes pleading "not guilty" to a crime that the defendant actually committed. What if instead a suspect tells a federal law enforcement officer that he did not engage in conduct that is criminal in nature? Is such a denial a federal crime? Until 1998, many circuit courts accepted the "exculpatory no" defense, saying that a simple denial of an accusation of criminal activity comes within a suspect's constitutional rights.[11] But that year, the Supreme Court put this practice to an end in *Brogan v. United States* (522 U.S. 398 (1998)).

James Brogan was a union leader who had illegally taken money on five occasions from a business that employed union members. The statute of limitations had run on four of the five.[12] One night federal agents knocked on Brogan's door and asked him whether he had accepted such funds. He answered "no," and was subsequently prosecuted for the false statement. Such a denial comes very close to simply saying, "I plead not guilty." Had Brogan said that, instead of "no," Justice Ginsburg observed in her concurring opinion (522 U.S. 411 (1998)), he would not have been prosecuted. Secondly, in cases like Brogan's, applying the statute to a situation in which the government already knows the truth, including situations in which the statute of limitations has already run, applying Section 1001 is an open invitation to law enforcement agents to create crimes when none that could be prosecuted has been committed.

Yet the language of Section 1001 makes no exception for exculpatory "no" cases, and the majority, in an opinion written by Justice Scalia, decided to follow the text as written. This drew sharp criticism from Justice Stevens' dissenting opinion, for courts routinely contextualize statutes to avoid having them apply to situations that were not intended to be covered. Judge Posner (2008) gives the example of a prosecutor who handles child pornography in the course of preparing a criminal trial. Surely the prosecutor does not become subject to

---

10 The documents are available at https://www.lawfareblog.com/george-papadopoulos-stipulation-and-plea-agreement (last visited October 15, 2021).
11 Moser v. United States, 18 F.3d 469, 473–474 (7th Cir. 1994); United States v. Taylor, 907 F.2d 801, 805 (8th Cir. 1990); United States v. Equihua-Juarez, 851 F.2d 1222, 1224 (9th Cir. 1988); United States v. Cogdell, 844 F.2d 179, 183 (4th Cir. 1988); United States v. Tabor, 788 F.2d 714, 717–719 (11th Cir. 1986); United States v. Fitzgibbon, 619 F.2d 874,880–881 (10th Cir. 1980); United States v. Chevoor, 526 F.2d 178, 183–184 (1st Cir. 1975), cert. denied, 425 U. S. 935 (1976).
12 Id. at 411 (Ginsburg, J., concurring).

criminal penalty even if the law does not carve out an exception for such a situation specifically.

In some respects, Brogan's denial and the denials of members of the Trump campaign share a common narrative. All of these individuals, when approached by law enforcement officers, could have asserted their rights under the Fifth Amendment, and not answered the questions. The biggest difference is that Brogan was caught by surprise in the night, whereas the Trump affiliates met with agents voluntarily and lied to them, perhaps assuming wrongly that there would be no independent record of what really happened. It is also possible that President Trump's affiliates did not commit a crime by meeting with the Russian representatives, and lied merely to protect the false story coming from the White House that there were no such contacts – criminal or not. Whatever their motives, it is hard to believe that people involved in a heavily-reported investigation of that sort were unaware that there might be consequences if they were caught lying to the FBI. This puts Brogan in a somewhat more sympathetic light since he may well have simply been pleading not guilty in his own way, but failed to use the acceptable language to do so.

We are thus left with four observations when it comes to how the law treats lies: First, making a truthful statement that is intended to lead the recipient to believing something false is not a lie, at least as far as the perjury statute is concerned (*Bronston*). Second, making an assertion one believes to be false is a lie, even if the assertion turns out to be true (*DeZarn*). Third, bald-faced lies are still lies, even if they do not fool anyone and were not intended to fool anyone (Sorensen, and examples of students lying to escape serious punishment). Fourth, pleading "not guilty" in court is not a lie, but saying "I didn't do it" to the police is a lie (*Brogan*). Philosophers are not in complete accord in drawing boundaries around the concept of lying.[13] Yet the illustrations in the literature suggest that the legal definition is in accord with the conclusions of many scholars who have taken positions on the definition of lying.

3 Deceit

Read carefully, Section 1001 does not require a lie as a premise for prosecution. Any effort to deceive, whether by making a false statement or otherwise, suffices.

---

13 For example, Jörg Meibauer (2005: 1373, 1382), takes the position that the deceit in cases like Bronston should be seen as falling within an extended definition of lying. Several chapters in this volume explore this question.

Michael Flynn pled guilty to lying to the FBI. But the kind of chicanery that Bronston had attempted would have also constituted a violation of the law.

3.1 Lying versus deception: Which is worse?

Samuel Bronston was not a perjurer, but that does not make him a paragon of virtue. His goal was to trick his creditors into thinking that he did not have assets that he actually did have, to prevent those assets being distributed among them by the Bankruptcy Court. Bronston engaged in an act of deception that apparently was thwarted as a result of the assets in question having been discovered independently.

People generally consider lying to be morally worse than deceiving by misdirection. Philosopher Jennifer Mather Saul (2012: 70) presents an experiment based on the following scenario to demonstrate the point:

> An elderly woman is dying. She asks if her son is well. You saw him yesterday (at which point he was happy and healthy), but you know that shortly after your meeting he was hit by a truck and killed. [Is it better] to utter (1) than (2) because (1) is merely misleading while (2) is a lie?
>
> (1) I saw him yesterday and he was happy and healthy.
>
> (2) He's happy and healthy.

Many people in Saul's study chose (1) over (2) because telling the truth is morally better than lying, even if the truth is intentionally misleading. But Saul (2002: 86) argues that there should be "no defensible moral preference" for deception through misdirection over lying. The result is the same. To Saul, the difficult issue is why so many of us feel better about ourselves uttering (1) rather than (2) if there is no moral basis for preferring one over the other. For the record, my views largely match those of Saul. If I tell my aunt that the meal she cooked for me was "unbelievable" in order to insincerely let her think that I thought it was good, I do not think I deserve much credit for being honest.

Others take the view that uttering a false statement is itself a moral wrong, which should be taken seriously in its own right. Seana Valentine Shiffrin (2015) presents strong argumentation in this direction, using Kant's "murderer at the door" thought experiment (Kant 1799) as a vehicle for analysis. (See Chapters 1–4 and references therein for additional commentary on the ethics of lying and misleading.)

3.2 What the law says about deceit

At this point, one may wonder why the legal system would create a safe harbor for fraudulent conduct in the court room whereas it is outlawed in everyday life. If anything, one might expect judicial proceedings to be a sanctuary for honesty and fair play. Stuart Green (2019: 7–8) explains the disparity this way:

> Why exactly should culpable deceit be easier to prove in cases of fraud than of perjury? The distinct contexts in which the two crimes are committed suggest a possible answer: As noted above, perjury involves statements made under oath, often in a formal, adversarial setting where the truth of the witness' statement can be tested through probing cross-examination. Fraud, by contrast, typically occurs in a commercial or regulatory setting, where the deceiver and deceived are engaged in an arm's length, often one-shot transaction. In such circumstances, there is no opportunity for careful fact-finding or cross-examination. Likely for this reason, the courts have tended to define deception more broadly in the fraud context than in that of perjury.

Green's explanation is consistent with that of the *Bronston* Court, which blamed the creditors' lawyer for not following up and asking the question that would have pinned Bronston down: "What about you personally?" (409 U.S. at 358). Indeed, the adversarial system does present the opportunity to probe further. But that does not really get to the heart of the matter. For one thing, in the world of business, at least in many transactional environments, both parties have ample opportunity to ask additional questions to undo the inferences drawn from misleading statements. While we may not wish to require those in the business world to be as distrustful as those in the world of adversarial litigation, the distinction between the two settings may not be adequate to justify such a sharp distinction in moral responsibility.

Deception, like lying, is generally disallowed in the business world, especially when a victim relies on a deceptive statement to his or her detriment. That is the classic definition of fraud. There are nuances, however. For one thing, the law forgives a certain amount of exaggeration in the law governing sales. This practice is called "puffery", and is allowed. The justification for this doctrinal permissiveness is that consumers are expected to be aware enough of what they are hearing so as not to be unduly influenced by the hyperbole. A current example is the television commercial advertising Mypillow.com. It comes with a jingle, "for the best night's sleep in the whole wide world, visit mypillow.com." When it comes to misrepresentations for which there are monetary or criminal sanctions, the conveyor's state of mind comes more into play. Consider Rule 10b-5 of the Securities and Exchange Commission (17 C.F.R. 240.10b-5):

It shall be unlawful for any person, directly or indirectly, by the use of any means or instrumentality of interstate commerce, or of the mails or of any facility of any national securities exchange, (a) To employ any device, scheme, or artifice to defraud, (b) To make any untrue statement of a material fact *or to omit to state a material fact necessary in order to make the statements made, in the light of the circumstances under which they were made, not misleading*, or (c) To engage in any act, practice, or course of business which operates or would operate as a fraud or deceit upon any person, in connection with the purchase or sale of any security.

Note that the rule specifically includes truthful statements that are designed to lead the reader or hearer to draw a false inference. This is exactly what Bronston did. It is not perjury, but it is an act of fraud.

In our book, *Speaking of Crime* (Solan and Tiersma 2005: 234–235), Peter Tiersma and I agree with the holdings in both *Bronston* and *DeZarn*, but find the justification not in the lawyer's responsibility to follow up as a matter of professional competence, but rather in the role morality of the lawyers. Lawyers are permitted to deceive in circumscribed ways that are defining features of the relationship between lawyer and adverse witness. To take two examples, lawyers are permitted, some say required, to produce false defenses. By "false defense" I mean a defense based on legitimate evidence that is likely to lead a trier of fact to an inference that the lawyer knows to be false. This license applies particularly to criminal defense lawyers. A lawyer who decides not to challenge the time of death in an autopsy report that contains errors in calculation would be remiss even if the lawyer knew from his own client that the estimated time of death is fairly accurate. Likewise, as Monroe Freedman has pointed out, a lawyer who fails to cross-examine a visually-impaired eyewitness on what she actually saw because he knows her account to have been accurate would be committing malpractice (Freedman 1975: 48).

Moreover, in the routine cross-examination of witnesses, it is the lawyer's job to persuade witnesses to agree to characterizations of uncontested events in ways that will help the lawyer's client. Witnesses need not agree to inaccurate characterizations, of course, but even such word choices as "smash" versus "hit" in a car accident case can have a profound effect on how a juror conceptualizes the event (Loftus and Palmer 1974). The moral issue arises when the lawyer knows that the characterization is sufficiently accurate so that the witness has an obligation to accept it, but that is not a fair characterization. That is, if the lawyer were speaking in casual conversation with a person she trusts, she would have used different language.

As Bradley Wendell points out, it is not enough to justify deceptive practices by lawyers as within the role of the lawyer in society unless we can justify the rules of the role itself on independent moral grounds. He writes (Wendell 2018: 154–155):

We tolerate lawyers engaging in these practices not because we are indifferent to lying, but because we recognize that bluffing in negotiations and arguing for false inferences are means to broader institutional ends such as protecting liberty and enabling citizens to have access to the rights allocated to them by law. The assessment of public actors as truthful or untruthful requires situating their conduct in context, including the expectations and beliefs of others who participate in the relevant social practices and institutions. This contextual, community-grounded evaluation also suggests that we may do better at realizing the value of truthfulness by instituting and reinforcing certain methodologies and practices that are adapted to the obstacles one is likely to encounter to the maintenance of truth.

Returning to *Bronston,* a witness does not answer questions in a vacuum. A witness answers questions that are often designed to elicit answers that will create a misimpression, at least from the witness's point of view. At the very least, the questions are intended to elicit answers that will serve the interest of the party the lawyer represents, even if neither the lawyer nor the witness would regard the exchange as producing a fair characterization from the perspective of a neutral observer.

This license for lawyers to produce a record that may go beyond the lopsided, even to the point of being deceptive, helps explain why Bronston should not go to prison for playing on the same field. The Cooperative Principle (Grice 1975) tells us that in ordinary conversation, we assume the other participant to be moving the discussion along in a cooperative manner, and we give the other individual the impression that we are doing the same.

In cross-examination, some of this cooperation holds. For example, Grice's maxim of relation ("Be relevant") is required of witnesses, although Bronston himself trickily flouted that maxim. Yet trial practice manuals encourage lawyers to be conversational in their cross-examination not to cooperate with the witness, but to lull the witness into being less guarded and more cooperative, increasing the likelihood of getting helpful responses (see Lubet 2013; Mauet 2013).

4 Bullshit

4.1 A brief note on President Trump

On December 30, 2017, the *Washington Post* published an article entitled, "In a 30-minute interview, President Trump made 24 false or misleading claims."[14] President Trump had been interviewed by the *New York Times* at one of his golf

---

14 https://www.msn.com/en-us/news/factcheck/in-a-30-minute-interview-president-trump-made -24-false-or-misleading-claims/ar-BBHuIgy?li=BBmkt5R&ocid=spartandhp. The *Post* maintained a

resorts, and was apparently not entirely truthful in his remarks. I will not summarize the details of the interview here, because this essay is focused on the legal system's handling of the various species of dishonesty. However, whether it is Bill Clinton talking about his sex life, George W. Bush talking about Iraq's efforts to acquire nuclear weapons, or Tony Blair's similar efforts, politicians are known to present information in a manner that is more concerned with the narrative they wish to create than with the truth of the matter. President Trump holds a special place in this succession. In the summer of 2017, the *New York Times* published a full-page list of what it called "Trump's Lies," updated later to include dates through November 11th.[15] Below are two examples:

> Feb. 18: "You look at what's happening in Germany, you look at what's happening last night in Sweden. Sweden, who would believe this?" (*Trump implied there was a terror attack in Sweden, but there was no such attack.*)

> March 17: "I was in Tennessee – I was just telling the folks – and half of the state has no insurance company, and the other half is going to lose the insurance company." (*There's at least one insurer in every Tennessee county.*)

In response to claims that Trump was no different from President Obama, the *Times* reported in late 2017 a comparative analysis showing that Trump had to date had produced more false statements in ten months than Obama had in his entire eight years in office.[16]

How many of President Trump's inaccurate statements are lies, how many are honest mistakes, and how many are bullshit is anyone's guess. Continuing to adopt Frankfurt's definition, bullshit is an assertion made without regard for whether the assertion is true or not. For the bullshitter, whether a statement is true or false is a matter of convenience. When the statement happens to be true, there will be less criticism, and accordingly, less inconvenience.

---

running list of Trump's false or misleading statements during his presidency, totaling them at more than 30,000 by the end of his term.

15 https://www.nytimes.com/interactive/2017/06/23/opinion/trumps-lies.html?_r=0.

16 The analysis (available at https://www.nytimes.com/interactive/2017/12/14/opinion/sunday/trump-lies-obama-who-is-worse.html), which claimed to use the same method to evaluate the truth of statements by both presidents that had been challenged as inaccurate, found that Trump had made 108 false statements in 10 months in office, whereas Obama had made 18 in his eight years in office.

4.2 How the law reacts to bullshit

Bullshit does not meet the criteria for either lying or deceiving because the requisite state of mind is absent. The person who neither knows nor cares about the truth cannot tell a lie. Moreover, bullshit may be the result of wishful thinking. People may have a general sense of a situation, and fill in the details without adequate evidence. The law is not consistent in its treatment of bullshit, but it is specifically disapproved in particular contexts.

4.2.1 Federal pleadings

Rule 11 of the Federal Rules of Civil Procedure requires that all court filings be signed, and that the signature is a certification of various representations, including (Rule 11(b), FRCP):

> Representations to the Court. By presenting to the court a pleading, written motion, or other paper – whether by signing, filing, submitting, or later advocating it – an attorney or unrepresented party certifies that to the best of the person's knowledge, information, and belief, formed after an inquiry reasonable under the circumstances:
>
> (3) the factual contentions have evidentiary support or, if specifically so identified, will likely have evidentiary support after a reasonable opportunity for further investigation or discovery; and
>
> (4) the denials of factual contentions are warranted on the evidence or, if specifically so identified, are reasonably based on belief or a lack of information.

This rule removes from lawyers (and pro se litigants) the right to make claims in court based on the hope that the evidence will later support the claim, unless it is specifically stated that the filer lacks evidence at the time to support the claim. In other words, it severely limits bullshit.

Added to this rule are cases decided by the US Supreme Court requiring detailed, factually-based pleadings in civil litigation. In *Ashcroft v. Iqbal* (556 U.S. 662 (2009)), decided in 2009, the Supreme Court set standards for a court's decision on whether to grant a motion to dismiss a complaint for failure to state a claim. Restating the test it had established in *Bell Atlantic Corp. v. Twombly* (550 U.S. 544 (2007)), the Court held (566 U.S. 662, 679 (2009)) that:

> only a complaint that states a plausible claim for relief survives a motion to dismiss. Determining whether a complaint states a plausible claim for relief will, as the Court of Appeals observed, be a context-specific task that requires the reviewing court to draw on its judicial experience and common sense. . . . In keeping with these principles a court considering a

motion to dismiss can choose to begin by identifying pleadings that, because they are no more than conclusions, are not entitled to the assumption of truth. While legal conclusions can provide the framework of a complaint, they must be supported by factual allegations. When there are well-pleaded factual allegations, a court should assume their veracity and then determine whether they plausibly give rise to an entitlement to relief.

Taken together, these cases require those who file civil cases in federal court have significant knowledge of facts, which are sometimes not in their control. When one adds to the pleading requirements the certifications under Rule 11, the likelihood of bullshit in federal pleadings has surely been reduced. I take no position here on concerns expressed that these cases have the effect of closing the court house door on many meritorious claims that require discovery to be adequately developed to meet the pleading standards.

4.2.2 Expanded definition of fraud

Recall that fraud requires an effort to lead someone to believe something that the speaker believes to be false. Yet some statutes, and many statements of the common law include "reckless disregard for the truth" as a substitute for knowingly making a false statement.[17] This standard requires somewhat more regard for the truth than does Frankfurt's bullshit, because the truth must be fairly overt for it to be recklessly disregarded. Nonetheless, the fact that an individual can commit fraud without knowing the truth and flouting it is a significant step away from classic definitions of deceit. (See also Chapters 1 and 9 for more on the reckless disregard criterion.)

By the same token, the Restatement (Second) of Contracts (§ 164(1)) makes a contract voidable for misrepresentation when the aggrieved party relies on a representation that is either fraudulent or material (or both). And a fraudulent misrepresentation can include bullshit (§ 162(1)):

> A misrepresentation is fraudulent if the maker intends his assertion to induce a party to manifest his assent and the maker (a) knows or believes that the assertion is not in accord with the facts, or (b) does not have the confidence that he states or implies in the truth of the assertion, or (c) knows that he does not have the basis that he states or implies for the assertion.

The broad definition of fraudulent misrepresentation makes sense in this context, where the remedy is rescission of a contract. If a person enters into an

---

17 See, e.g., *Pace v. Parrish*, 247 P.2d 273, 274–75 (Utah 1952); Eurycleia Partners, LP v. Seward & Kissel, LLP, 12 N.Y.3d 553 (2009).

agreement because the other party misinformed her, that party should not be bound as long as the misinformation was of a material fact, regardless of the state of mind of the purveyor of falsity.

5 Conclusion

This essay has attempted to demonstrate differential tolerance for various forms of dishonest conduct in legal contexts. Lying is never allowed as a formal matter, but it is tolerated in courtrooms when offered by law enforcement agents (see also Chapter 11). Deception short of lying is permitted by witnesses in court, but not by people engaged in commercial life. Bullshit, the bread and butter of political life, is outlawed as a species of fraud in some circumstances, tolerated in others. When we add to this set of facts the materiality requirement in both perjury and fraud cases, the law appears to recognize the fact that people do not always tell the truth, while also seeking to ensure that the legal system operates with sufficient integrity so that dishonesty does not compromise the integrity of business interactions or the truth-seeking function of the courtroom.

References

Armstrong, Sharon Lee, Lila R. Gleitman & Henry Gleitman. 1983. What some concepts might not be. *Cognition* 13: 263–308.

Augustine. 420. Against lying. http://www.newadvent.org/fathers/1313.htm.

Austin, J.L. 1962. *How To Do Things With Words*. Cambridge, MA: Harvard University Press.

Coleman, Linda and Paul Kay. 1981. Prototype semantics: the English word *lie*. *Language* 57: 26–44.

Fallis, Donald. 2009. What is lying? *Journal of Philosophy* 106: 29–56.

Frankfurt, Harry. 2005. *On Bullshit*. Princeton: Princeton University Press.

Freedman, Monroe. 1975. *Lawyers' Ethics in an Adversary System*. New York: Bobbs-Merrill.

Gaines, Philip. 2015. Toward a communicative approach to law- and rule-making. In Lawrence Solan, Janet Ainsworth & Roger Shuy (eds.), *Speaking of Language and Law: Conversations on the Work of Peter Tiersma*, Chapter 38. Oxford: Oxford University Press.

Green, Stuart. 2019. Lying and the law. In Jörg Meibauer (ed.), *The Oxford Handbook of Lying*, 483–494. Oxford: Oxford University Press.

Grice, H.P. 1975. Logic and conversation. In Peter Cole & Jerry Morgan (eds.), *Syntax and Semantics 3: Speech Acts*, 41–58. New York: Academic Press.

Hampton, James A., Zachary Estes & Sabrina Simmons. 2007. Metamorphosis: essence, appearance, and behavior in the categorization of natural kinds. *Memory & Cognition* 35(7): 1785–1800. DOI: 10.3758/bf03193510.

Horn, Laurence. 2017. Telling it slant: Toward a taxonomy of deception. In Dieter Stein and Janet Giltrow (eds.), *The Pragmatic Turn in Law*, 23–55. Berlin: De Gruyter.

Kant, Immanuel. 1799. On the supposed right to die from benevolent motives. Available at http://www.sophia-project.org/uploads/1/3/9/5/13955288/kant_lying.pdf.

Kuntz, Phil (ed.). 1998. *The Starr Report: The Evidence*. Darby, PA: Diane.

Loftus, Elizabeth F. & John C. Palmer. 1974. Reconstruction of automobile destruction: An example of the interaction between language and memory. *Journal of Verbal Learning and Verbal Behavior* 13: 585–589.

Lubet, Steven. 2013. *Modern Trial Advocacy: Analysis and Practice*, 4th ed. Louisville, CO: National Institute for Trial Advocacy.

Mauet, Thomas A. 2013. *Trial Techniques and Trials*, 9th Edition. Philadelphia: Wolters Kluwer.

Meibauer, Jörg. 2005. Lying and falsely implicating. *Journal of Pragmatics* 37: 1373–99.

Posner, Richard A. 1999. *An Affair of State: The Investigation, Impeachment and Trial of President Clinton*. Cambridge, MA: Harvard University Press.

Posner, Richard A. 2008. *How Judges Think*. Cambridge, MA: Harvard University Press.

Rosch, Eleanor. 1975. Cognitive representations of semantic categories. *Journal of Experimental Psychology: General* 104(3): 192–233. https://doi.org/10.1037/0096-3445.104.3.192.

Saul, Jennifer. 2012. *Lying, Misleading, and What is Said*. Oxford: Oxford University Press.

Shargel, Gerald L. 2007. Federal Evidence Rule 608(b): Gateway to the minefield of witness preparation. *Fordham Law Review* 76: 1263–1294.

Shiffrin, Seana Valentine. 2015. *Speech Matters: On Lying, Morality, and the Law*. Princeton: Princeton University Press.

Solan, Lawrence. 2002. The Clinton scandal: Some legal lessons from linguistics. In Janet Cotterill (ed.), *Language in the Legal Process*, 180–195. Berlin: Springer.

Solan, Lawrence. 2012. Lawyers as insincere (but truthful) actors. *Journal of the Legal Profession* 36: 487–527.

Solan, Lawrence. 2018. Lies, deceit, and bullshit in law. *Duquesne University Law Review* 56: 73–104.

Solan, Lawrence and Peter Tiersma. 2005. *Speaking of Crime: The Language of Criminal Justice*. Chicago: University of Chicago Press.

Sorensen, Roy. 2007. Bald-faced lies! Lying without the intent to deceive. *Pacific Philosophical Quarterly* 88: 251–264.

Sperber, Dan & Deirdre Wilson. 1995. *Relevance: Communication and Cognition*, 2nd Edition. Oxford: Blackwell.

Tiersma, Peter. 2004. Did Clinton lie? Defining "sexual relations". 79*Chi.-Kent L. Rev.* 927 (2004). Available at http://papers.ssrn.com/sol3/papers.cfm?abstract_id=470645.

Wendell, Bradley. 2018. Truthfulness as an ethical form of life. *Duquesne University Law Review* 56: 141–167.

Winter, Steven L. 2001. *A Clearing in the Forest: Life, Law, and Mind*. Chicago: University of Chicago Press.

Index

https://doi.org/10.1515/9783110733730-016

Printed in the USA
CPSIA information can be obtained
at www.ICGtesting.com
LVHW010200181223
766737LV00017B/1721